Groundwater

Resource Evaluation, Augmentation, Contamination, Restoration, Modeling and Management

Groundwater

Resource Evaluation, Augmentation, Contamination, Restoration, Modeling and Management

Edited by

M. Thangarajan

Senior Deputy Director (Retd.)
National Geophysical Research Institute
Hyderabad, India

Springer

CP

A C.I.P. Catalogue record for this book is available from the Library of Congress.

ISBN 978-90-481-7435-5
ISBN 978-1-4020-5729-8 (e-book)

Copublished by Springer,
P.O. Box 17, 3300 AA Dordrecht, The Netherlands
with Capital Publishing Company, New Delhi, India.

Sold and distributed in North, Central and South America by Springer,
233 Spring Street, New York 10013, USA.

In all other countries, except India, sold and distributed by Springer, Haberstrasse 7, D-69126 Heidelberg, Germany.

In India, sold and distributed by Capital Publishing Company,
7/28, Mahaveer Street, Ansari Road, Daryaganj, New Delhi, 110 002, India.

www.springer.com

Cover Illustration: Crystal clear water in a highly weathered large well at Kannivadi village, Kodaganar river basin, Tamil Nadu, India. Photograph taken by editor.

Printed on acid-free paper

Foreword

The demand for water resources is increasing day by day due to ever increasing population, mostly from developing countries. This has resulted in abstracting more water from the subsurface stratum and forcing the water managers to manage the limited groundwater resources in a more scientific way, which in turn needs a more sophisticated way of assessing the underground resource and manage it optimally. There is an urgent need to locate high yielding boreholes in the hard rock region by using geophysical methods. Electrical imaging technique in conjunction with remote sensing and geographical information system (GIS) technique has proved to be a potential tool for the purpose. Hydrodynamics of fractured aquifer system in hard rock region is not yet fully understood. The understanding of the groundwater pollution migration in porous and fractured aquifer system and the seawater intrusion in the coastal aquifer has to be improved further. Various aspects of groundwater modeling and in particular issues related to model calibration, validation and prediction has to be understood in much better way. One should integrate all the above issues for effective understanding of the assessment and management of groundwater resources.

There is a need to have a comprehensive book to deal with all the above. My former colleague, Dr. M. Thangarajan, Retired Scientist-G, NGRI, Hyderabad, India has successfully edited a book on GROUNDWATER (Resource Evaluation, Augmentation, Contamination, Restoration, Modeling and Management) by inviting topics from various experts across the globe.

It is my firm view that this book will form a comprehensive text for those engaged in the resource evaluation, augmentation, contamination modeling and management.

I wish to congratulate Dr. M. Thangarajan to bring out this book in an excellent manner.

Dr. V.P. Dimri
Director
National Geophysical Research Institute
Hyderabad, India

Preface

Groundwater plays a major role in the livelihood of mankind by providing water for drinking, irrigation and industrial purposes. The rapid population growth in the last three decades all over the globe resulted in exploiting more groundwater. The distribution of groundwater—both in space and time—is more erratic as it depends on the subsurface geological and climatic conditions. In many countries, the decline of water level indicates that the resources are depleted very fast. It is, therefore, necessary to assess the available subsurface resource in a more judicious scientific manner and then apply it for evolving optimal utilization purposes. There is an urgent need to have a comprehensive book, which contains the entire spectrum of groundwater assessment and management aspects. I had seen many books, which cover only the specific aspects on groundwater exploration, exploitation, augmentation, pollution and remediation and mathematical modeling but not many books on the integrated aspects of all. It was, therefore, planned to bring a book, which covers the abovesaid aspects by inviting specific topics from various experts of the globe.

The aim of the book is to provide theoretical background on the application of remote sensing and GIS techniques in the delineation of subsurface resources as described in Chapter 1. Chapter 2 discusses the principles of electrical resistivity and imaging (tomography) techniques in the identification of potential boreholes in hard rock region. Principles of pumping test analysis and interpretation of data to evolve aquifer parameters transmissivity 'T' and storativity 'S' through numerical technique have been brought out in Chapter 3. Basic principles and the application of theory of regionalized variables (Geostatistics-Kriging) for the interpolation of sparse hydrogeological data and its estimation error have been stressed in Chapter 4. Augmentation of groundwater resources through aquifer storage and recovery (ASR) along with the quality problem due to reaction of rocks and injection fluid is dealt in Chapter 5. Chapter 6 deals with the environmental problem. The sources and the process of fluoride and arsenic contamination have been brought out clearly in this article.

Chapter 7 deals with the characterization of fracture properties in hard rock areas through hydrogeological investigation at different scales. A methodology was evolved to delineate the vertical distribution of conductive fracture zones and their permeability through flowmeter vertical profiles during fluid injection and evolving the spatial distribution of permeability by making use of slug test data. The above methodology was tested in Maheswaram watershed, a hard rock region in Andhra Pradesh (India). Groundwater flow and mass transport modeling play a major role in the assessment and management of groundwater resources. Modeling principles and various types of models, which are in vogue for various applications are presented in Chapter 8. An exclusive chapter on the model calibration and issues related to validation, sensitivity analysis, post-audit, uncertainty evaluation and assessment of prediction data needs is also presented. Groundwater development and management of coastal and island aquifers through field investigations and mathematical modeling is brought out in Chapter 10. Basic principles of SUTRA (USGS) finite element model are also highlighted. The management of groundwater resources through community participation approach and some aspects of remedial measures of contaminated aquifer may provide some insight to the groundwater professionals, which form the Chapter 11.

Groundwater management needs assessment, which in turn needs a model. A model needs a set of mathematical equations to describe the system. The equations have to be solved through a set of characteristic parameters, initial and boundary conditions of the aquifer system, which in turn have to be obtained through field investigations. Field investigations need a set of procedures, which in turn needs guidelines to carry out field investigations. I believe and hope that this book may provide the needed guidelines and answers for all the above.

I am thankful to Dr. V.P. Dimri, Director, National Geophysical Research Institute (NGRI), Hyderabad, India for providing infrastructure facilities at NGRI to complete this book. I extend my gratitude to all the contributors for this book who are well known experts in their respective fields. The following have contributed: F.M. Howari, Mohsen M. Sherif and Mohamed S. Al Asam from UAE University, Al Ain, UAE; V.P. Singh, Louisiana State University, Baton Rouge, USA; Ron D. Barker, University of Birmingham, Birmingham, UK; V.S. Singh, NGRI, Hyderabad, India; Shakeel Ahmed, NGRI, Hyderabad, India; Chris Barber, Center for Groundwater Studies, Adelaide, Australia; N. Madhavan and Prof. V. Subramanian from the School of Environmental Sciences, JNU, New Delhi, India; J.C. Maréchal and B. Dewandel from BRGM, France; K. Subrahmanyam, NGRI, Hyderabad, India; Claire R. Tiedeman and Mary C. Hill from USGS, USA; and A. Ghosh Bobba, National Water Research Institute, Burlington, Canada.

My colleagues at NGRI, Hyderabad viz. Dr. S. Thiagarajan, Mr. G. Ramandha Babu and Y.S.N. Murthy are thanked for their support in preparation of the manuscript. I take this opportunity to thank both Prof. V.P. Singh and Prof. Mary C. Hill who inspired me to bring out this book. Finally, Capital Publishing Company, New Delhi, India is thanked for their keen interest in publishing this book.

M. Thangarajan
Former Scientist-G & Head, Groundwater Modeling Group
National Geophysical Research Institute
Hyderabad, India

Contents

Foreword v

Preface vii

1. Application of GIS and Remote Sensing Techniques in Identification, Assessment and Development of Groundwater Resources
Fares M. Howari, Mohsen M. Sherif, Vijay P. Singh and Mohamed S. Al Asam 1

2. Electrical Resistivity Methods for Borehole Siting in Hardrock Region
Ron D. Barker 26

3. Parameterization of Groundwater Aquifer System
V.S. Singh 61

4. Application of Geo-statistics in Hydrosciences
Shakeel Ahmed 78

5. Augmentation of Groundwater Resources through Aquifer Storage and Recovery (ASR) Method
C. Barber 112

6. Environmental Impact Assessment, Remediation and Evolution of Fluoride and Arsenic Contamination Process in Groundwater
N. Madhavan and V. Subramanian 128

7. Characterization of Fracture Properties in Hard Rock Aquifer System
J.C. Maréchal, B. Dewandel and K. Subrahmanyam 156

8. Groundwater Models and Their Role in Assessment and Management of Groundwater Resources and Pollution
M. Thangarajan 189

9. Model Calibration and Issues Related to Validation, Sensitivity Analysis, Post-audit, Uncertainty Evaluation and Assessment of Prediction Data Needs
Claire R. Tiedeman and Mary C. Hill 237

10. Groundwater Development and Management of Coastal Aquifers (including Island Aquifers) through Monitoring and Modeling Approaches
A. Ghosh Bobba 283

11. Management of Groundwater Resources
M. Thangarajan 334

Index 357

1

Application of GIS and Remote Sensing Techniques in Identification, Assessment and Development of Groundwater Resources

Fares M. Howari, Mohsen M. Sherif[1], Vijay P. Singh[2] and Mohamed S. Al Asam[3]

Geology Dept., College of Science, UAE University, Al Ain, UAE
[1]Civil and Environmental Engineering Dept., College of Engineering, UAE University Al Ain, UAE
[2]Department of Civil and Environmental Engineering, Louisiana State University Baton Rouge, LA 70803-6405, USA
[3]Ministry of Agriculture and Fisheries, Dubai, UAE

PREAMBLE

Geographical Information Systems (GIS) and Remote Sensing (RS) techniques have emerged as efficient and powerful tools in different fields of science over the last two decades. The GIS has the ability to store, arrange, retrieve, classify, manipulate, analyze and present huge spatial data and information in a simple manner. The RS technique is used to collect detailed information in space and time even from inaccessible areas. Nowadays, both GIS and RS are regarded as essential tools for groundwater studies especially for extended and complex systems.

This article reviews practices, problems and prospects of using Geographic Information Systems and Remote Sensing in the exploration, assessment, analysis and development of groundwater resources. Review of literature indicates successful attempts that have used GIS and RS, and studies that have included conceptualizations of space and time embedded in the current generation of GIS and RS. However, there are limitations to applications of these techniques. For example, in many cases, incompatibility between hydrological and hydrogeological models and GIS and RS tools exist. By reframing the future research agenda from the emerging geographic information science perspective, it is believed that integration of hydrological/ hydrogeological modeling with GIS should proceed with the development of a high-level common platform that is compatible with both GIS and hydrological/hydrogeological models. More research to better handle, communicate and reduce the uncertainties in the process of GI Science-based hydrological and hydrogeological modeling is needed. On the other hand, it is evident to water resources specialists that remote sensing provides efficient means of observing hydrological variables over large and remote areas.

This article also discusses issues and information that need to be considered while dealing with major hydrological/hydrogeological investigations, such as those dealing with land surface temperature from thermal infrared data, surface soil moisture from passive microwave data, selecting groundwater recharge sites, simulation of groundwater systems, water quality using visible and near-infrared data, and estimating landscape surface roughness using radar. Methods for estimating related variables of interest, such as fluxes, evapotranspiration and runoff, using state variables are also described. The article is concluded with an example for the employment of RS and GIS techniques in groundwater studies.

INTRODUCTION

Satellite remote sensing has become a common tool to investigate different fields of earth and natural sciences (Barret and Kidd, 1987; Barret and Curtis, 1982). The progress of the performance and capabilities of the optoelectronic and radar devices mounted on-board remote sensing platforms have further improved the capability of instruments to acquire information about water resources and hydrological/hydrogeological systems. With the advent of new high-spatial and spectral resolution satellite and aircraft imagery new applications for large-scale mapping and monitoring have become possible (e.g., Fortin and Bernier, 1991; Jensen, 1986). Integration with Geographic Information Systems (GIS) allows a synergistic processing of multi-source spatial data. Using of GIS in hydrogeology is only at its beginning, but there have been successful applications that started to develop (Barton, 1987; Bhasker et al., 1992; Clark, 1998; Gossel et al., 2004). Groundwater resources are dynamic in nature as they are affected by various human activities, including the expansion of the cultivated and irrigated lands, industrialization, urbanization, and others. Because it represents the largest available source of fresh water lying beneath the ground it has become crucial not only for targeting of groundwater potential zones, but also monitoring and conserving this important resource. In addition to targeting the groundwater potential zones, it is also important to identify suitable sites for artificial recharge to sustain groundwater systems and avoid their depletion due to over-pumping and/or insufficient natural recharge. When natural recharge from rainfall events cannot meet groundwater demands, the balance is disturbed and hence it calls for artificial recharge on a countrywise basis (Saswosky and Gardner, 1991; Shulz, 1994; More, 1991; Sameena et al., 2000). To improve the information and data feeding the GIS, pressure is put to advance the remote sensing platforms. The development and advancement of remote sensing platforms is crucially needed to improve our knowledge and monitoring ability on natural resources.

On the other hand, remote sensing with its advantages of spatial, spectral and temporal availability of data covering large and inaccessible areas within a short time has become a very handy tool in assessing, monitoring and conserving groundwater resources. Satellite data provides quick and useful baseline information on the parameters controlling the occurrence and movement of groundwater, such as geology, lithology, stratigraphy, structural controls, geomorphology, soils, land-use/cover and lineaments (Das, 1994, 1990). However, all the controlling parameters have rarely been studied together because of the non-availability of data, integrating tools and modeling techniques. Hence a systematic study of these factors leads to a better delineation of prospective zones in an area, which is then followed up on the ground through detailed hydrogeological and geophysical investigations.

Fotheringham and Rogerson (1994) indicated that for almost two decades in the 1960s and 1970s, geographic information systems (GIS) and hydrological modeling developed in parallel with a few interactions. Major research efforts toward the integration of GIS with hydrological modeling did not take place until the late 1980s, as part of the GIS community's efforts to improve the analytical capabilities of GIS (e.g., Goodchild et al., 1992) and hydrologists' new demand for accurate digital representations of the terrain (Clark, 1998; Singh & Fiorentino, 1996). Nowadays, both GIS users and hydrologists have increasingly recognized the mutual benefits of such integration from the successes of the last few years (Gossel et al., 2004).

Remote sensing images have been used successfully to identify boundaries of inaccessible watersheds and determine the order and slopes of channels. Outcropping and recharge areas of aquifers, soil and rock types, land cover, land erosion in coastal zones, movement of sand dunes, and development of agricultural, industrial, urbanization and tourism activities have been successfully studied using RS (Lanza et al., 1997; Sui and Maggio, 1999; Schmugge et al., 2000; Su Z.B. Troch P.A. 2003). Nevertheless, many studies require supporting field information to ensure the accuracy of the RS images. This information can be introduced to hydrological models, such as HEC-HMS, to estimate the surface water runoff and base flow. Information about permeable (active) and impermeable (inactive) areas (cells or elements) can also be imported to groundwater models, such as FEFLOW, MODFLOW and SUTRA, to simulate groundwater flow and solute transport in aquifers.

The most important feature of the RS images is that all information and data collected are geo-referenced and can be easily integrated in GIS databases and hydrological/hydrogeological models. In other words, study domains and other parameters and boundaries can be imported/exported among different surface and groundwater simulation packages with little or no errors.

INTEGRATING GIS AND REMOTE SENSING

Visual interpretation has been the main tool for evaluation of groundwater prospective zones for over two decades (Beven and Moor, 1992; Blyth, 1993; Das, 1990). It has also been found that remote sensing, besides helping in targeting potential zones for groundwater exploration, provides inputs towards estimation of the total groundwater resources in an area, the selection of appropriate sites for drilling or artificial recharge and the thickness of unconsolidated deposits. By combining the remote sensing information with adequate field data, particularly well inventory and yield data, it is possible to arrive at prognostic models to predict the ranges of depth, yield, success rate and types of wells suited to various terrains under different hydrogeological domains. Based on the status of groundwater development and groundwater irrigated areas (through remote sensing), artificial recharge structures, such as percolation tanks, check dams and subsurface dykes, can be recommended upstream of groundwater irrigated areas to recharge the wells in the downstream areas so as to augment groundwater resources. Examples of successful studies include the work of Engman et al. (1989), Gupta et al. (1996) and Hannaford and Hall (1980).

Sherif et al. (2004) conducted a comprehensive study to assess the effectiveness of dams in recharging aquifers in three selected Wadis in the United Arab Emirates. Remote Sensing images were used to identify watersheds contributing to the water storage in the ponding areas of different dams. Recharge areas and active (permeable) and inactive (impermeable) zones were also identified using the RS images. All information, including locations of pumping and observation wells, pumping rates, lithological cross sections, geologic and hydraulic parameters and climatic data, were integrated into a GIS database. Such information was in several layers that could be imported by most of the groundwater simulation models. The RS technique and GIS were also employed in another study for groundwater management in two coastal aquifers of the United Arab Emirates and Sultanate of Oman (Sherif et al., 2004b). Figure 1 presents a RS image for Wadi Ham in the UAE including the main hydrogeological/geological features of the Wadi.

Many researchers are now using digital techniques to derive the geological, structural and geomorphological details (Humes et al., 1994; Jensen, 1986; Jakson, 1977) to understand the aquifer system. The various thematic layers generated using remote sensing data like borehole lithology/structural, geomorphology, landuse/cover and lineaments can be integrated with slope, drainage density and other collateral data in a Geographic Information System (GIS) framework and analyzed using a model developed with logical conditions associated with groundwater zones as well as artificial recharge sites (Das, 1990; Peck, 1981; Perry et al., 1988; Rango et al., 2000; Gossel et al., 2004). Digital enhancement techniques are found to be suitable since they improve the feature sharpness and contrast for simple interpretation (Fig. 2).

Field data can also be integrated into GIS, since most of the groundwater related data would be available from pumping wells/bore hole logs, would be point information, and could be interpolated to get spatial data, i.e., each grid cell would have valid data without any gaps (e.g., Goodchild et al., 1993). Point data could be water level, water quality, weathered zone thickness, saturated zone thickness, elevation, yield of wells, rainfall at various rain gauge stations, porosity of soil, hydraulic conductivity, K, transmissivity (T) and storage coefficient (S), and others. Depending on the objectives of the study and the porous media under consideration the important layers of information data and other related variables that would be stored in different layers could be identified. A GIS database including various layers should be developed in such a manner as to allow for an efficient access, retrieval, organization, manipulation and analysis of the available information. On the other hand, contour or zonation maps can be developed from point information to provide a better understanding of the areal distribution of variables in the study domain (Fig. 2).

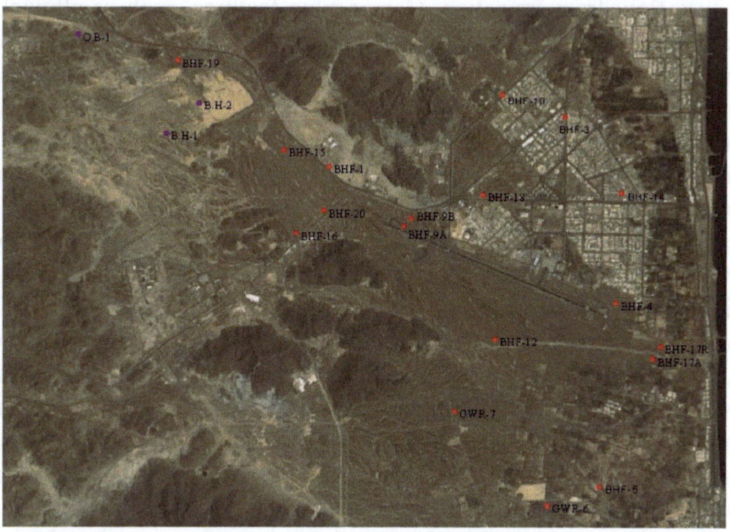

Figure 1. RS image of Wadi Ham in the UAE (after Sherif et al., 2004). (Colour reproduction on Plate 1)

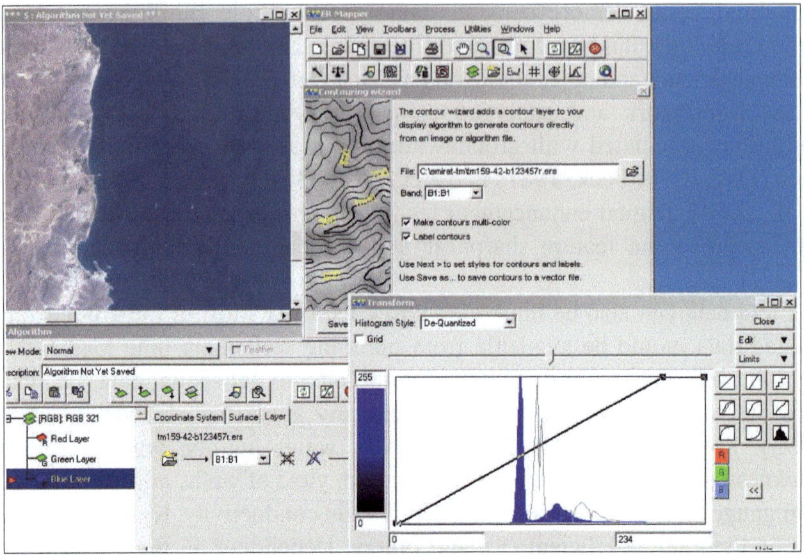

Figure 2. Image processing activities to improve feature sharpness and contrasts. (Colour reproduction on Plate 1)

HYDROLOGICAL APPLICATIONS AND ASSOCIATE VARIABLES

According to Chow (1986), aquifer recharge occurs in nature by rainfall, seepage from canals and reservoirs, lagoons, forested lands, irrigated fields, and return flow from irrigation. Geomorphic features, such as alluvial fans buried pediments, old stream channels and the deep-seated interconnected fractures, are indicators of subsurface water accumulation. These features represent natural recharge sites due to their high permeability and water holding capacity. Howari et al. (2002) and Howari (2003) studied soil permeability using remote sensing techniques. Moreover, it is clear that the higher the permeability the lower the drainage density and the higher the drainage density the higher the surface water runoff. Identification of lineaments has an immense importance in hard rock hydrogeology as they can identify rock fractures that localize groundwater (Das, 1990). Hydrogeologists usually infer subsurface hydrological conditions through surface indicators, such as aerial geological features, linear structures and others. Most of the geological linear features are assumed to be the zone of fractured bedrocks and the position of porous and permeable state where enhanced well yields can be expected (Das, 1997). The advent of increasing computing power and GIS technique, physically-based hydrologic modeling has become important in contemporary hydrology for assessing the impact of human intervention and/or possible climatic change on basin hydrology and water resources (Lanza et al., 1997; Sui and Maggio, 1999; Schmugge et al., 2000; Su Z.B. Troch P.A., 2003). The use of distributed physically based (conceptual) hydrological modeling with prudent simplification is appropriate for providing a reasonable solution to large scale hydrological problems associated with planning and optimal allocation of resources.

HYDROLOGICAL MODELING AND GIS

GIS has influenced the development and implementation of hydrologic models at several different levels, and it provides representations of these spatial features of the Earth, while hydrologic modeling is concerned with the flow of water and its constituents over the land surface and in the subsurface environment. Various hydrological modeling techniques have enabled GIS users to go beyond the data inventory and management stage to conduct sophisticated modeling and simulation. For hydrological modeling, GIS, especially through their powerful capabilities to process Digital Elevation Models (DEM), have provided modelers with new platforms for data management and visualization. The rapid diffusion of GIS has the potential to make various hydrological models more powerful, accurate, and transparent and can enable the communication of their operations and results to a larger group of users. The growing literature on the integration of GIS with hydrological modeling attests to the recognition of such benefits (DeVantier & Feldman, 1993; Maidment, 1993; McDonnell, 1996; Moore, 1996).

GIS researchers and hydrologists have recognized that lack of sophisticated analytical and modeling capabilities is one of the major deficiencies of GIS technology (Maidment, 1993). Several research initiatives worldwide have focused on the improvement of spatial analytical and modeling capabilities of GIS technology during the past 10 years (Goodchild et al., 1992). Integration of GIS with hydrological modeling was part of these broad research efforts to link spatial analysis and modeling with GIS. Although overlapping with many other GIS modeling efforts in terms of general methodology (e.g., Goodchild et al., 1993), integration of GIS with hydrological modeling has a set of different issues from other kinds of GIS-based environmental modeling (Goodchild, Parks & Steyaert, 1993). For example, unlike most other kinds of environmental modeling, hydrological modeling has a set of well-established practices and standards widely accepted by hydrologists and hydraulic engineers, and modeling results are sometimes used for regulatory purposes. Current practices of integrating GIS with hydrological modeling thus deserve a separate scrutiny (Lanza et al., 1997; Sui and Maggio, 1999; Schmugge et al., 2000; Su Z. B Troch P. A., 2003).

Improving Hydrological Modeling by Including GIS Functionalities

According to Sui and Maggio (1999), embedding GIS functionalities in hydrological modeling packages has been adopted primarily by hydrological modelers who think of GIS essentially as a mapping tool and conceptually irrelevant to the fundamentals of hydrological modeling. This approach usually gives system developers maximum freedom for system design. Implementation is not constrained by any existing GIS data structures, and usually this approach is capable of incorporating the latest development in hydrological modeling (Brimicombe and Bartlett, 1996; Clark, 1996). The downside of this approach is that the data management and visualization capabilities of these hydrological modeling software packages are, in no way, comparable to those available in commercial GIS software packages, and programming efforts also tend to be intensive and sometimes redundant. The developers of the latest version of River-CAD, HEC-HMS 2.0, River-Tools, MODFLOW and SUTRA have basically taken this approach (Djokic et al., 1995).

On the other hand, a few leading GIS software vendors in recent years have made extra efforts to improve the analytical and modeling capabilities of their products. Pioneered by HEC-SAS developed by the Army Corps of Engineers (Davis, 1978), several commercial software vendors have developed stand-alone GIS modules with functions that can be used for a variety of hydrological modeling needs. Certain hydrological modeling functions have been embedded in leading generic GIS software packages, such as ESRI's ArcStorm and ArcGrid, Integraph's InRoads, etc. This approach builds on the top of a commercial GIS software package and takes full advantage of built-in GIS functionalities, but the modeling capabilities are usually simplistic and calibrations must take place outside of the package. Also, these models tend not to be industry standard and/or have not been validated (Shamsi and Fletcher, 1994; Shuman, 1993).

Existing Problems

GIS have provided hydrologists and hydraulic engineers with the ideal computing platform for data inventory, parameter estimation, mapping and visualizing results for hydrological/hydraulic modeling, thus greatly facilitating the design, calibration, and implementation of various hydrological/hydraulic models (Lanza et al., 1997; Sui and Maggio, 1999; Schmugge et al., 2000; Su Z. B Troch P. A., 2003). However, we should not let the fancy maps and graphics of GIS blind us from the real issues in hydrological modeling. The current practices are dominated by technical concerns without adequately addressing some of the important conceptual issues involved in the integration of GIS with hydrological modeling. While the technical problems related to the database integration are well documented (Adam & Gangopadhyay, 1997; Buogo & Chevallier, 1995), only a few papers in the literature have discussed the broad conceptual issues involved in the integration of GIS with hydrological modeling. Several problems in both hydrological models and the current generation of GIS are usually observed. These problems must be addressed before we can make the integration of GIS with hydrological modeling theoretically consistent, scientifically rigorous, and technologically interoperable (Lanza et al., 1997; Sui and Maggio, 1999; Schmugge et al., 2000; Su Z. B Troch P. A., 2003).

Inherited Problems

According to Chow, Maidment and Mays (1988), hydrological models can be classified according to their conceptualization and assumptions of three key parameters—randomness, space and time. The models widely used in the current practices of GIS-based hydrological modeling are dominated by deterministic lumped models because of the availability of various modeling packages, such as the US Army Corps of Engineers HEC-HMS and HEC-RAS, the US Soil and Conservation Service's TR-20 and TR-40, USDA's SWAT, DoT's WSPRO, EPA's WASP and BASINS, and USGS's DRM3 and PRMS, etc. (Singh and Frevert, 2002a, b). Several researchers have challenged the foundations of these models (Grayson, Moore & McMahon, 1992; Smith & Goodrich, 1996) and many researchers have been active to develop spatially distributed and stochastic models (Beven & Moore, 1992; Romanowicz, Beven & Moore, 1993; Wheater, Jakeman & Beven, 1995). However, these newly developed models are still mostly confined to research laboratories and not widely used in practice. Obviously, our primary objective should not only focus on doing the thing right in the technical sense but also, perhaps more importantly, challenge whether it is the right thing to do in a scientific sense.

Computational Problem

With its historical roots in computer cartography and digital image processing, the development of GIS to date has relied upon a limited map metaphor (Burrough & Frank, 1995). The majority of GIS datasets are currently represented in vector format, which is convenient due to storage efficiency but can be difficult to manipulate analytically. To achieve tractability, vector format GIS data is often rasterized (e.g., conformed to a grid-cell representation). The processes involved in vectorization of map data as well as rasterization of vector data have indiosyncratic errors that manifest as representational error in a given GIS system. Consequently, the representation schemes and analytical functionalities in GIS are geared toward map layers and geometric transformations. The layer approach implicitly forces a segmentation of geographic features (Raper and Livingstone, 1995). This representation scheme is not only temporally fixed but is also incapable of handling overlapping features (Hazelton, Leahy & Williamson, 1992). The absolute conceptualization of space has forced space into a geometrically indexed representation scheme via planar enforcement, and time is conceptualized as discrete slices. In contrast, depending on how randomness is handled, space and time in hydrological models can be conceptualized quite differently. Although technically we can plug in various hydrological models into GIS through the strategies outlined in the previous section, GIS and hydrological models are just used together, not really integrated because of the fundamentally different spatial data representation schemes involved

(Abel, Kilby & Davis, 1994). Therefore, in order to accomplish the seamless integration of GIS and hydrological models, more research is needed at a higher level to develop and incorporate novel approaches to conceptualizing space and time that is interoperable both within GIS and hydrological models (Lanza et al., 1997; Sui and Maggio, 1999; Schmugge et al., 2000; Su Z. B Troch P. A., 2003).

Obviously, the current practices of integrating GIS and hydrological modeling are essentially technical in nature and have not touched upon the more fundamental issues in either hydrological models or GIS. Simply being able to run a HEC-HMS or HEC-RAS model in Arc/Info or a CAD system improves neither the theoretical foundation nor the performance of the model. GIS-based hydrological modeling has resulted in a tremendous amount of representational compromise (Gan, Dlamini & Biftu, 1997). Such problems call for a fresh look at the integration of GIS with hydrological modeling.

Uncertainties of GIS as Related to Hydrological Investigations

In spite of well-known quality issues in spatial data handling, most applied modeling in practice neglects error and uncertainty. If we succeed in developing GIScience-based hydrological modeling interoperable across computing platforms, hydrological modeling will be more widely accessed and used to address water-related issues of great societal concerns. Robust hydrological modeling will be an integral part of management tools for sustainable watershed development. The interoperable GIScience-based hydrological model will also greatly empower citizens and communities in flood-prone areas to play an important role in flood control, flood mitigation, floodplain mapping, and flood insurance studies. However, every single step in the process of integrating GIS with hydrological modeling is full of uncertainties, ranging from data acquisition to model calibration to results visualization. Currently, we still lack effective methods to handle and communicate these uncertainties (Heuvelink, 1998; Tung, 1996; Kundzewicz, 1995). Second, new methods of communicating results of uncertainty measures need to be developed. Both visual and non-visual methods of communicating uncertainties are worth exploring (Lanza et al., 1997; Sui and Maggio, 1999; Schmugge et al., 2000; Su & Troch, 2003). It has long been recognized by the GIS community that the ability to communicate to users the uncertainty of the digital databases is a critical step in maintaining professional integrity in applications of great societal concerns, such as floodplain mapping, flood insurance rating, and watershed development planning (Clark, 1996; Hwang, Karimi & Byun, 1998; Klinkenberg & Joy, 1994). Although there is no general agreement or professional protocols, the methods of communicating uncertainties of spatial databases reviewed by Hunter and Goodchild (1996) can be applied to convey the uncertainties of integrating GIS with hydrological modeling.

REMOTE SENSING AS A SOURCE OF INFORMATION TO GIS: CAPABILITIES AND LIMITATIONS

Remote sensing is the process of inferring surface parameters from measurements of the upwelling electromagnetic radiation from the land surface. This radiation is both reflected and emitted by the land. The former is usually the reflected solar radiation while the latter is in both the thermal infrared (TIR) and microwave portions of the spectrum (Jensen, 1986; Sabins, 2002). There is also reflected microwave radiation as in imaging radars. The reflected solar radiation is used in hydrology for snow mapping vegetation/ land cover and water quality studies. Active microwave or radar has promise because of the possibility of high spatial resolution. However, surface roughness effects can make it difficult to extract soil moisture information. Remotely sensed observations can contribute to our knowledge of these quantities and, especially, their spatial variation. With remote sensing we not only observe the surface but can also obtain the spatial variability and if observations are made repeatedly the temporal variability can be obtained. In this section we will concentrate on those applications of remote sensing which are most promising in hydrology (e.g., Shultze, 1988).

Major focus of remote sensing research in hydrology has been to develop approaches for estimating hydrometeorological states and fluxes. The primary set of state variables include land surface temperature, near-surface soil moisture, snow cover/water equivalent, water quality, landscape roughness, land use and vegetation cover. The hydrometeorological fluxes are primarily soil evaporation and plant transpiration or evapotranspiration, and snowmelt runoff. We will describe methods which have been used to quantify the components of the water and energy balance equation using remote sensing methods as described by Sui and Maggio, 1999; Schmugge et al., 2000 and Su and Troch, 2003. The water balance is commonly expressed as:

$$\frac{\Delta S}{\Delta t} = P - ET - Q \qquad (1)$$

where $\Delta S/\Delta t$ is the change in storage in the soil and/or snow layer, P is the precipitation, ET is the evapo-transpiration, and Q is the runoff. Because the energy and water balance at the land surface are closely linked we also need to consider the energy balance equation which is typically written as:

$$R_N - G = H + LE \qquad (2)$$

where R_N is the net radiation, G is the soil heat flux, H is the sensible heat flux and LE is the latent heat flux, all in W m^{-2}. The quantity $R_N - G$ is commonly referred to as the available energy, and ET and LE represent the same water vapour exchange rate across the surface–atmosphere interface, except that ET is usually expressed in terms of the depth of water over daily and longer time scales, namely mm/day.

REMOTE SENSING TECHNIQUES TO ASSESS WATER QUALITY

Water quality is a general descriptor of water properties in terms of physical, chemical, thermal, and/or biological characteristics. In-situ measurements and collection of water samples for subsequent laboratory analyses provide accurate measurements for a point in time and space but do not give either the spatial or temporal view of water quality needed for accurate assessment or management of water bodies. Substances in surface water can significantly change the backscattering characteristics of surface water (Forster and Wittman, 1983). Remote sensing techniques for monitoring water quality depend on the ability to measure these changes in the spectral signature backscattered from water and relate these measured changes by empirical or analytical models to water quality parameters. The optimal wavelength used to measure a water quality parameter is dependent on the substance being measured, its concentration, and the sensor characteristics (Howari et al., 2002).

Major factors affecting water quality in water bodies across the landscape are suspended sediments (turbidity), algae (i.e., chlorophylls, carotenoids), chemicals (i.e., nutrients, pesticides, metals), dissolved organic matter (DOM), thermal releases, aquatic vascular plants, pathogens, and oils (Sui and Maggio, 1999; Schmugge et al., 2000; Su and Troch, 2003). Suspended sediments, algae, DOM, oils, aquatic vascular plants, and thermal releases change the energy spectra of reflected solar and/or emitting thermal radiation from surface waters, which can be measured by remote sensing techniques. Most chemicals and pathogens do not directly affect or change the spectral or thermal properties of surface water; they can therefore only be inferred indirectly from the measurements of other water quality parameters affected by these chemicals (Shultze, 1988).

Empirical or analytical relationships between spectral properties and water quality parameters are established. The general forms of these empirical equations are:

$$Y = AX^{B} \tag{3}$$

where Y is the remote sensing measurement (i.e., radiance, reflectance, energy), X is the water quality parameter of interest (i.e., suspended sediment, chlorophyll), and A and B are empirically derived factors. In empirical approaches statistical relationships are determined between measured spectral/thermal properties and measured water quality parameters. Often information about the spectral/optical characteristic of the water quality parameter is used to aid in the selection of best wave length(s) or best model in this empirical approach. The empirical characteristics of these relationships limit their application to the condition for which the data were collected.

Suspended sediments are the most common pollutant both in weight and volume in surface waters of freshwater systems. Significant relationships between suspended sediments and radiance or reflectance from spectral wavebands or combinations of wavebands on satellite and aircraft sensors have been shown. Ritchie and Cooper (1991) and Ritchie et al. (2000) showed that an algorithm for relating remotely sensed data to the sediment load developed for one year was applicable for several years.

Direct measurement of rainfall from satellites for operational purposes has not been generally feasible because the presence of clouds prevents observation of the precipitation directly with visible, near infrared and thermal infrared sensors (Sui and Maggio, 1999; Howari et al., 2002; Schmugge et al., 2000; Su and Troch, 2003). However, improved analysis of rainfall can be achieved by combining satellite and conventional gauge data. Useful data can be derived from satellites used primarily for meteorological purposes, including polar orbiters, such as the U.S. National Oceanographic and Atmospheric Administration (NOAA) series and the Defense Meteorological Satellite Program (DMSP), and from geostationary satellites, such as Global Operational Environmental Satellite (GOES), Geosynchronous Meteorological Satellite (GMS) and Meteosat, but their visible and infrared images can provide information only about the cloud tops rather than cloud bases or interiors. However, since these satellites provide frequent observations (even at night with thermal sensors), the characteristics of potentially precipitating clouds and the rates of changes in the cloud area and shape can be observed. From these observations, estimates of rainfall can be made which relate cloud characteristics to instantaneous rainfall rates and cumulative rainfall over time. For the practicing hydrologist, satellite rainfall methods are most valuable when there are no or very few surface gauges for measuring rainfall (Sui and Maggio, 1999; Schmugge et al., 2000; Su and Troch, 2003).

There are, of course, many types of satellite in operation. Here, we will look at some important satellites that provide a background to the dynamical data produced by the meteorological satellites of hydrological applications; LANDSAT, SPOT and ERS are known as the 'Earth Resources' satellites. The availability of meteorological and Landsat satellite data has produced a number of techniques for inferring precipitation from the visible and/or infrared (VIS/IR) imagery of clouds. The GOES Precipitation Index (Arkin, 1979) derived from thresholding the infrared brightness temperature of cloud tops has been used to study the distribution of tropical rainfall. The University of Bristol (Barrett and Martin, 1981; D'Souza and Barrett, 1988) has led the development of a cloud indexing (Moses and Barrett, 1986) approach and a thresholding (Barrett et al., 1988) approach. A life-history approach, developed by Scofield and Oliver (1977), considers the rates of change in individual convective clouds or clusters of convective clouds. This approach is the basis of a flash flood system that assimilates GOES data with ground-based and atmospheric data to forecast precipitation amounts for use in a flood forecast model (Clark and Morris, 1986).

Evapotranspiration: In general, remote-sensing techniques cannot measure evaporation or evapotranspiration (ET) directly. However, remote sensing does have two potentially very important roles in estimating evapotranspiration. First, remotely sensed measurements offer methods for extending point measurements or empirical relationships, such as the Thornthwaite (1948), Penman (1948) and Jensen-Haise (1963) methods, to much larger areas, including those areas where measured meteorological data may be sparse. Secondly, remotely sensed measurements may be used to measure variables in the energy and moisture balance models of ET. Although, there has been progress made in the direct remote sensing of the atmospheric parameters which affect ET, such as the Raman Light Detection And Ranging (LIDAR). This is essentially a ground-based, point measurement and will not be covered in this report. A common approach in estimating *ET*, as described by Schmugge et al., 2000; Sui and Maggio, 1999 and Su and Troch, 2003, is to solve for the latent heat flux, *LE*, as a residual in the energy balance equation for the land surface, namely,

$$LE = R_N - G - H \tag{4}$$

Remote sensing methods for estimating these components are described in various places in literature. Typically, with reliable estimates of solar radiation, differences between remote sensing estimates and observed $R_N - G$ are within 10%. The largest uncertainty in estimating *LE* comes from computing *H*.

Additional approaches for estimating *ET* from remote sensing data are being explored. Barton (1978) and Davies and Allen (1973) have modified this formula for evaporation from an unsaturated land surface by the surface layer soil moisture. Barton used air-borne microwave radiometers to sense soil moisture remotely in his study of evaporation from bare soils and grasslands. Soares et al (1988) demonstrated how thermal infrared and C-band radar could be used to estimate bare soil evaporation. Choudhury et al., (1994) have shown strong relationships between evaporation coefficients and vegetative indices. Another approach being pursued is the development of numerical models that simulate the heat and water in the soil and drive it with the energy balance at the surface (Camillo et al., 1983, Taconet et al., 1986). Taconet and Vidal-Madjar (1988) have used this approach with Advanced Very High Resolution Radiometer (AVHRR) and Meteosat data.

Radar remote sensing has been used as a potential tool for soil moisture measurement since the pioneering work in the 1970s and 1980s (Ulaby, 1974; Ulaby et al., 1978; Ulaby et al., 1986; Toll, 1985; Schmugge, 1983). Radar is sensitive to variations in soil moisture because of the changes in the dielectric properties of soil produced by changes in water content. The sensitivity of radar to variations in moisture content potentially results in contrasting backscatter. The experimental bases of the dielectric properties of natural materials are provided by Ulaby et al. (1986) and Soulis et al. (1998). Figure 3 shows the relationship between the dielectric constant of a typical soil and its volumetric moisture content. As moisture content increases, the real component of the dielectric constant increases and on reaching a transition value of moisture, it increases more rapidly (Ulbay et al., 1986; Soulis et al., 1998; Tansey et al., 1999). Tansey et al. (1999) indicated that for particular volumetric soil moisture, the real component of the dielectric constant is greater for sand-rich soils than for clay-rich soils. This is due to the greater availability of free water that can interact with the radar signal in sand-textured soils. Consequently, the strength of the signal returned to the satellite or the backscatter is less for a dry soil than for a wet soil.

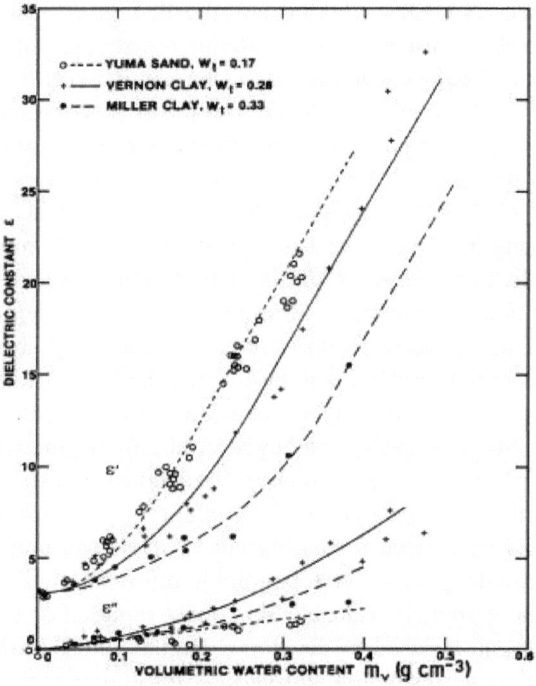

Figure 3. The complex dielectric constant, comprising real (') and imaginary (") parts, shown as a function of volumetric moisture content for three soil types at 5 GHz. Also shown are the approximate transition volumetric soil moisture values (W_t) discussed in the text (after Wang & Schmugge, 1980).

Generally, field soil moisture measurements are used to calibrate and establish a relationship between in-situ and remote sensing estimates. Relating point measurements with radar responses over an aerially averaged pixel scale can be difficult. Glenn and James (2003) presented a geostatistical method to tackle this problem which is described below. Van Oevelen (1998) was able to downscale high-resolution aerial soil moisture estimates to low-resolution soil moisture estimates from the HAPEX-Sahel 1992 (Goutorbe et al., 1994). Other studies incorporating geostatistics in radar remote sensing of soil moisture include the works of Dempsey et al. (1998a, b). These studies use kriging and principal component analyses to discriminate variations in soil texture and soil moisture over time and space (Glenn and James, 2003).

Other previous studies on synthetic aperture radar (SAR) imagery have shown qualitative relationships between radar backscatter and soil moisture. Following the launch of the European Space Agency (ESA). European Remote Sensing Satellite (ERS-1) in 1991 and the Japanese Earth Resources Satellite (JERS-1) (1992), there has been a multitude of SAR sensors in space during the 1990s including the Shuttle Imaging Radar (SIR-C) (1994), Radarsat (1995), ESA ERS-2 (1995) and the Radarsat 2 launched in 2004. However, to be able to use airborne or satellite based radar data in operational programmes it will be necessary to establish quantitatively how the returned signal of radar is related to soil moisture and the effects of surface roughness, soil type, and vegetation cover and growth stage, as a function of frequency and polarization.

Runoff and Hydrologic Modeling: Remote sensing data can be used to obtain almost any information that is typically obtained from maps or aerial photography. In many regions of the world, remotely sensed data, and particularly Landsat, Thematic Mapper (TM) or Systeme Probatoire, d'Observation de la Terre (SPOT) data, are the only sources of good cartographic information. Drainage basin areas and the stream network are easily obtained from good imagery, even in remote regions (Lanza et al. 1997; Sui and Maggio, 1999; Schmugge et al., 2000; Su and Troch P. A., 2003). There have also been a number of studies to extract quantitative geomorphic information from Landsat imagery (Haralick et al., 1985). Topography is a basic need for hydrologic analysis and modeling. Remote sensing can provide quantitative topographic information of suitable spatial resolution to be extremely valuable for model inputs. For example, stereo SPOT imagery can be used to develop a Digital Elevation Model (DEM) with 10 m horizontal resolution and vertical resolution approaching 5 m in ideal cases (Case, 1989). A new technology using interferometric SAR has been used to demonstrate similar horizontal resolutions with approximately 2 m vertical resolution (Zebker et al., 1992).

One of the first applications of remote sensing data in hydrologic models used Landsat data to determine both urban and rural land use for estimating runoff coefficients (Jackson et al., 1976). Land use is an important characteristic of the runoff process that affects infiltration, erosion, and evapo-transpiration. Distributed models, in particular, need specific data on land use and its location within the basin. Most of the work on adapting remote sensing to hydrologic modeling has involved the Soil Conservation Service (SCS) runoff curve number model (U.S. Department of Agriculture, 1972) for which remote sensing data are used as a substitute for land cover maps obtained by conventional means (Jackson et al., 1977, Bondelid et al., 1982). In remote sensing applications, one seldom duplicates detailed land use statistics exactly. For example, a study by the Corps of Engineers (Rango et al., 1983) estimated that an individual pixel may be incorrectly classified about one-third of the time. However, by aggregating land use over a significant area, the misclassification of land use can be reduced to about two percent, which is too small to affect the runoff coefficient or the resulting flood statistics. Investigations have shown (Jackson et al., 1977) that for planning studies the Landsat approach is cost effective. The authors estimated that the cost benefits were on the order of 2.5 to 1 and can be as high as 6 to 1 in favour of the Landsat approach. These benefits increase for larger basins or for multiple basins in the same general hydrological area. Mettel et al. (1994) demonstrated that the recomputation of pmf's for the Au Sable River using Hydrologic Engineering Center (HEC-HMS) and updated and detailed land use data from Landsat TM resulted in 90% cost cuts in upgrading dams and spillways in the basin.

Other types of runoff models that are not based only on land use are beginning to be developed. For example, Strubing and Schultz (1983) have developed a runoff regression model that is based on Barrett's (1970) indexing technique. The cloud area and temperature are the satellite variables used to develop a temperature weighted cloud cover index. This index is then transformed linearly to mean monthly runoff. Rott (1986) also developed a daily runoff model using Meteosat data for a cloud index. Recently, Papadaakis et al. (1993) have used a cloud cover index from satellite imagery to estimate monthly area precipitation. A series of non-linear reservoirs then transforms the precipitation into monthly runoff values. This approach was successfully demonstrated for the large (16,000 sq. km) Tano River Basin in Africa and illustrates the value of remote sensing data when conventional data are not readily available. Ottle et al. (1989) have shown how satellite derived surface temperature can be used to estimate ET and soil moisture in a model that has been modified to use these data. Duchon et al. (1992) have used Landsat to identify uniform land cover areas and GOES data for input insulation for a monthly water balance model. RADARSAT-1 is sensitive to median soil moisture levels, returning low brightness values for low moisture contents and high brightness values for high moisture contents (Fig. 4).

APPLICATIONS OF GIS AND RS IN GROUNDWATER STUDIES

Remote sensing and GIS techniques are regarded as powerful tools for presentation and analysis of data and information. RS images provide important information related to the soil type, land use, orientation and intensity of fractures, identification of recharge and discharge areas, outcropping areas, coordinates of pumping and observation wells, and others. Depending on the soil texture, radar images can also provide useful information about the subsurface formation. Digital elevation models can be used to acquire the topography of the land surface from RS images. Topographic (contour) maps can thus be developed and exported directly as shape files or others to groundwater simulation packages. On the other

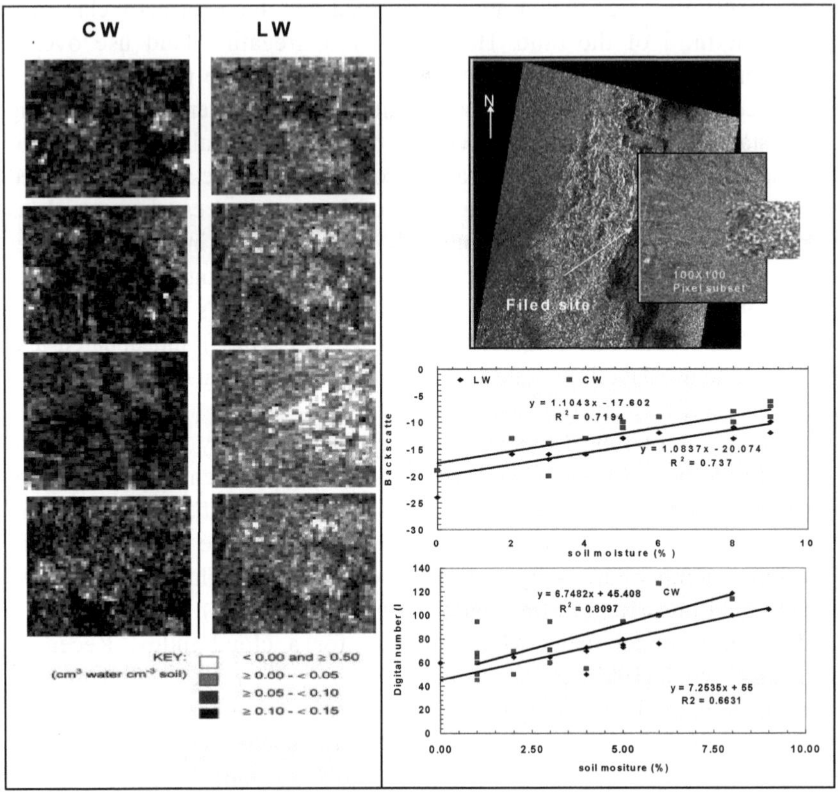

Figure 4. Relationship between soil moisture, backscattered coefficients and digital numbers in Fujairah, United Arab Emirates (after Howari, 2004)

hand, all data including the RS images can be organized and stored in different layers using ARC-GIS or any other GIS system. Contour maps for groundwater levels, aquifer bottom, thicknesses of the different layers, hydraulic parameters (porosity, permeability, transmissivity, storativity, and others), water quality, and other types of analyses can be done using GIS. Such information can also be exported directly as input files for groundwater simulation packages. An example for the application of GIS and RS in groundwater studies is elaborated hereafter.

Sherif et al. (2004), in collaboration with the Ministry of Agriculture and Fisheries, UAE, conducted a project to assess the groundwater recharge from three recharge dams in Wadi Bih, Wadi Tawiyean, and Wadi Ham, UAE. Figure 5 presents a remote sensing image for the location of these three wadis. Rain gauge stations are also located on the images. The sub-basins, boundaries and the order of all tributaries were identified and delineated using ARC-View. The total catchment area and the areas of all sub-basins were calculated by the same GIS system. All other information including the length and average slope of streams, soil type, and land cover/ land use were deduced from the RS image. Vegetation areas in the three wadis were also calculated.

This information was introduced to the HEC-HMS and simulation runs were conducted for calibration. A comparison between the simulated and observed water storage values of the three dams was made. The difference between the observed and simulated storage values was on the order of 5% for most events. For some small events, the difference was on the order of 20%. A sensitivity analysis was also performed to reduce uncertainties. The model was then used to assess the surface water runoff and water storage that would result from rainfall events of different intensities and different durations in the three wadis.

Figure 6 presents a RS image for the selected study domain of groundwater flow in the area of Wadi Ham. The study area extends from the coastline of the Gulf of Oman in the east to the upper catchment area of the dam of Wadi Ham in the west. Within the study domain, two ophiolite outcrops (mostly impermeable layers) are observed and are regarded as inactive or no-flow zones. The dam of Wadi Ham and the ponding area are also marked on the RS image, as shown in Fig. 6. Two well fields, Saharaah and Khalba, are also located within the study area and are regarded as discharge zones. The exact locations of existing observation wells were determined in the field using a GPS and were introduced to the RS image. The pumping schedule and rates were provided for the purpose of this study. Several boreholes and wells were drilled and pumping tests were conducted to identify the hydrogeological parameters of the aquifer system in the area. In addition, intensive geophysical investigations were conducted and several geological profiles were outlined. Figure 7 presents a typical geological cross section along the main course of Wadi Ham.

A constant head boundary condition was assumed at the coastline of the Gulf of Oman. A specified flux (inflow boundary) was also considered wherever a base flow was expected. Otherwise, hard rocks surround the domain and the boundaries were considered impermeable.

The RS image and all other information were introduced to ARC-View. The main advantage of the establishment of such a GIS database is that all the information and data were geo-referenced. In addition, important data that are needed for any groundwater model could be directly extracted or evaluated from the GIS database. For example, the total area of the ophiolite outcropping (inactive impermeable zones) within the selected domain was calculated from the ARC-View tools as 6.65 km^2. Likewise, the ponding area at the flooding level was estimated as 0.4 km^2. In other words, the statistical, contouring, mapping, tabulating and all other powerful tools of GIS systems can be easily and efficiently integrated into groundwater simulation models, especially at the stage of preparation of input data (pre-processing).

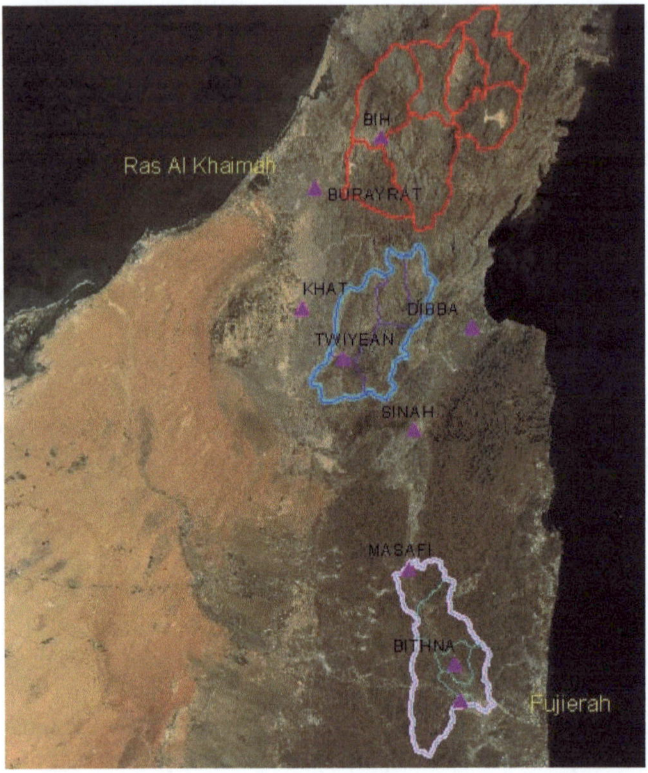

Figure 5. Remote sensing image for three locations in Wadi along with rain gauge stations. (Colour reproduction on Plate 2)

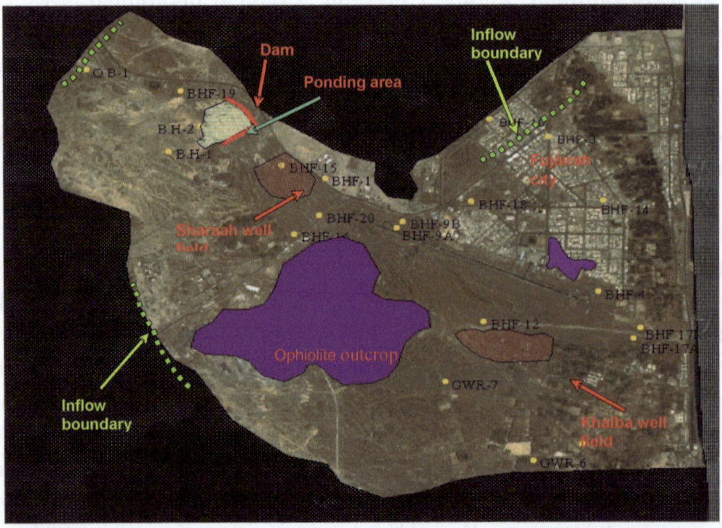

Figure 6. RS image including the main features of the study area in Wadi Ham, UAE (Sherif et al., 2004). (Colour reproduction on Plate 2)

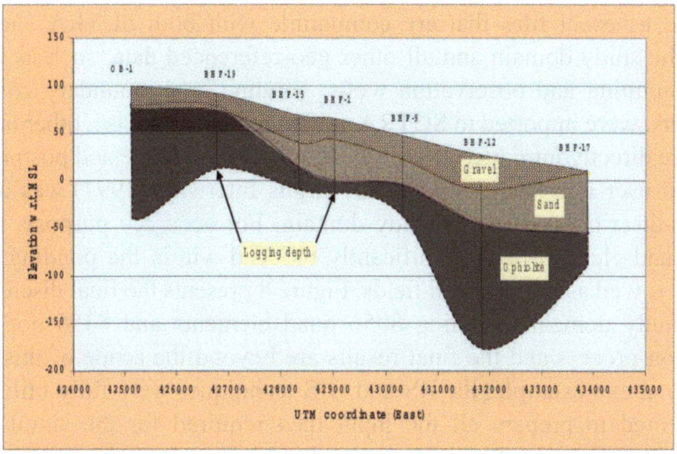

Figure 7. Typical geological cross-section along the course of Wadi Ham (Sherif et al., 2004). (Colour reproduction on Plate 2)

Sherif et al. (2004) employed the USGS model "SUTRA, a model for saturated-unsaturated variable-density ground-water flow with solute or energy transport (Voss, 1984; Voss and Provost, 2002)" to evaluate the recharge from the dam of Wadi Ham. SUTRA is a computer programme that simulates the groundwater flow in porous media and the transport of either energy or dissolved substances in a subsurface environment. The code employs a two or three dimensional finite element and finite difference method to approximate the governing equations that describe the two interdependent processes that are simulated:

1. Fluid density dependent saturated or unsaturated groundwater flow
2. Either (a) transport of a solute in the groundwater, or (b) transport of thermal energy in the groundwater

SUTRA-GUI (Winston and Voss, 2003) may be employed in one, two or three–dimensional simulation problems. Flow and transport simulation may be either steady state which requires a single solution step, or transient, which requires a series of time steps in the numerical solution. SUTRA may be employed for two-dimensional areal, cross sectional, and fully three-dimensional modeling of saturated groundwater flow systems and unsaturated zone flow. Although the simulation was limited to the groundwater flow, SUTRA-GUI, a flow and solute/energy transport model, was selected because of the vicinity of the study area from the Gulf of Oman. Simulation runs were conducted using the calibrated flow model to study the seawater intrusion problem upon availability of proper data.

All the required input data files were prepared as layers of information using ARC-View. In addition, stress periods from rainfall events and water storage in the ponding area as well as the pumping rates and schedules were tabulated as excel files that are compatible with both SUTRA and ARC-View. The study domain and all other geo-referenced data, such as inactive zones, pumping and observation wells, ponding area, boundary conditions and others, were imported to SUTRA as shape files. Likewise, other tabulated data were directly introduced to SUTRA. Argus-One (a pre and postprocessor graphical user interface environment, Argus Interware, 1997) was used for the auto-discritization of the study domain. For accuracy purpose, the size of the quad elements was significantly reduced within the ponding area of the dam as well as the two well fields. Figure 8 presents the final discretization of the study domain including 8056 quad elements and 8356 nodes. The simulation process and the final results are beyond the scope of this article.

In the above example, the RS and GIS techniques were fully utilized and implemented to prepare all the input data required for the simulation of groundwater flow in Wadi Ham. It should be noted the stage of data preparation for groundwater simulation studies is not only difficult and time consuming but also represents a major source of errors and uncertainties, especially when dealing with geo-referenced data of a huge size. The employment of available GIS systems in the organization and preparation of data for groundwater models would certainly simplify the process, minimize the time needed for preparation of input files and increase the assurance of data accuracy. The outputs can also be presented in an informative and clear manner.

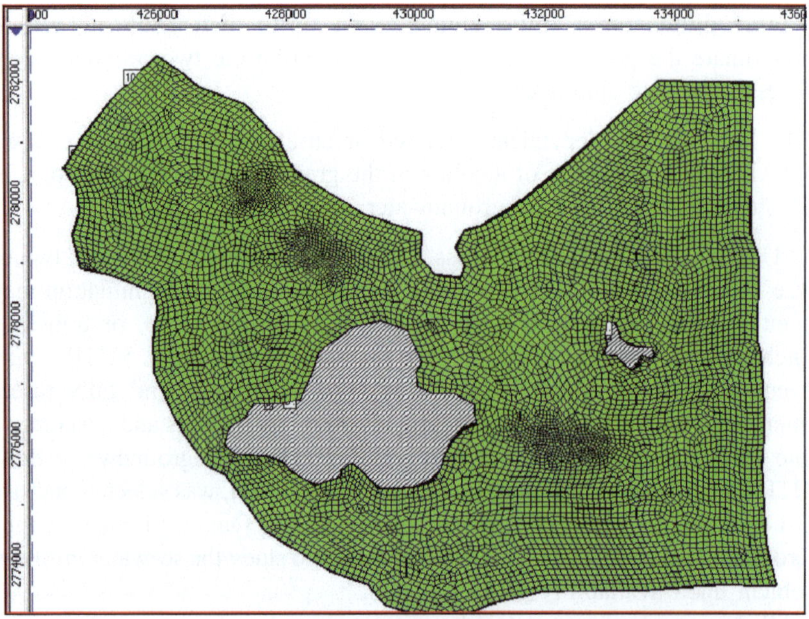

Figure 8. Initial discritization of the study domain in Wadi Ham using Argus-One.

REFERENCES

Albertson, J.D., Kustas, W.P. and Scanlon, T.M. (2001). Large eddy simulation over heterogeneous terrain with remotely sensed land surface conditions. *Water Resour. Res*. **377**: 1939-1953.

Allord, G.J. and Scarpace, F.L. (1979). Improving Streamflow Estimates through Use of Landsat. *In:* Satellite Hydrology, 5th Annual William T. Pecora Memorial Symposium on Remote Sensing: Sioux Falls, SD. pp. 284-291.

Argus Interware (1997). Argus Numerical Environment and MeshMaker User's Guide, 510 p.

Barrett, E.C. and Kidd, C. (1987). The use of SMMR data in support of a VIR/IR satellite rainfall monitoring technique in highly-contrasting climatic environments. *In:* Passive Microwave Observing from Environmental Satellites, a Status Report (ed. J.C. Fischer), NOAA Tech. Rep. NESDIS 35, Washington DC, pp. 109-123.

Barrett, E.C. and Curtis, L.F. (1982). Introduction to Environmental Remote Sensing. Chapman and Hall, London.

Barrett, E.C. and Martin, D.W. (1981). The Use of Satellite Data in Rainfall Monitoring. Academic Press, London, 340 pp.

Barrett, E.C. (1970). The estimation of monthly rainfall from satellite data. *Mon. Weather. Rev.*, **98:** 322-327.

Barton, I.J. (1978). A case study comparison of microwave radiometer measurements over bare and vegetated surfaces. *J. Geophy. Res.*, **83:** 3513-3517.

Beven, K.J. and Moore, I.D. (Eds.) (1992). Terrain analysis and distributed modelling in hydrology. Chichester, UK: Wiley & Sons.

Bhaskar, N.R., Wesley, P.J. and Devulapalli, R.S. (1992). Hydrologic parameter estimation using geographic information systems. *Journal of Water Resources Planning and Management*, **118:** 492-512.

Blyth, K. (1993). The use of microwave remote sensing to improve spatial parameterization of hydrological models. *J. of Hydrology*, **152:** 103-129.

Brimicombe, A.J. and Bartlett, J.M. (1996). Linking GIS with hydraulic modeling for flood risk assessment: The Hong Kong approach. *In:* M.F. Goodchild, L.T. Steyaert and B.O. Parks (Eds.), GIS and environmental modeling: Progress and research issues (pp. 165-168). Fort Collins Co: GIS World Books.

Camillo, P.J., Gurney, R.J. and Schmugge, T.J. (1983). A soil and atmospheric boundary layer model for evapotranspiration and soil moisture studies. *Water Resources Research*, **19:** 371-380.

Carlson, T.N. and Buffum, M.J. (1989). On estimating total daily evapotranspiration from remote surface measurements. *Remote Sens. Environ.*, **29:** 197-207.

Carroll, S.S. and Carroll, T.R. (1989). Effect of forest biomass on airborne snow water equivalent estimates obtained by measuring terrestrial gamma radiation. *Remote Sensing Environment*, **7:** 313-320.

Carroll, T.R. and Vadnais, K.G. (1980). Operational airborne measurement of snow water equivalent using natural terrestrial gamma radiation. *In:* Proc. 48th Annual Western Snow Conf., Laramie, WY: 97-106.

Chow, V.T., Maidment, D.R. and Mays, L.W. (1988). Applied hydrology. New York: McGraw-Hill.

Clark, M.J. (1996). Professional integrity and the social role of hydro-GIS. *In:* K. Kovar and H.P. Nachtnebel (Eds.), HydroGIS: Applications of geographic information systems in hydrology and water resources management (IAHS

Publication, **235:** 279-287). Wallingford, CT: International Association of Hydrological Sciences.

Clark, M.J. (1998). Putting water in its place: A perspective on GIS in hydrology and water management. *Hydrological Processes*, **126:** 823-834.

Das, D. (1990). Satellite remote sensing in subsurface water targeting. Proceeding ACSM-ASPRS annual convention, 99-103.

Das, D. (1996). Environmental appraisal for water resource development proceedings. International conference on Disaster Management (ICODIM) 499-507 Guwahati: India, Organised by Tejpur University.

DeVantier, B.A. and Feldman, A.D. (1993). Review of GIS applications in hydrologic modeling. *Journal of Water Resources Planning and Management*, **119:** 246-261.

DeVries, J.J. and Hromadka, T.V. (1993). Computer models for surface water. *In:* D.R. Maidment (Ed.), Handbook of hydrology (pp. 23.1-24). New York: McGraw.

Dewey, K.F. and Heim, R. Jr. (1981). Satellite observations of variations in northern hemisphere seasonal snow cover. NOAA Technical Report NESS 87, Washington DC, 83 pp.

Dillard, J.P. and Orwig, C.E. (1979). Use of satellite data in runoff forecasting in the heavily forested, cloud covered Pacific northwest. Proc. Final Workshop on Operational Applications of Satellite Snow Cover Observations. NASA CP-2116, pp. 127-150.

Djokic, D., Coates, A. and Ball, J.E. (1995). GIS as integration tool for hydrologic modeling: A need for generic hydrologic data exchange format. Paper presented in the 1995 ESRI User Conference. Redlords, CA, 22–26 May. Available: *http://www.esri.com/base/common/userconf/proc95/to250/* p245.html

Dozier, J. (1984). Snow reflectance from Landsat-4 thematic mapper. IEEE *Trans. Geosci. and Rem. Sens*. GE-**22(3):** 323-328.

Engman, E.T. and Gurney, R.J. (1991). Remote Sensing in Hydrology. Chapman and Hall, London, 225 pp.

Engman, E.T., Angus, G. and Kustas, W.P. (1989). Relationship between the hydrologic balance of a small watershed and remotely sensed soil moisture. Proc. IAHS Third Intl. Assembly, Baltimore, IAHS Publ., **186:** 75-84.

Gossel, W., Ebraheem, A.M. and Wycisk, P. (2004). A very large scale GIS-based groundwater flow model for the Nubian sandstone aquifer in Eastern Sahara (Egypt, northen Sudan and Eastern Libya). Report—*Hydrogeology Journal.*

Howari, F.M., Goodell, P.C. and Miyamoto, S. (2002). Spectral properties of salt crusts formed on saline soils. *J. Environ. Qual.,* **31:** 1453-1461.

Howari, F.M. (2003). The use of remote sensing data to extract information from agricultural land with emphasis on soil salinity. *Australian Journal of Soil Research,* **41(7):** 1243-1253.

Lanza, L.G., Schultz, G.A. and Barrett, E.C. (1997). Remote sensing in hydrology: Some downscaling and uncertainty issues. *Physics and Chemistry of The Earth*, **22(3-4):** 215-219.

Schmugge, T.J., William, P. Kustas, Jerry, C. Ritchie, Thomas J. Jackson and Al Rango (2002). Remote sensing in hydrology. *Advances in Water Resources*, **25(8-12):** 1367-1385.

Sherif, M.M. and others (2004). Assessment of the Effectiveness of Al Bih, Al Tawiyean and Ham Dams in Groundwater Recharge using Numerical Models. Progress Report, Ministry of Agriculture and Fisheries, Dubai, UAE.

Su, Z.B. Troch P.A. (2003). Applications of quantitative remote sensing to hydrology. *Physics and Chemistry of the Earth*, **28(1-3):** 1-2.

Sui, D.Z. and Maggio, R.C. (1999). Integrating GIS with hydrological modeling: practices, problems, and prospects. *Computers, Environment and Urban Systems*, **23(1):** 33-51.

Voss, C.I. and Provost, A.M. (2002). SUTRA: a model for saturated-unsaturated variable-density ground-water flow with solute or energy transport. US. Geological Survey Water Resources Investigation Report WRIR 02-4231, 250 pp.

Voss, C.I. (1984). SUTRA—Saturated-Unsaturated Transport: A Finite Element Simulation Model for Saturated-Unsaturated, Fluid-Density-Dependent Groundwater Flow with Energy Transport or Chemically-Reactive Single Solute Transport: US. Geological Survey Water Resources Investigation Report WRIR 84-4369 (revised 1990, 1997), 409 pp.

Winston, R.B. and Voss, C.I. (2003). SutraGUI, a graphical-user interface for SUTRA, a model for saturated- unsaturated variable-density ground-water flow with solute or energy transport. US. Geological Survey Open-File report 03-285, 114 pp.

2

Electrical Resistivity Methods for Borehole Siting in Hardrock Region

Ron D. Barker

The School of Earth Sciences
University of Birmingham, Birmingham, UK

INTRODUCTION

The measurement of earth resistivity is the most commonly employed geophysical technique in hydrogeological investigations but is perhaps one of the least well understood. The ease with which electrical measurements can now normally be made has led many non-geophysicists to believe, quite falsely, that resistivity is simple. Unfortunately this has led to many common errors in field practice and interpretation and ultimately to the production of ubiquitous, poor quality, survey reports. Inevitably this has resulted in the often poor reputation of this technique which is unfortunate because used correctly the resistivity technique is one of the most cost-efficient and powerful of the geophysical tools. This article outlines the principles of the techniques now in common use and shows how the results should be interpreted to obtain the most useful subsurface information

PRINCIPLES

Current flow from electrodes

The main aim of a resistivity survey is the measurement of the electrical resistivity of the earth, knowledge of which enables the estimation of its hydrogeological properties. The first step towards this is the measurement of the resistance of the ground by passing an electrical current into it through a metal stake (electrode), which acts as the current source. Of course, current can only pass into the ground if it can also exit through a second electrode (or current sink) so that the circuit is completed. In the immediate vicinity of the source electrode the current flows into the ground in all directions with the pattern shown in Fig. 1.

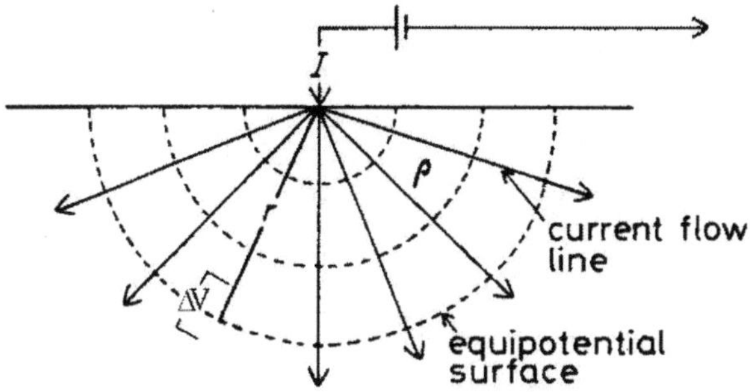

Figure 1. Current flow from a single electrode on the ground surface

Here the radial lines are representative lines of current flow. The concentric lines are lines of equal potential. The electrical potential at a point is the work done in bringing a unit electrical charge from infinity to that point against the flow of current. It can be seen, therefore, that the potential increases towards the current source (Fig. 1). The expression for the electrical potential is probably the most important in resistivity surveying and that on which the few equations presented below are based. The potential, V, at a distance, r, from a point current source, I, on the surface of the earth of resistivity, ρ, is given by:

$$V = \frac{\rho I}{2\pi r} \tag{1}$$

The pattern of current flow in a homogeneous earth when two electrodes are considered is shown in Fig. 2. Normally it is the difference in potential, ΔV, between two metal stakes (potential electrodes) at points on the surface, which is measured using electric voltmeter and this is referred to as the voltage. The measured voltage between two points M and N is:

$$\Delta V = V_M - V_N \tag{2}$$

where V_M is the potential at electrode M.

The potential at M is the sum of potentials due to any current sources and sinks. In the case of there being only a single current source and a single current sink, the potential difference is:

$$\Delta V = (V_{MA} - V_{MB}) - (V_{NA} - V_{NB}) \tag{3}$$

where V_{MA} is the potential at M due to current entering at A. As B is a current sink where current leaves the earth, V_{MB} and V_{NB} are negative.

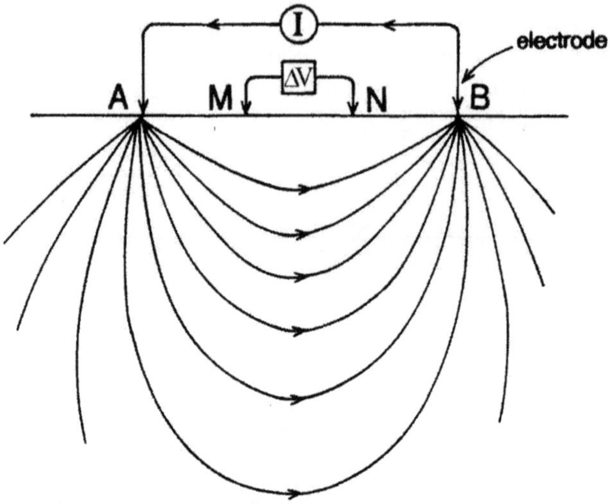

Figure 2. Current flow in the subsurface. *A* and *B* are current electrodes and *M* and *N* are potential electrodes. The resistance is given by the ratio Δ*V*/*I*.

If the distances between different electrodes are defined as in Fig. 3, equation (3) can be rewritten by substituting in equation (1) with appropriate distances. Then

$$\Delta V = \frac{\rho I}{2\pi}\left\{\frac{1}{r_1} - \frac{1}{r_2} - \frac{1}{R_1} + \frac{1}{R_2}\right\} \tag{4}$$

or

$$\Delta V = \frac{\rho I}{2\pi} K$$

where K is known as the geometric factor and depends on the spacing between electrodes and their arrangement and ρ is the resistivity of a homogeneous earth.

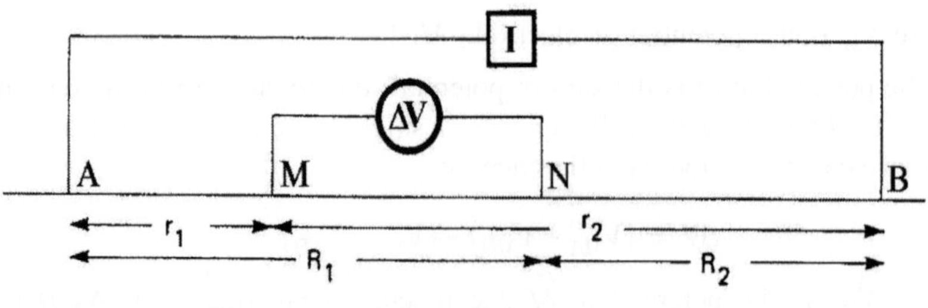

Figure 3. A generalized four-electrode array

Over a normal heterogeneous subsurface, the measured resistivity is a weighted mean of the resistivities of all the individual bodies of rock, which make up the earth and is termed the "apparent resistivity", ρ_a.

The arrangement of electrodes is referred to as the "array" or "configuration" and of the many which have been tried and tested only two have attracted significant attention in hydrogeological investigations: the Wenner and the Schlumberger arrays. Each of these has particular advantages under different conditions. A third electrode arrangement, the bipole-bipole array, is also used occasionally.

Electrode Arrays

Wenner Array

In the Wenner array (Fig. 4a), the electrodes are inserted into the ground in a line with equal spacing between adjacent electrodes. The apparent resistivity of the earth is then obtained from equation (4) as:

$$\rho_a = 2\pi a R \tag{5}$$

where a is the distance between adjacent electrodes and R is the resistance measurement, $\Delta V/I$.

Schlumberger Array

The Schlumberger arrangement is a colinear array in which the inner two potential electrodes (Fig. 4b) are closely spaced. The expression for the apparent resistivity assumes that the electrode spacing, b, is infinitely small so that the potential gradient at the centre of the array is measured. Then

$$\rho_a = \frac{\pi L^2 \Delta V}{4bI} \tag{6}$$

In practice the potential electrode spacing, b, must be finite so that a resistance can be measured but as long as b is always less than $L/5$, any errors introduced will be less than 3% (Keller and Frischknecht, 1966). The following equation for apparent resistivity is derived from equation (4) and has the advantage that no assumptions concerning b need be made.

$$\rho_a = \pi \frac{(L^2 - b^2)}{4b} R \tag{7}$$

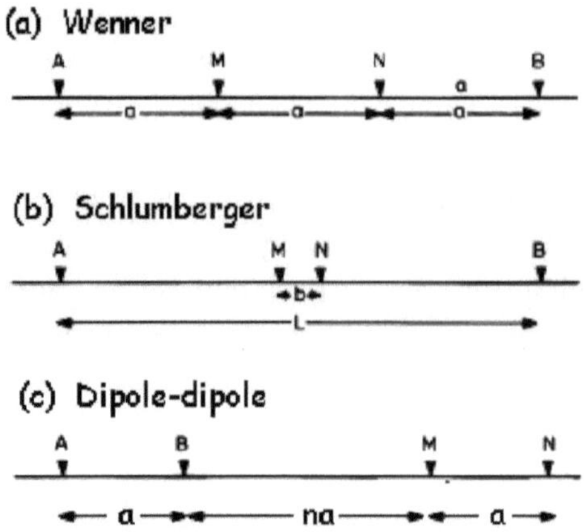

Figure 4. Common electrode arrangements. A and B are current electrodes, M and N are potential electrodes

Dipole-Dipole Array

The dipole-dipole colinear electrode arrangement (Fig. 4c) passes current across a closely spaced pair of electrodes (a dipole) with the voltage being measured across a second dipole some distance away. In this case the apparent resistivity is defined by:

$$\rho_a = \pi R n(n+1)(n+2)a. \tag{8}$$

Strictly speaking, if the dipole lengths are not small compared to the distance between them, the array is described as a bipole-bipole array.

Whichever electrode array is used, a single measurement of apparent resistivity gives very little information about the structure of the subsurface. To provide this a series of measurements of apparent resistivity must be made in one of a number of types of resistivity survey: a resistivity traverse, a vertical electrical sounding or an electrical imaging survey. However, before discussing these different forms of data collection, some other important aspects of resistivity measurement must be understood.

Four-Electrode Measurement

So far we have explained how measurements of earth resistance are made with a set of four electrodes. This contrasts with the conventional laboratory measurement of resistivity made by an electronics engineer who normally uses just two electrodes. The reason for this is that if a current is passed into the earth through two contact points made by pushing metal stakes into the ground, the current is found to pass not only through the earth resistance, but also through two high resistances which occur at the points of contact. These contact resistances (R_{C1} and R_{C2} in Fig. 5) can be several orders of magnitude higher than the earth resistance (R_e) of interest. If the voltage is measured across the same two points (a two-electrode measurement), the measured resistance is the sum of the earth resistance and the two contact resistances (Fig. 5a). To obtain a measure of the earth resistance only, the voltage must be measured across two additional electrodes (Fig. 5b). This is referred to as a 4-electrode measurement. The voltmeter must have high input impedance so that it draws negligible current from the circuit.

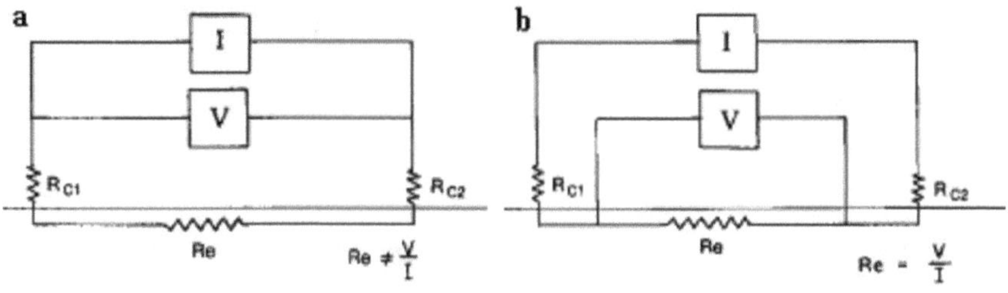

Figure 5. Earth resistance measurement using (a) a 2-electrode system and (b) a 4-electrode system.

Current Flow and the Electrical Properties of Rocks

Charge Transfer

The final aim of a resistivity survey is the determination of geological structure. To do this we need to understand electrical current passes through rocks. Transport of electrical current or charge transfer takes place in the following ways.

1. Electronic Conduction. Here current flow is by movement of electrons in a conductive medium. It is the normal means by which current flow occurs in the laboratory in wires and in the field in conductive minerals such as magnetite and various sulphide ores. The main rock forming minerals such as quartz, calcite and feldspars are non-conductors.

2. Ionic (Electrolytic) Conduction. In porous, water-filled rocks, electrical charge can be carried by the ions in solution in the electrolyte. The most commonly occurring salt is sodium chloride and so the ions bale to carry the charge are Cl^- and OH^- in one direction and Na^+ and H^+ in the other. The resistivity of the rock decreases with increasing porosity and also with increasing electrolyte salinity.

3. Clays. In clay minerals, electrical flow takes place by means of surface charge built up at the surface of each individual clay mineral. Soft young clays may have a surface area as much as several hundred square metres per gram of material and the resulting surface charge provides an easy path for electrical flow with the result that clays normally exhibit a very low resistivity.

The bulk resistivity of a rock will depend, therefore, on:

(a) the amount of conductive minerals present,
(b) the amount of water,
(c) the salinity of the water, and
(d) the amount of clay.

Example resistivities for common rocks are shown in Fig. 6.

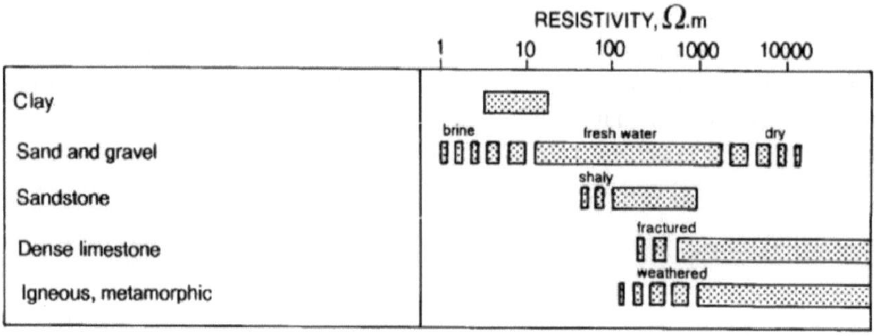

Figure 6. Range of resistivity values of some common rock materials

INSTRUMENTATION

Resistance meters

Most resistivity soundings are now measured using pulsed DC earth resistance meters of which there are several in the market. The meter applies a square wave current pulse to the ground with the polarity being reversed on every alternate pulse. The observed voltage is processed to remove inductive coupling and the effects of slowly varying earth currents. Modern instruments also allow the observed accuracy to be improved by averaging over several resistance measurements.

Electrodes

Electrodes are normally mild steel, stainless steel or copper rods inserted into the ground to a depth of a few centimetres. The depth of insertion should be enough to achieve a good contact with the earth but should not be such that the theoretical requirement of point electrodes is invalidated. Generally speaking the depth of insertion should be less than a twentieth of the total electrode array length. In moderately dry conditions remarkable improvement in ground contact can be achieved by pouring a small amount of water into the hole into which the electrode is inserted. In extreme conditions electrode positions will need to be watered copiously and left over night. At large spacings, multiple stakes inserted to considerable depth may be used as electrodes.

Cables

Normally single core PVC-covered cables are used to connect the electrodes to the resistance meter, although if polyurethane covered cables can be obtained these are far more durable. Multicore cables can also be employed since inductive coupling effects, a problem with older meters, are largely removed with pulsed DC instruments.

Field tests

It is possible to demonstrate that if the roles of the current and potential electrodes are exchanged, the measured resistance remains the same, i.e.

$$R_{AMNB} = R_{MABN} \tag{9}$$

This is the reciprocal relationship and holds whatever the electrode array and spacing and whatever the geological structure. If a test is carried out and the two resistances are not equal, an instrumental problem must be anticipated. This is a useful field test and should be checked if poor data are recorded in the field.

Depth of Investigation

It is often useful to have some idea of the depth being investigated by any particular arrangement and spacing of electrodes. Conventionally the probing depth of the Wenner array is given as equal to the electrode spacing, a, and for the Schlumberger as $L/2$, where L is the spacing between current electrodes AB. In fact, both these rules of thumb considerably overestimate the general depth of investigation. The problem can be best understood by studying the simple case of a subsurface, which consists of a material of a single, constant resistivity. This "homogeneous half-space" can be considered to consist of an infinite number of thin horizontal sheets of infinitesimally small thickness. The contribution that each sheet makes to the measured resistance (termed the normalized depth of investigation characteristic, NDIC) may be plotted as shown in Fig. 7 (Roy and Apparao, 1971; Banerjee and Pal, 1986).

The NDIC curve shows that most of the resistance measured with the Wenner array originates from the depth range 0.5L to 2.5L. Although there is clearly no single depth of investigation it is often useful to use a single value for reference. In this case it is found that the most practically useful value is the median depth (Edwards, 1966; Barker, 1989), or the depth from below which and from above which 50% of the signal originates. In this way the depths of investigation for the common electrode arrays can be calculated. Of course much of the signal originates from greater depths so it is also useful to define the maximum effective depth of investigation as that depth from below which only 10% of the signal originates. These parameters are listed below for the two most popular electrode arrays.

Configuration	*Depth of investigation*	*Maximum effective depth of investigation*
Wenner	0.17L (i.e. ≈ $a/2$)	0.41L
Schlumberger	0.19L (i.e. ≈ AB/5)	0.42L

So using a Wenner array with a = 100 m, the maximum signal can be expected to originate from around a depth of 20 m although some significant response from all depths down to 40 m is likely.

It is important to remember that these values pertain to a homogeneous earth. Depth of investigation is decreased if the resistivity increases rapidly with depth, such as in hardrock areas, and is increased if resistivity decreases with depth, such as in areas where groundwater salinity increases with depth.

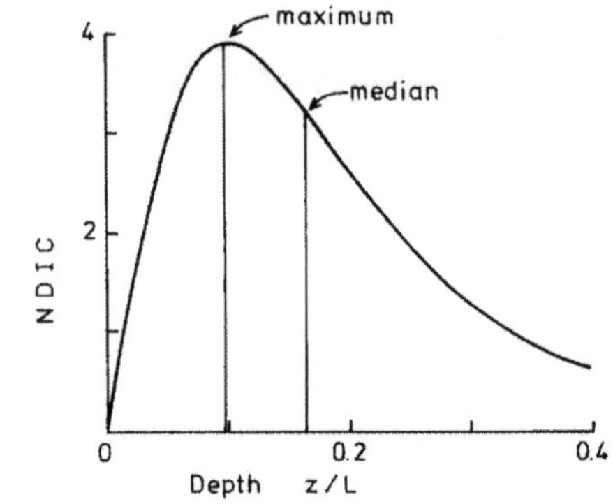

Figure 7. Normalized depth of investigation characteristic curve for the Wenner array (L is the distance between the outer current electrodes)

Horizontal Traversing and Mapping

Horizontal traversing (or profiling) is the survey technique employed to locate lateral changes in resistivity such as those caused by faults, vertical fissure zones, narrow buried valleys and plumes of contaminated groundwater. Any electrode configuration may be employed. The spacing is fixed and the whole electrode arrangement is moved progressively along the traverse line after each measurement of resistance. Each result is plotted at the geometric centre of the electrode array to produce a graph of apparent resistivity against distance. Several parallel traverses may be combined to produce a map of the survey area.

The interpretation of this type of resistivity data is normally qualitative, because anomalies over even the simplest geological structure can be quite complex. In practice even a thin layer of overburden obscuring the structure is enough to remove much of the complexity, which will anyway likely be masked by general background noise caused by near-surface geological variation.

Figure 9 is an example of a horizontal profile across a fissure zone in a hardrock area. The lower resistivity of the fractured water-bearing rock is clearly observed. This type of result is sometimes used as the only criterion for siting water supply wells although the method has been largely replaced by electromagnetic profiling.

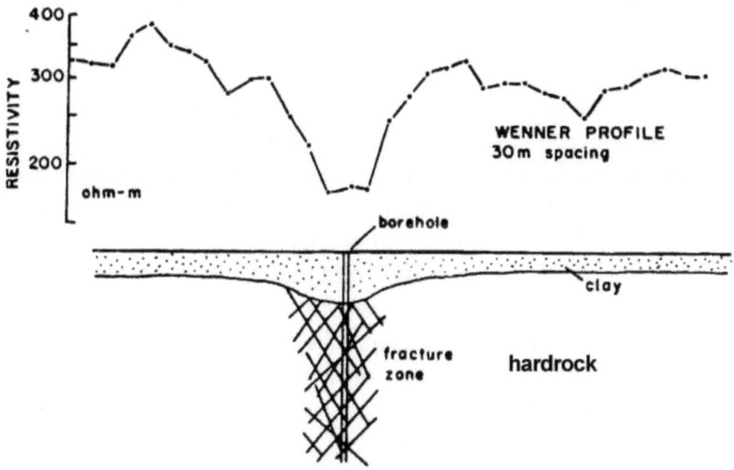

Figure 8. Offset Wenner traverse across a fracture zone in hardrock.

Although most electrode configurations may be employed in horizontal traversing, the Wenner and bipole-bipole arrays have the practical advantage of requiring equally spaced electrode positions along the profile. The Schlumberger array, however, gives anomalies of greater amplitude.

In carrying out a resistivity traverse, it is important to use an electrode spacing, which gives the required depth of penetration. Generally the values given above can be used. However, in hardrock and other areas, where the deeper geological formation has a very high resistivity, the depth of investigation of an electrode array is considerably reduced, as the electrical current prefers to flow through the low resistivity layer at the surface. In these cases larger electrode spacings should be employed in order to achieve the required depth of investigation. If possible the choice of electrode spacing should be made after studying a few resistivity soundings measured over the area to be profiled.

Vertical Sounding

Principles

This is an important resistivity survey technique and is used to examine vertical variations in resistivity. The technique is useful for investigating the subsurface structure, which is thought to be approximately horizontally layered one, although, as we shall see, quite complex structures may also be studied.

Each of the electrode arrangements already discussed may be employed in sounding although the field techniques are often quite different. Normally the centre of the electrode arrangement remains fixed while the inter-electrode spacing is increased between resistance measurements. Information relating to greater depths is obtained as the spacing is increased from a minimum of around 1 m to a maximum of more than 100 m. The measurements are plotted on double logarithmic paper with electrode spacing as abscissa and apparent resistivity as ordinate. The technique is illustrated in Fig. 10 for the case of a two-layer earth. At a very small electrode spacing (Fig. 9a) the current flows mainly in the top layer and the measured apparent resistivity is close in value to the true resistivity of this layer. As the spacing is increased (Fig. 9b) the current flows to greater depths and the measurement is intermediate between the resistivities of the two layers. At very large spacings (Fig. 9c) the current flows mainly in the deeper layer and the measured apparent resistivity approaches the true resistivity of this layer.

The resulting sounding curve must theoretically be smooth if it is to be correctly interpreted in terms of horizontal layers. In the first instance, if the abscissa is read as a depth scale instead of electrode spacing, the sounding curve can be considered as a much smoothed plot of the subsurface resistivity variation with depth. The curve of Fig. 10 is recognised as a two-layer curve because at small spacings the curve is asymptotic to a single resistivity (in this case 70 Ωm) and at larger spacings the resistivity increases and approaches a second resistivity (360 Ωm).

A variety of interpretation techniques enable the thicknesses and true resistivities of the main layers to be determined from this curve. Clearly a good interpretation will depend on the quality of the measured sounding curve so before discussing interpretation in detail we will consider some other important aspects of resistivity sounding, which depend on the electrode arrangement employed in the survey.

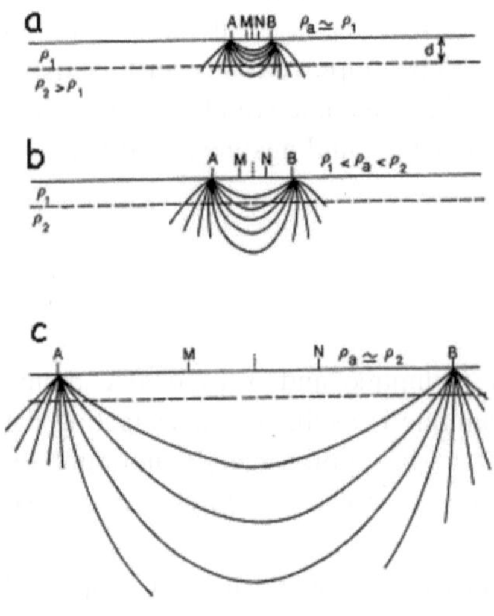

Figure 9. The principle of vertical electrical sounding

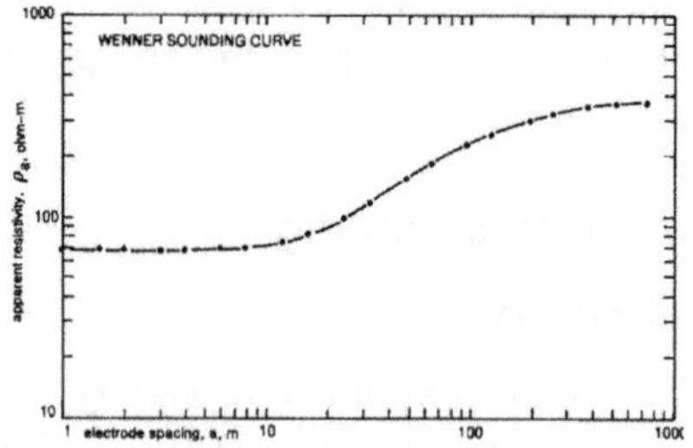

Figure 10. An example of two-layer sounding curve

Wenner Sounding

The Wenner is a simple arrangement to employ in sounding as all four electrodes are moved between resistance measurements. The electrode spacing may be increased as required although a commonly used spacing series is the following:

1, 1½, 2, 3, 4, 6, 8, 12, ...2^n, $3(2^n)$.

When plotted on double logarithmic paper the data points are approximately equally spaced along the curve.

Offset Sounding

Near-Surface Lateral Resistivity Effects

Vertical soundings are carried out in order to study a subsurface which is essentially horizontally layered. If lateral changes in resistivity occur in some of the layers especially in the topmost layer, sharp distortions to the sounding curve will result (Fig. 11). It has been a common error in the past to misinterpret these distortions as due to thin horizontal layers. However, computer studies will normally quickly confirm that it is impossible to model sharp irregularities with horizontal layers.

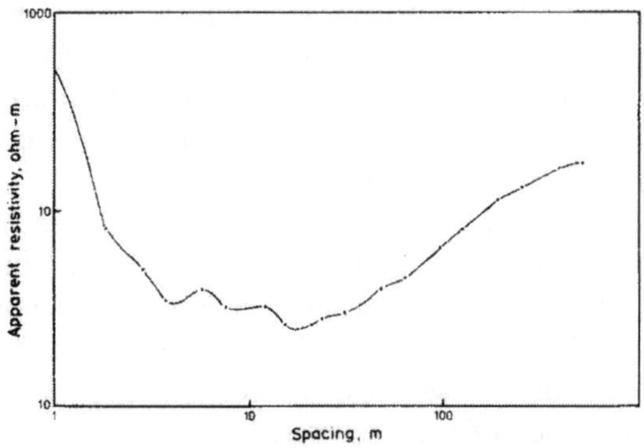

Figure 11. Wenner sounding showing irregularities caused by lateral resistivity variations

The origin of the spurious values is easily appreciated from a study of signal contribution sections such as those shown in Fig. 12. Each section shows the relative contribution, to the total measured resistance, of a unit volume of earth at any point on a plane through the line of the four electrodes (Barker, 1979). We see that the greatest contribution to the signal comes from material near the surface with very little contribution to the measurement originating from depths below *L*/3. Although the earth around each electrode gives the greatest contribution, negative areas normally cancel with positive areas so that little signal actually comes from this region. If the values of the sections are averaged over thin horizontal layers, the NDIC curve of Fig. 7 is obtained.

When a strong lateral change in resistivity (e.g a boulder) occurs to one side of an electrode, the negative zone will no longer cancel with a positive zone and an erroneous measurement will be made. Many of the irregularities observed on sounding curves and horizontal traverses are caused by the presence of boulders, ditches, tracks, fences and other features close to the electrodes. For this reason it is always wise to position the sounding lines away from tracks or similar features and in the middle of fields or other relatively undisturbed ground.

However, where near-surface variations in resistivity do occur, their effect on the sounding curve can be reduced using the Offset Wenner technique (Barker, 1981), a modification to the Wenner technique. The technique is illustrated in Fig. 13, which shows the signal contribution section for the Wenner electrode array with a large boulder near one of the electrodes in an otherwise homogeneous medium of 100 Ωm resistivity. The boulder, which is represented here by a fully conductive sphere, is situated in a positive zone (Fig. 13a) and the measured apparent resistivity is calculated as 91.9 Ωm. If the four electrodes are now "offset" to one side by a distance equal to the electrode spacing (Fig. 13b), negative regions now fall where positive regions were previously and vice versa. This is true except for the region at each end of the electrode line. The resistivity measured in this case is 107.8 Ωm. The average of these two measurements is the "Offset Wenner resistivity" and in this case is 99.85 Ωm, the resistivity of the subsurface in the absence of the boulder. An error due to lateral resistivity effects of around 8% in each case has thus been reduced to less than 0.2%.

Comparisons of this type depend on the position of the anomalous body relative to the electrodes and its dimensions. However, studies show that improvement in the resistance measurement is best for anomalous bodies, which have dimension less than half the electrode spacing. Although it is less effective with deeper sources, the technique provides a means of identifying the presence of such sources.

This principle can be incorporated into sounding by using a five electrodes system. This is no more difficult than normal sounding for the additional sounding remains fixed at the centre of the spread (Fig. 14). A switching device is used to change from one adjacent set of four electrodes to the other. Two resistances (R_{D1} and R_{D2}) are obtained which are averaged to provide the Offset Wenner resistance, R_D, in which the spurious effects of near-surface lateral resistivity variations are greatly reduced or even eliminated.

Calculation Of Lateral Resistivity Effects

One important advantage of the Offset sounding system is the ability it provides to estimate the magnitude of lateral resistivity effects. This estimate is provided at any spacing, *a*, by the "offset error", $e_{off}(a)$, defined by the following expression:

$$e_{off}(a) = \frac{R_{D1}(a) - R_{D2}(a)}{R_D(a)} \times 100\% \tag{10}$$

where R_D, R_{D1} and R_{D2} are defined in Fig. 14. Near-surface lateral effects cause offset errors which change randomly in magnitude and sign with increasing electrode spacing. The Offset technique substantially reduces these effects. If $e_{off}(a)$ is consistently greater than 20% and of the same sign, deeper and larger scale, lateral variations such as dipping beds and faults, are likely to be present. Although smooth curves may still be recorded, an interpretation in terms of horizontal layers could be invalid. The Offset error will indicate when such situations are encountered.

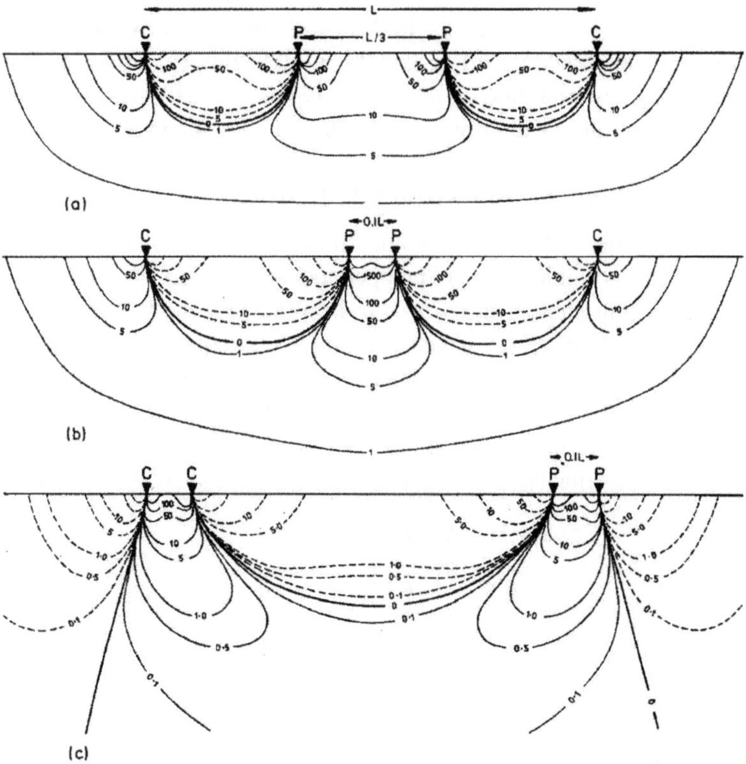

Figure 12. Signal contributions for: (a) Wenner, (b) Schlumberger and (c) dipole-dipole configurations. Contours show the relative contribution made by individual volume elements of earth to the total potential difference measured between the two potential electrodes; broken lines are negative contours.

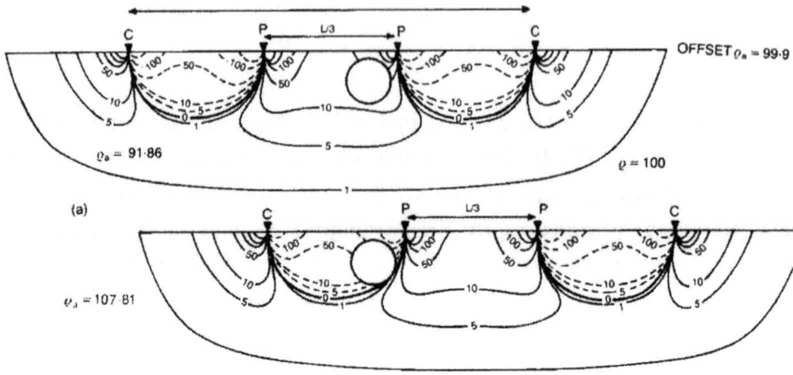

Figure 13. Signal contribution sections for the Wenner array with conducting sphere in positive region – broken lines show negative values of signal contribution and background resistivity is 100 Ωm; (b) offset position of electrodes – sphere now falls in a negative region. The distortion of contours due to the presence of sphere is not shown.

Instrumentation for Offset Sounding

A multicore cable system enabling measurements to a maximum spacing of 256 m is easily portable and may be operated by one man. To avoid coupling effects the system should be used with a pulsed DC earth resistance meter. Although the number of resistance measurements required is increased with the Offset system—28 measurements to produce a fourteen point Wenner curve—the small amount of extra time that this involves with a modern instrument is more than offset by the ease of operation of the technique so that one man can measure a single sounding in much less than one hour.

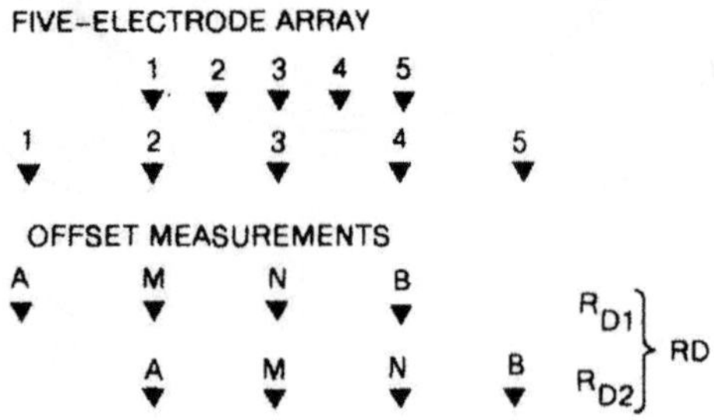

Figure 14. Principle of Offset Wenner sounding

Schlumberger Sounding

The Schlumberger configuration is in widespread use in many parts of the world. The field technique is different from that used with the Wenner array for the aim in using this array is to position the potential electrodes close enough together so that an approximate measurement of the potential gradient at the centre of the array is achieved. As long as the potential electrode separation, *b*, does not exceed *L*/5, errors in this approximation should not exceed 3%. As long as this criterion is met it would not be necessary to move the potential electrodes; if it were not that the measured voltages are very small and decrease still further as the current electrodes are moved apart. The voltages may fall close to noise levels in some instruments and there normally comes a point when *b* must be increased in order to permit an accurate resistance measurement.

In sounding with the Schlumberger array it is normal to move the potential electrodes, *M* and *N*, once for every five or six changes of the current electrodes. This type of survey procedure results in the typical segmented form of the Schlumberger sounding curve shown in Fig. 15. The effect of increasing *b* is always to give a positive error to the apparent resistivity. Therefore, the right hand end of each segment of a Schlumberger curve will represent lower (and more accurate) apparent resistivities than the beginning of the following segment. A smooth sounding curve can be constructed by drawing a curve passing through the last points of each segment.

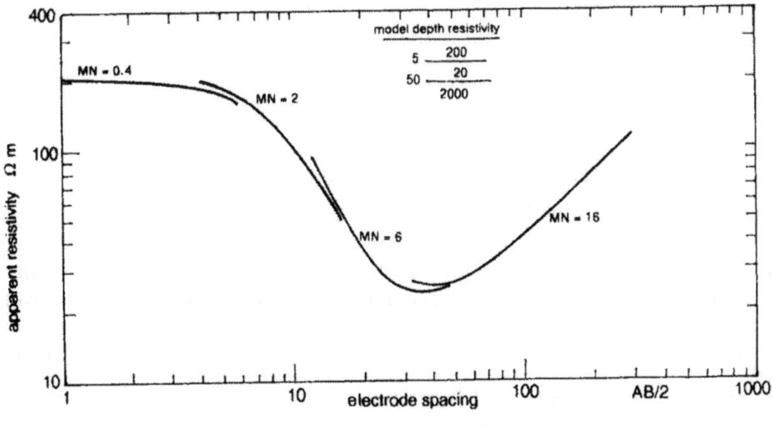

Figure 15. Schlumberger sounding curve showing segments measured with different MN spacings

The main advantage claimed for the Schlumberger array is the possibility it provides for removing lateral resistivity effects. Conventional argument claims that the potential electrodes mostly feel electrode effects, which causes a vertical shift of the curve segment recorded for the single *b* spacing affected by the inhomogeneity. In order to eliminate the lateral effect, it is an easy matter to move the displaced segment to its correct position as illustrated in Fig. 16. However, this will only allow correction of potential electrode effects; current electrode effects will still give some irregularity to the curve. Sometimes if a large number of the segments is affected it is impossible to determine which segments have been shifted.

Comparison of Wenner and Schlumberger Arrays

Geophysicists involved in resistivity sounding generally belong to one of the two main faiths: Wenner or Schlumberger. Advocates of one electrode arrangement will often vehemently oppose the use of the other and conversion from one faith to the other is rare. Much of the argument for and against the two arrays is based on traditional reasoning, which has since been superceded by modern ideas, and on published errors of judgement, which have never been repudiated. A summary of the pros and cons is given in Table 1.

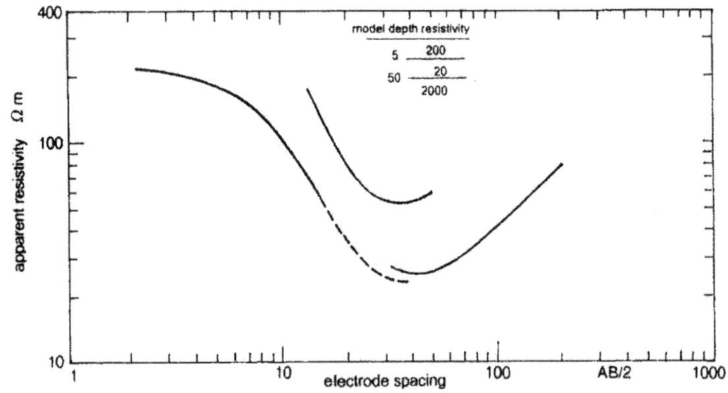

Figure 16. Effect on Schlumberger sounding curve of near-surface change in resistivity

Table 1. Summary comparison of the relative advantages and disadvantages of using the Wenner and Schlumberger arrays in vertical resistivity sounding

	Argument	*Winner*
1	The measured voltage (and hence resistance) is much smaller with the Schlumberger array. This results in greater noise and so better and more powerful equipment will be needed.	W
2	Shorter cable lengths are used with the Schlumberger array. This can be important when carrying out very deep sounding.	S
3	The Wenner array gives an exact curve, the Schlumberger a segmented curve.	W
4	The total number of electrode positions is smaller with the Schlumberger array (but is smaller still with the Offset Wenner).	S
5	Lateral effects can be corrected using the Schlumberger array. This is a conventional argument, which in fact is only partially valid. The argument is that a near-surface variation in resistivity close to one of the electrodes will cause one of the curve	W

(*Contd.*)

Table 1. (*Contd.*)

	segments to be displaced vertically. To correct for the effect, the segment is simply moved into its correct position. However, it is possible to demonstrate that this is only possible with lateral effects, which affect the potential electrodes, but not those which affect the current electrodes. In this respect the Offset Wenner technique is superior.	
6	The Schlumberger array has a greater depth of investigation. For a homogeneous earth, the depth of investigation for the Schlumberger is indeed slightly greater than for the Wenner array, but note that the difference is slight.	S
7	The Schlumberger array is more sensitive to lateral effects. Figure 17 shows traverses across a buried sphere for the Wenner, Schlumberger and Offset Wenner arrays, where the Schlumberger array responds best to the sphere and would be the best array for traversing across such features. However, this is a disadvantage when it comes to sounding.	W

It is important to note that there is little difference between the two arrays and that correctly carried out; surveys with either array must end up giving the same interpretation. This means that the choice of array may be purely personal.

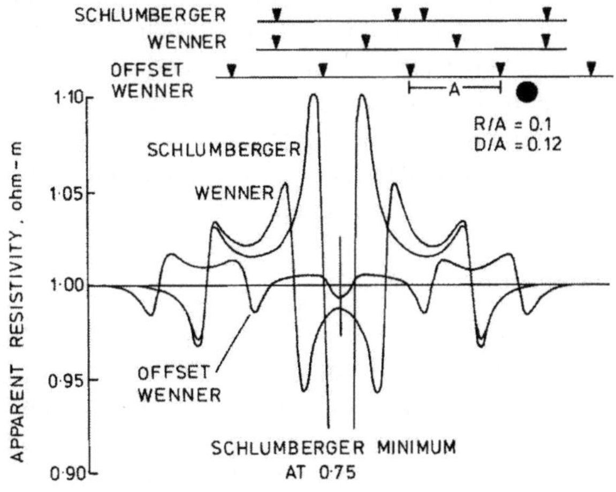

Figure 17. Traverses across a buried sphere

SOUNDING CURVE INTERPRETATION

Interpretation Techniques

Whichever field technique or electrode arrangement is employed, it is important to ensure that the end result is a sounding curve which is smooth and which can therefore be interpreted in terms of horizontal layers. The interpretation techniques, which are generally available to the hydro-geologist can be grouped as follows:

1. Approximate empirical techniques
2. Use of master curves
3. Interactive computer modelling
4. Automatic computer inversion.

It is tempting to think that automatic computer interpretation is the most desirable technique to employ simply because of its high technology. This would be so if the computer could be relied upon to yield exactly that interpretation that the hydrogeologist would expect given certain geological knowledge and constraints. Sadly this is not so. The computer will normally throw out only one of a wide range of possible interpretations, which naturally result from the significant ambiguity inherent in the technique. There is then a danger that the inexperienced worker will accept the computer result as the only correct solution when another might be preferred.

The advantage of interactive computer interpretation is that the hydrogeologist can work to a certain extent from a geological basis and knowledge of the problem being studied. By testing a variety of models he learns rapidly what alternatives are possible and what are not. He can then refine the interpretation by using an automatic inversion algorithm as the final stage. This is a very powerful approach to sounding curve interpretation.

At the other end of the spectrum, there are also many approximate methods for interpreting sounding curves. Some of these, such as the Barnes method and the Moore cumulative method, are still occasionally used. However, very large errors can be introduced into the interpretation, especially if more than two layers are considered, and better results will be obtained with the more soundly based and accurate techniques discussed below. For this reason we will consider approximate techniques no further.

Total Curve Matching

The simplest accurate method for interpreting sounding curves is by comparing the field curve with a family of theoretical curves until one is found which has the same shape as the field curve. Depths and resistivities may then be read from the theoretical curve. The curve fitting method is applicable to two-, three- and four-layer problems. A set of 2400 theoretical Wenner type curves has been published by Mooney and Wetzel (1956) and a selection of three-layer curves for use with the Schlumberger array by Compagnie Générale de Géophysique (1955).

Although the search through a large number of curves may be a formidable task, there is certain logic to the search, which is limited by the parameters of the first layer. With practice an interpretation can be made quickly and efficiently. The method has the disadvantage that a theoretical curve, which fits the data, may not be available. For more than three layers the technique can become time-consuming. The advantage of the method is that if a precise curve match can be achieved, the resulting model will be an accurate interpretation of the field curve; no computer technique will significantly improve the interpretation, as the master curves are themselves computer generated.

The technique is illustrated with reference to the field curve of Fig. 10, which is judged to represent a two-layer subsurface, and a set of two-layer master curves (Fig. 18). To carry out an interpretation, the field resistivity curve is plotted on transparent paper on a double logarithmic scale with the same modulus as that used for the master curves. If the field resistivity curve is a two-layer curve it must have the same form as one of the type curves or be of a form that can be interpolated between two adjacent ones. The field curve is therefore superimposed on the appropriate set of type curves and, as shown in Fig. 18, keeping the two sets of axes parallel, it is moved about until a fit is obtained, or until it appears to be interpolated between adjacent type curves in the correct place. Now the $\rho_a/\rho_1 = 1$ axis of the master curves cuts the ordinate of the field curve at $\rho_a = \rho_1$. Similarly the $a/h = 1$ axis of the type curves must cut the abscissa of the field curve where $a = h$. The resistivity contrast for the field curve is that of the type curve to which it is the closest fit, or if the curve is interpolated it is intermediate between those of the nearest type curves. In the case of Fig. 18, $\rho_1 = 68\ \Omega$m and the thickness, h, or depth to base, d_1, of the first layer is 20 m. The resistivity contrast, ρ_2/ρ_1, is 6, so the second layer resistivity is approximately 400 Ωm.

The curve fitting method can be extended directly to three and four layer problems, although the sets of available master curves become more extensive.

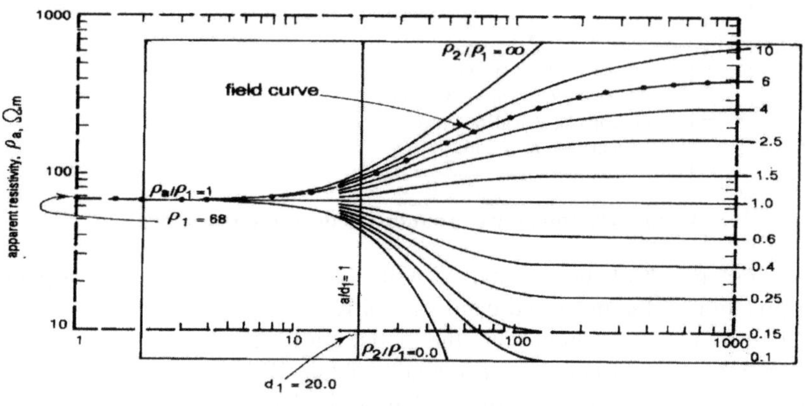

Figure 18. Interpretation of field curve using a set of two-layer master curves

Interactive Computer Interpretation

A very popular technique of resistivity sounding curve interpretation involves interaction with the computer as shown in Fig. 19. In this technique, the field curve is displayed on the VDU of a computer. An initial model estimate is provided to the computer, which then calculates the theoretical curve, which the model would produce. Both the theoretical and field curves are then displayed on the VDU on the same double log scale.

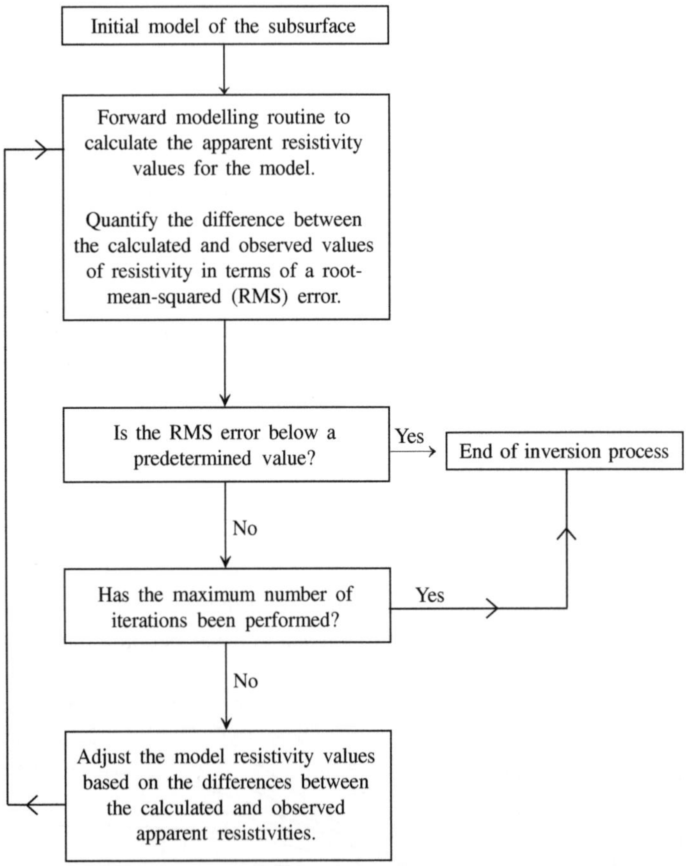

Figure 19. Interactive sounding curve interpretation

A close match of the two curves will indicate that the model is a good interpretation of the measured data. Any differences between the two curves are used as a basis for modifying the model until a good match between the theoretical and field curves is eventually obtained.

One of the problems with this technique is that a starting model has to be entered on the computer. A very approximate model may be obtained by visually fitting two layer curves to the field-sounding curve as shown in Fig. 20. First the curve is divided into segments each of which can be approximated by a two-layer master curve. A two-layer curve is then fitted to each segment and starting from the left side of the curve the depths and resistivities of the layers are read off. With experience this can be done visually and very quickly. It does not matter if errors are made, as these will be removed in the interactive interpretation carried out on the computer.

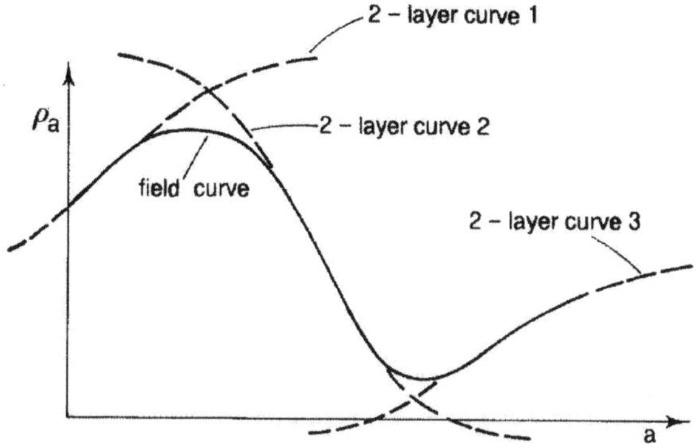

Figure 20. Visually fitting two-layer curves to provide a starting model for a computer interpretation

Accuracy of Sounding Curve Interpretation

An agreement of ±10% between the depths indicated by a sounding interpretation and the depths of the same interfaces observed in an adjacent borehole is considered very good; disagreement of far more than this is common. The reasons why this may be so are from the many sources of error, which can enter the process at any stage from the survey to the final interpretation. Instrumental errors are normally small, as modern meters are capable of measuring resistance under ideal conditions with an accuracy of better than 1%. However, there are many errors, which originate from survey problems such as having the wrong spacing between electrodes, observer error, topographic and environmental effects and lateral resistivity effects. Together these might contribute a total error of 5% to 10% to the determined apparent resistivity. Of course some of these will be random errors and within limits will not be a problem, while others which stem from topographic and environmental variations may be systematic and will lead to faulty interpretation of the curves. In addition the sounding technique has a poor resolving power and considerable inherent ambiguity; these problems are discussed next.

Resolution

Generally speaking, the vertical electrical sounding technique cannot recognize layers whose thickness is less than 10% of their depth. This is illustrated in Fig. 21, which shows representing sounding curves obtained over a three-layered earth in which the second layer has a variable thickness.

Where the second layer is thick compared to the depth to its top surface, it is clearly manifest on the sounding curve and so should appear in the interpretation. However, in the case of a layer, which is thin, the distortion it produces on the sounding curve is less than what might be considered within the 10% accuracy of the technique. This provides the limit of detection for the technique.

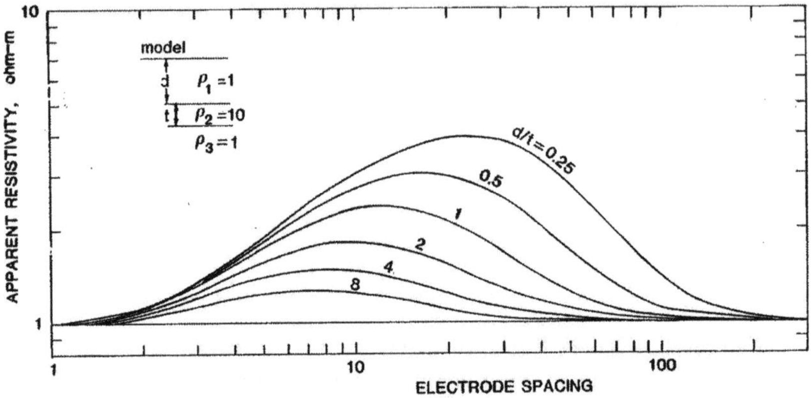

Figure 21. Sounding curves over a three-layer model in which the second layer varies in depth

Equivalence

The problem of equivalence is encountered with three-layer curves where the second layer has a much higher (or lower) resistivity than the layers above and below. In this case a variety of equivalent interpretations may be determined in which the middle layer takes different thicknesses and resistivities. The different interpretations for the case of a high resistivity middle layer have one feature in common; the transverse resistance, T, of the middle layer, the product of the layer thickness and its resistivity, is a constant, i.e

$$T = h\rho = \text{constant} \tag{11}$$

Although the transverse resistance of the middle layer can be unambiguously determined from the sounding curve, the layer thickness and resistivity cannot be so interpreted without some controlling information from other sources. This type of ambiguity can only be resolved with additional geological information.

When the middle layer has a lower resistivity than the layers above and below, it is the longitudinal conductance S, the product of the layer thickness and its conductivity, which governs the equivalence, i.e.

$$S = \frac{h}{\rho} = h\sigma = \text{constant} \tag{12}$$

Equivalence of this type is a serious problem in borehole siting in basement areas where the water saturated aquifer has a lower resistivity than both the overlying alluvium and the underlying unfractured bedrock.

Suppression

Suppression is the problem encountered in three-layer curves where the second layer has a resistivity intermediate to that of the layers above and below. In this situation the second layer will be suppressed, i.e. not manifest in the shape of the sounding curve, unless it has a large enough thickness. Then a three-layer curve can adequately be interpreted with two layers except when the thickness of the second layer increases beyond about three times the depth to its top.

The Two-stage Approach to Sounding Interpretation

The penultimate stage in the interpretation of a series of resistivity soundings is the construction of a vertical section through a line of the soundings. This is termed a geoelectrical section (Fig. 22a) and in the first instance displays the resistivities of the interpreted layers and their thicknesses correctly plotted in relation to the topography. Correlation between layers of similar resistivity is attempted and geological information is also included.

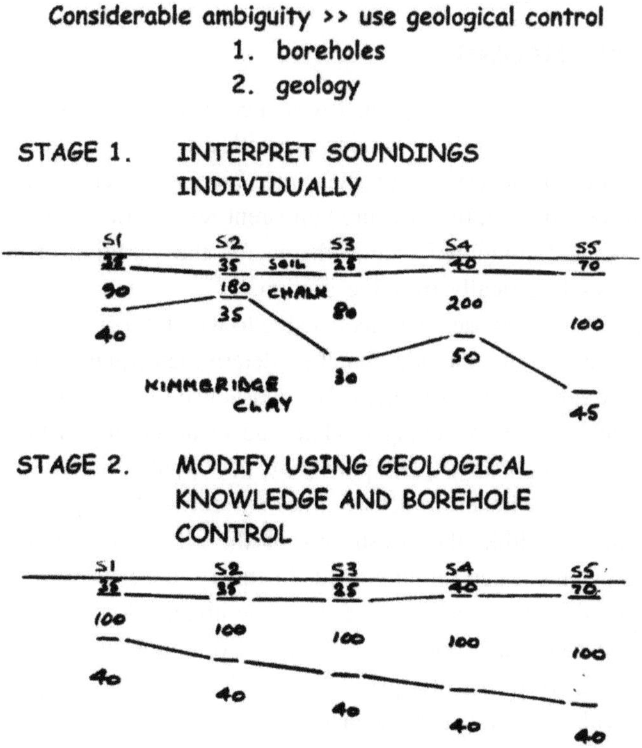

Figure 22. The two-stage approach to vertical sounding interpretation

At this stage the section will probably not bear much resemblance to real geology and this is because each sounding has been individually interpreted with no reference to adjacent soundings. In Fig. 22a, for example, the initial construction of the geoelectrical section along a line of six soundings has resulted in considerable variability in the formation resistivities and interface depths. However, in this region the bedrock and the overlying clay are both known to have consistent properties and so are unlikely to show variable resistivity.

This then leads to the next and very important stage in the interpretation—the reappraisal and reinterpretation of each sounding to test for geological consistency. It is at this stage that the effects of equivalence and suppression are considered, and borehole and other information are introduced. For example, in Fig. 22a, the soundings suggest that an average resistivity of the hardrock might be 4000 Ωm. The soundings are then interpreted but fixing the resistivity of the bedrock at 4000 Ωm. A single average resistivity for the regolith is also tested. In this case, good interpretations are possible and the geoelectrical section of Fig. 22b can be constructed. We find that by bringing consistency to the formation resistivities, the formation thicknesses have also adjusted to give a structure, which is also more realistic geologically.

ELECTRICAL IMAGING

Electrical imaging (tomography) involves measuring a series of constant separation traverses along the same line but with the electrode spacing being increased with each successive traverse. Since increasing separation leads to greater depth penetration, the measured apparent resistivities may be used to construct a vertical contoured section displaying the variation of resistivity both laterally and vertically over the section.

Modern field systems use a multicore cable to which 50 or more electrodes are connected at takeouts moulded on at predetermined equal intervals. Such a cable is very much like a seismic cable and is used in a similar way. The cable is connected to a switching module and to an earth resistance meter and computer through an RS232 port (Fig. 23). With these systems, any electrodes may be switched to act as either current A, B or potential M, N electrodes and so within the constraints of the electrodes emplaced, any electrode arrangement can be employed. In theory, the measurements from any array should be processed to give the same image. However, in practice either the two-electrode (pole-pole), Wenner, or dipole-dipole arrays are employed. For a fixed line of equally spaced electrodes the arrays have the following advantages and disadvantages:

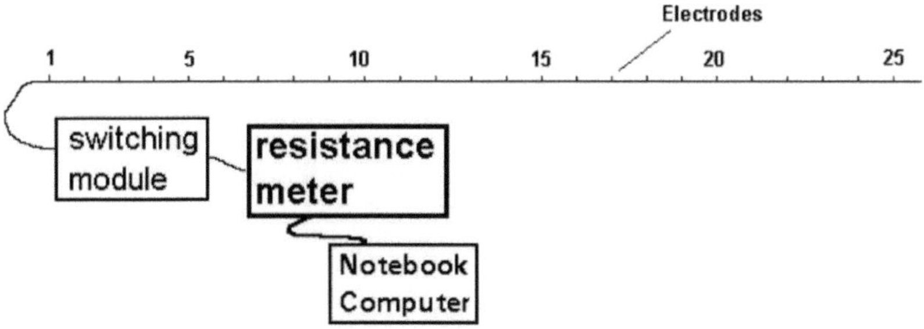

Figure 23. The basic instrument system employed in computer controlled electrical imaging

1. The ***two-electrode*** array has the greatest depth range, lowest resolution and least sensitivity to geological noise.
2. ***Wenner*** covers an intermediate depth range, has intermediate resolution and shows moderate sensitivity to geological noise,
3. ***Dipole-dipole*** has the smallest depth range, highest resolution and greatest sensitivity to noise.

In addition there are practical problems to be overcome with using the two-electrode array as this requires the use of electrodes placed at considerable distances from the imaging line. Also an important advantage of the Wenner array is that the number of measurements required to construct a pseudosection is generally much smaller than with the other arrays.

The first stage in the production of an electrical image is the construction of a pseudosection, an initial very approximate 'image' produced by plotting each apparent resistivity on a vertical section at a point below the centre of the four measuring electrodes and at a depth that is equivalent to the median depth of investigation (Barker, 1989; Edwards, 1977) of the array employed. This depth is referred to as the pseudo-depth. The data are contoured to form a pseudo depth-section that qualitatively reflects the spatial variation of resistivity in cross-section (Fig. 24).

The contoured pseudo-section contains three types of information. It clearly will contain considerable subsurface geological information that is reflected in the general form of the pseudosection. However, the pseudosection also contains geometrical information so that the anomaly across a simple structure will appear quite different for each of the different electrode arrays; what appears a simple structure on a two-electrode pseudosection may be quite complex on a dipole-dipole pseudosection. In addition the pseudosection will contain a certain amount of geological noise, the distorting effects of near-surface lateral changes in resistivity that occur close to the electrodes (often referred to as 'electrode effects'). Although shallower than the depth range of the image they can nevertheless produce spurious resistance readings at any spacing (Barker, 1979). Their extreme effect is to produce inverted

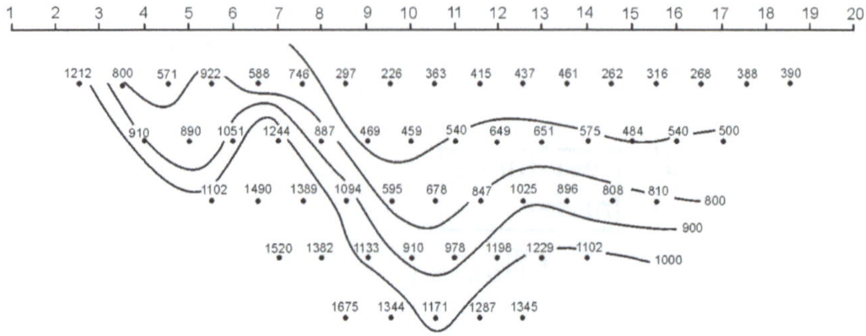

Figure 24. An example of Wenner pseudosection

'V' anomalies across the pseudosection. The reason why the dipole-dipole array is more sensitive to geological noise is its more complex response to any sub-surface structure.

In order to remove geometrical effects from the pseudosection and produce an image of true depth and true formation resistivity, the observed data must undergo a form of processing known as inversion. The inversion technique recently described by Loke and Barker (1995, 1996) appears to be a powerful and effective means of processing pseudosection data to provide a contoured

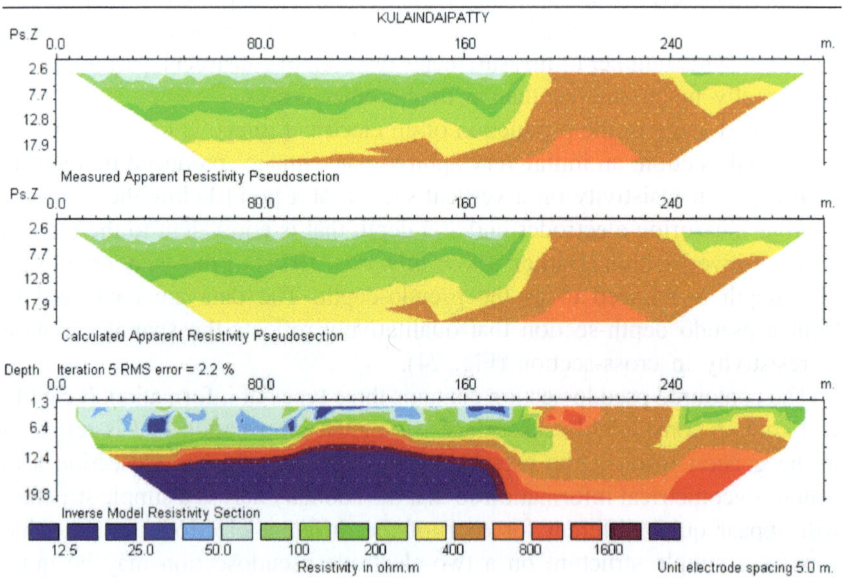

Figure 25. Electrical image over weathered basement in India. (a) is observed data plotted as a coloured pseudosection (Ps.Z = pseudo-depth), (b) is the pseudosection computed from the model and (c) is the image or model showing true depth and true formation resistivity. (Colour reproduction on Plate 3)

image of true depth and true formation resistivity. It uses a sensitivity matrix of coefficients based on the signal contribution section (Barker, 1979) to form the first model. Then, using a finite difference forward modelling algorithm modified from Dey and Morrison (1979), an iterative least squares optimisation technique is applied. An acceptable model is normally arrived at within five iterations and the whole process can be carried out in the field on a modern notebook computer in less than a couple of minutes.

Figure 25 shows the results of such an inversion. Note that the resulting image (Fig. 25c) shows boundaries, which are much sharper than those indicated by the observed data plotted as a pseudosection (Fig. 25a). Also the disturbing effects of near-surface changes in resistivity have been migrated back to near their correct locations. Such images consistently show very good agreement with drilling results, particularly with electrical logs obtained from boreholes on the section. Electrical imaging is rapidly replacing electrical sounding as the preferred electrical survey technique.

CASE HISTORY: ELECTRICAL IMAGING FOR BOREHOLE SITING NEAR DINDIGUL, TAMIL NADU, INDIA

This example describes the first use of electrical imaging in borehole siting in hardrock areas of India. The material is described more fully in Barker et al. (2003).

Water Supply in Hardrock Areas of India

The geological structure normally encountered in hardrock areas of India is granite or granite gneiss overlain by a variable thickness of weathered material, a regolith normally produced by the in-situ weathering of the basement rock (Acworth, 1987). The regolith normally grades into solid unfractured basement over several tens of metres, although often the boundary between the two may be fairly sharp. Hydrogeologically the weathered overburden has a high porosity and contains a significant amount of water, but, because of its relatively high clay content, it has a low permeability. The bedrock on the other hand is fresh but frequently fractured, which gives it a high permeability. But as fractures do not constitute a significant volume of the rock, fractured basement has a low porosity. For this reason a good borehole, providing long term high yields, is one which penetrates a large thickness of regolith, which acts as a reservoir, and also intersects fractures in the underlying bedrock, the fractures providing the rapid transport mechanism from the reservoir and so hence the high yield. Boreholes which intersect fractures, but which are not overlain by thick saturated regolith, cannot be expected to provide high yields in the long term. Boreholes which penetrate saturated regolith but which find no fractures in the bedrock are likely to provide sufficient yield for a handpump only.

Location of Fractures

The short-term success of borehole siting is clearly dependent on the borehole intersecting some fractures in the bedrock. However, once the bedrock is covered by any thickness of weathering, the fractures are notoriously difficult to find and geophysics provides no direct solution to the problem. Indeed, the belief that geophysical techniques can find fractures below overburden is a commonly held myth.

Although it is easily demonstrated that fractures of a few centimetres thickness, which may be very important hydrogeologically, cannot normally be located by geophysics once they are buried below a few metres of overburden, nevertheless, it is possible to use geophysics to ***increase the chances*** of intersecting fractures while drilling.

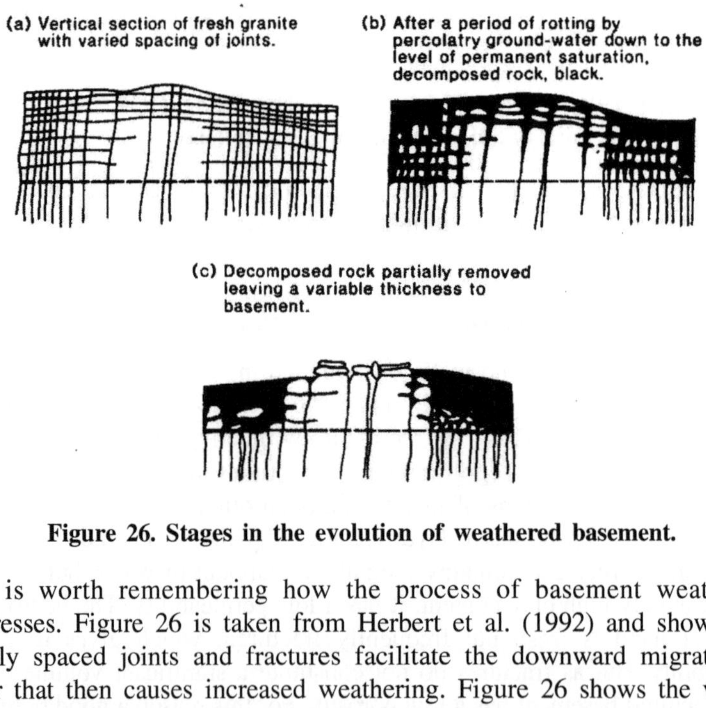

Figure 26. Stages in the evolution of weathered basement.

It is worth remembering how the process of basement weathering progresses. Figure 26 is taken from Herbert et al. (1992) and shows how closely spaced joints and fractures facilitate the downward migration of water that then causes increased weathering. Figure 26 shows the various stages of weathering and how irregular the change from fresh to weathered bedrock can be. The important feature is that, where fractures are present, the bedrock is expected to be more strongly weathered to a greater depth than where it is unfractured. For this reason it is best to drill at points where the bedrock reaches its greatest depth. This has a double advantage. Firstly the greater depth ensures a thick reservoir of water in the overburden, and secondly, the reservoir is more likely to be underlain by bedrock that is fractured.

This knowledge has led to the application of combinations of geophysical methods, normally electromagnetic profiling, followed by resistivity sounding, in successful borehole siting programmes in Africa (Barker et al., 1992; Carruthers and Smith, 1992; Olayinka and Barker, 1994). Now however, there is a move towards the use of electrical imaging as a replacement to resistivity sounding in many applications of resistivity sounding.

Field Surveys

To test electrical imaging in a typical hardrock environment, surveys were carried out at sites around Dindigul, in Tamil Nadu. The surveys formed part of a groundwater modelling study of the Basin of the River Kodaganar undertaken by National Geophysical Research Institute (NGRI), Hyderabad. The basin covers an area of 2300 km^2 and is a drought-prone hard-rock (crystalline basement) region. The River Kodaganar is effluent, with water flowing only during the rainy season (October – December) for between 10 and 20 days. Because of the lack of surface water, the rural communities depend for their drinking water on groundwater supplies from boreholes and dug wells. The groundwater occurs in both shallow and fractured aquifer systems to depths of 100 m, although most abstraction is very much less than this.

In order to examine potential borehole sites, a manual imaging system was employed consisting of a single cable with 25 takeouts at 5 m intervals was used together with a simple manual switchbox. In order to speed up the data collection and reduce the number of required measurements, all surveys were done using the Wenner array. The cable and switchbox were connected to an SSR MPL resistance meter, manufactured in India by IGIS Ltd, Hyderabad. Electrodes were 0.5 metre lengths of stainless steel, which were planted to a depth of 0.4 m. Each electrode was watered to ensure good contact with the ground (this is done most effectively by withdrawing the electrode from the ground, filling the hole with water and replanting the electrode).

Five images were measured in areas where the basement was shallow and where it was necessary to locate fracture zones in the bedrock. These fracture zones, where associated with a reasonable thickness of weathered overburden, might be considered suitable locations for boreholes. The lines varied in length from 110 m to 300 m to give an effective maximum depth of imaging of around 20 m. A summary of the lines is given in Table 1 and their locations are shown in Fig. 27.

Table 2. Imaging lines and their locations

Line	Number of Electrodes	Location	Length (m)
Line 1	61	Kulaindaipatty	300
Line 2	34	Mathinipatty	165
Line 3	23	Virudalaipatty	110
Line 4	34	Rajagoundanoor	165
Line 5	65	Giriyappa Naiyaknoor	320

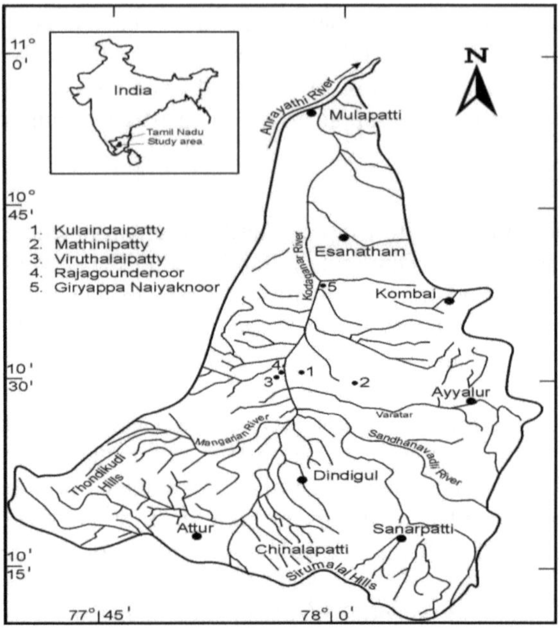

Figure 27. Study area and location of survey lines.

Results

Line 1 at Kulandaipatty

An image line was measured across a shallow dry river valley at Kulandaipatty. The site was chosen as it was thought by hydrogeologists working in the area to contain a geological lineament, although it was not one suggested by satellite imagery. If the lineament was a fracture zone associated with thick overburden, it would provide a zone of high hydraulic conductivity.

The first electrode of the image line was planted next to a dug well. The line went immediately across the dry riverbed about 1-2 metres in depth and then across open fields. A patch of outcrop was observed near electrodes 37-40 (180-200 m).

Figure 25a shows observed apparent resistivity data plotted as a conventional pseudosection, with apparent resistivity plotted against pseudo-depth, but with colour infill instead of line contours. This suggests that low resistivity material overlies higher resistivity bedrock. These data have been inverted to produce the image of Fig. 25c, which shows high resistivity (>300 Ωm) bedrock overlain by low resistivity (<100 Ωm) weathered material. The bedrock topography is variable with the thickness of overburden between 5 m and 10 m, except near electrodes 37-40 where the bedrock rises to the surface. A patch of unweathered bedrock, possibly a corestone, clearly shows on the image at a profile distance of 200 m. Unfortunately, there is no obvious evidence for the presence of a fracture zone. The best position for a borehole here would be around the present location of the dug well, where the bedrock is deepest, although other sites nearby could be investigated as they might show deeper weathering.

The model of Fig. 25c is used to compute a model dataset, which is shown in Fig. 25b. The close agreement between Figs. 25a and 25b indicate that the inversion has provided an accurate model of the subsurface.

Line 2 at Mathinipatty

This image was measured across an area that had previously been investigated with resistivity sounding and where indications of a strong change in apparent resistivity had been observed. The image started not far from dug wells near a road and went across uncropped dry sandy fields.

The image (Fig. 28) clearly shows a strong change in depth to bedrock, which is greater than 20 m at the start of the line, but which rises sharply to approach the surface near the 80 m electrodes. Again the thickest overburden and most suitable location of a borehole would be near the present dug wells. Although fractures are not directly indicated, the increased thickness of weathering here provides a good reservoir with a higher probability of bedrock fracturing.

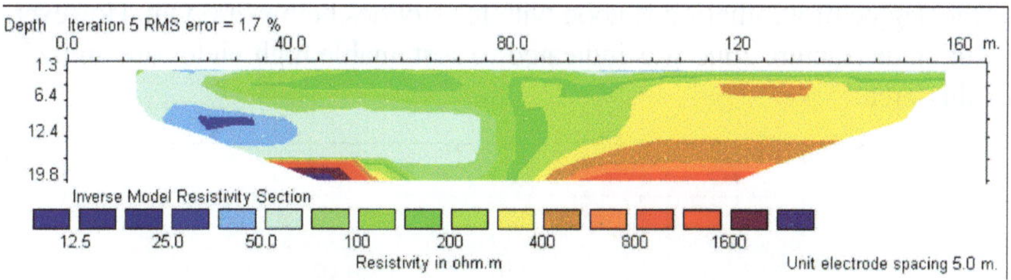

Figure 28. Electrical image along Line 2 at Mathinipatty (Colour reproduction on Plate 3)

Line 3 at Virudalaipatty

Borehole siting teams are frequently asked to indicate a suitable location for a well within the narrow confines of a village boundary. In this case we were asked to investigate the possibility of locating an additional well in Virudalaipatty and so an image line was measured along the middle of the high street. Electrode 1 was planted a couple of metres from a junction with the main road and the 23rd electrode was a short distance (1 m) from a house at the other end of the street. A dug well was located on the side of the street near electrode 16.

The image (Fig. 29) shows shallow bedrock covered by about 15 m of weathered rock, which increases in degree of weathering away from the road junction. The best place for a borehole would be at the furthest point from the road junction. However, there was no obvious sign of a fracture zone and so yields are likely to be only good enough for handpump supplies and not significantly better than existing dug well supplies. Where fractures are located in the bedrock and high yields obtained, these are unlikely to be sustainable over long periods.

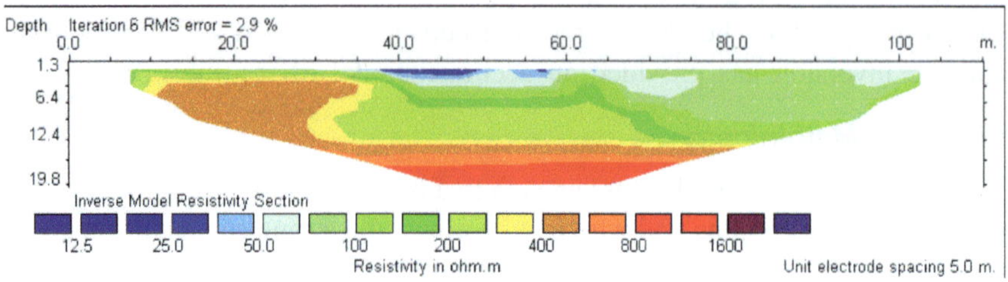

Figure 29. Electrical image along Line 3 at Virudalaipatty (Colour reproduction on Plate 3)

Line 4 at Rajagoundanoor

An image line was measured at the small village of Rajagoundanoor across an open bare sandy soil field with patches of quartzitic outcrop. The image (Fig. 30) shows fairly level bedrock with a slight dip away from the village. A borehole for a handpump supply could be located anywhere around here as the degree of weathering is good with resistivities below 200 Ωm. However, no obvious fracture zone was indicated so sustainable high yields are unlikely in this area.

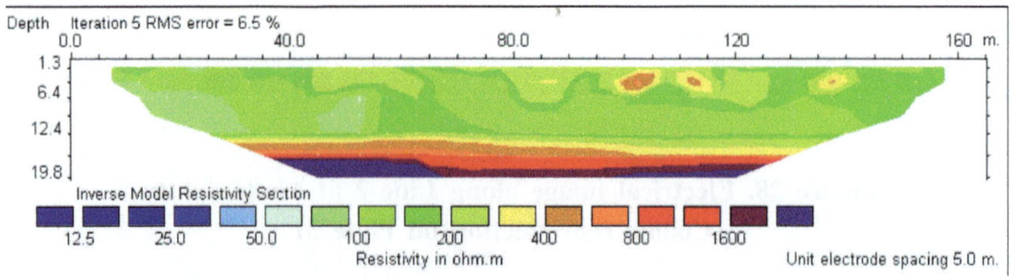

Figure 30. Electrical image along Line 4 at Rajagoundanoor (Colour reproduction on Plate 3)

Line 5 at Giriyappa Naiyakanoor

This image was carried out to investigate a lineament indicated by satellite imagery. Line 5 was an extended line of 65 electrodes positioned across irrigated paddy fields, a road and some uncropped dry fields. Electrode 35 was in a water-filled drain along the side of the road and electrodes 36 and 37 were on either side of the road.

The image (Fig. 31) shows that the lineament is probably a bedrock ridge 190 m along the line. Linear outcrops of granite were observed running parallel to the road for several kilometres. The bedrock is fairly shallow (12

m) along the first part of the image and has a lower resistivity that might represent the effects of recharge due to irrigation. To the north of the bedrock ridge, the bedrock drops to much more than 20 m in depth but here has a higher resistivity, which possibly correlates with the lack of irrigation, but which might also represent a lower grade of weathering.

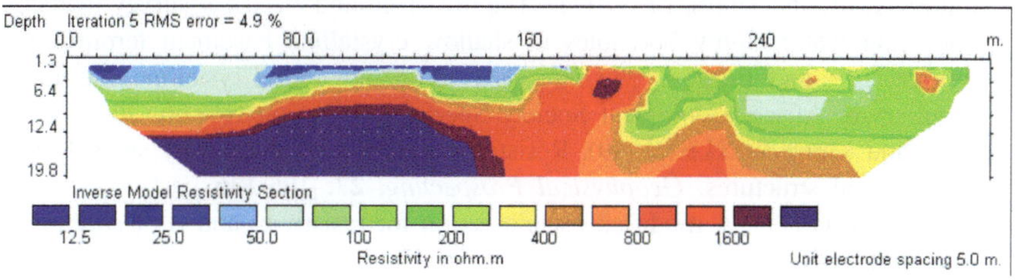

Figure 31. Electrical image along Line 5 at Giriyappa Naiyakanoor. (Colour reproduction on Plate 4)

Conclusions

Resistivity techniques are a powerful borehole-siting tool and, although there are many other geophysical techniques, which may be used, the rapid increase in the popularity of electrical imaging has brought the focus back onto electrical methods. Electrical images provide a more detailed view of subsurface structure than can be obtained using other geophysical techniques and may therefore lead to a better understanding of the local hydrogeology. Siting boreholes on the basis of increased thickness of overburden should provide better yielding boreholes with a greater degree of sustainability, although it should be remembered that very often the geological situation might preclude any high yielding boreholes.

It is important to remember that electrical imaging is not a reconnaissance tool. It is too slow and more suitable for the detailed study of problem areas that have been selected using other techniques, such as EM or resistivity traversing, satellite imagery, or the restricted demands of a local community. A rapid resistivity or electromagnetic traversing technique and an electrical imaging system would be a powerful combination of tools for borehole siting in hard rock regime.

REFERENCES

Acworth, R.I. (1987). The development of crystalline basement aquifers in a tropical environment. *Quarterly Journal of Engineering Geology*, **20:** 265-272.

Banerjee, B. and Pal, B.P. (1986). A simple method for determining the depth of investigation characteristics in resitivity prospecting. *Exploration Geophysics*, **17:** 93-95.

Barker, R.D. (1979). Signal contribution sections and their use in resistivity studies. *Geophysical Journal of the Royal Astronomical Society*, **59:** 123-129.

Barker, R.D. (1989). Depth of investigation of a generalised colinear 4-electrode array. *Geophysics*, **54:** 1031-1037.

Barker, R.D., White, C.C. and Houston, J.F.T. (1992). Borehole siting in an African accelerated drought relief project. *In:* Wright, E.P. and Burgess, W.G. (eds), Hydrogeology of Crystalline Basement Aquifers in Africa. Geological Society Special Publication No. 66, 183-201.

Caruthers, R.M. and Smith, I.F. (1992). The use of ground electrical survey methods for siting water-supply boreholes in shallow crystalline basement terrains. *In:* Wright, E.P. and Burgess, W.G. (eds), Hydrogeology of Crystalline Basement Aquifers in Africa. Geological Society Special Publication No. 66, 203-220.

Dey, A. and Morrison, H.F. (1979). Resistivity modelling for arbitrary shaped two-dimensional structures. *Geophysical Prospecting*, **27:** 106-136.

Edwards, L.S. (1977). A modified pseudosection for resistivity and IP. *Geophysics*, **42:** 1020-1036.

Herbert, R., Barker, J.A. and Kitching, R. (1992). New approaches to pumping test interpretation for dug wells constructed on hard rock aquifers. *In:* Wright, E.P. and Burgess, W.G. (eds), Hydrogeology of Crystalline Basement Aquifers in Africa. Geological Society Special Publication No. 66, 221-242.

Keller and Frischknecht (1966). Electrical Methods in Geophysical Prospecting; Pergamon Press, Oxford, p. 501.

Loke, M.H. and Barker, R.D. (1995). Least-squares deconvolution of apparent resistivity pseudosections. *Geophysics*, **60:** 1682-1690.

Loke, M.H. and Barker, R.D. (1996). Rapid least squares inversion of apparent resistivity pseudosections by a quasi-Newton method. *Geophysical Prospecting*, **44:** 131-152.

Money, H.M. and Wetzel, W.E. (1956). The potential about a point electrode and apparent resistivity curves for a two, three- and four-layered earths; University of Minnesota Press, Minnespolie, Minn; p. 146.

Olayinka, A.I. and Barker, R.D. (1990). Borehole siting in crystalline basement areas of Nigeria with a microprocessor-controlled resistivity traversing system. *Ground Water*, **28:** 178-183.

Roy, A. and Apparao, A. (1971). Depth of Investigation in Direct Current Methods. *Geophysics*. **36:** 943-959.

3

Parameterization of Groundwater Aquifer System

V.S. Singh

Scientist, National Geophysical Research Institute
Hyderabad-500007, India

PREAMBLE

In order to assess groundwater potential in any area, and/or to evaluate the impact of pumpage on the groundwater regime, it is essential to know the aquifer parameters. These are chiefly Transmissivity (T) and Storativity (S). These parameters are also vital for the management of the groundwater resources through the use of groundwater flow model. These parameters are estimated either by means of in-situ test or performing test on aquifer samples brought in the laboratory. The applicability of the result from the laboratory test has limitations while in-situ tests give representative aquifer parameters. The most common in-situ test is pumping test performed on wells, which involves the measurement of the fall and rise of water level with respect to time. The change in water level (drawdown/recovery) is caused due to pumping of water from the well. The change in water level with time is then interpreted to arrive at aquifer parameters. Theis (1935) was first to propose a method to evaluate aquifer parameters from the pumping test on a bore well in a confined aquifer. Since then, several methods have been developed to analyze the pumping test data (time-drawdown) under different conditions. Before we describe these methods in detail, let us recall some of the definitions, which are frequently used.

WATER BEARING FORMATION

A geological formation, which can yield water in sufficient quantity to be of consequence as a source of supply is termed as aquifer. Such formations are porous and pores are interconnected (eg. sands). Some formations are porous but the interconnections between the pores are poor. These are termed as aquiclude (such as clay). The formations which have interconnected pores but not sufficient enough to allow significant horizontal flow through them, are called aquitard (eg. silt). Although significant groundwater does not flow through them but under certain conditions induced vertical flow (leakage) can take place across them. The formations, which do not have any pores or voids, are termed as aquifuge for example compact granite. Such rocks are impervious. The saturated rock formations in the nature form the aquifer under different conditions. For example, deposition of weathered material forms an aquifer. Under various conditions, various types of aquifers are formed which are described below.

Confined Aquifer

The completely saturated permeable formations when occur in between two impervious layers, the aquifer is termed as confined aquifer as shown in Fig. 1. In such aquifers the pressure of water is usually higher than that of the atmospheric pressure and therefore the water in the well stands above the top of the aquifer. This water level is called piezometric surface.

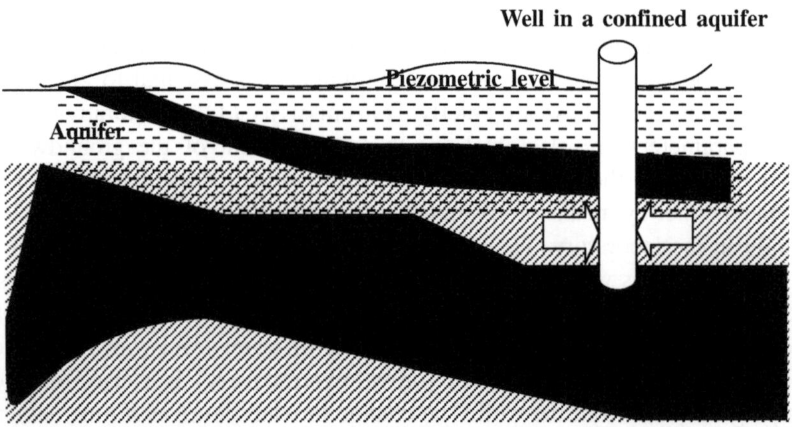

Figure 1. Illustration of confined aquifer condition.

Unconfined Aquifer

The permeable formation partly saturated with water, when rests over an impervious bed is called an unconfined aquifer (Fig. 2). The upper part of saturated zone is bounded by free water level called water table or phreatic level. The water level at the upper surface is at atmospheric pressure.

Semi-Confined Aquifer

In confined aquifer, when one of the confining layers is semi-pervious such as aquitard, the aquifer is known as semiconfined or leaky aquifer (Fig. 3).

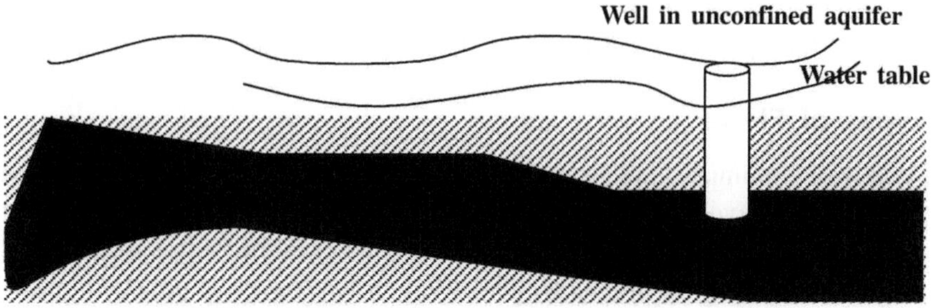

Figure 2. Illustration of an unconfined aquifer condition.

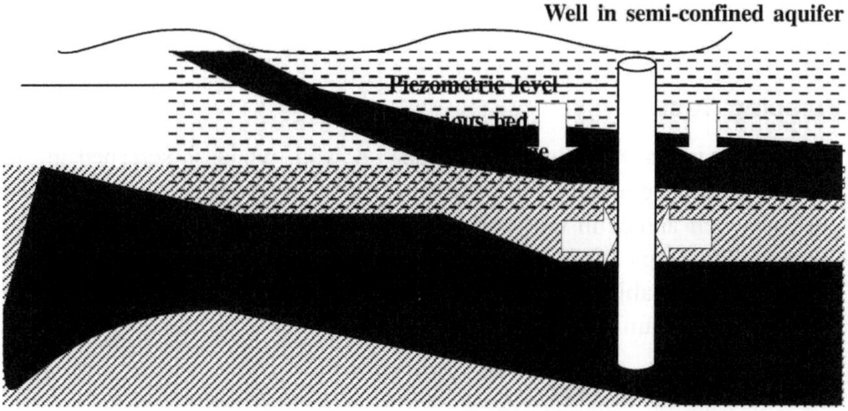

Figure 3. Illustration of a semi-confined aquifer condition.

As the well tapping the confined aquifer is pumped, the vertical flow (leakage) across the semiconfining layer is generated.

Hydraulic Properties of Aquifer

The occurrence and movement of groundwater in the aquifer are characteristically defined by two parameters namely transmissivity and storativity. Permeability is the property of the medium, which determines the ease of flow in the aquifer. It is often expressed as the quantity of water, which flows across the unit cross-section of the aquifer under unit hydraulic gradient per unit time. The unit is metre per day. Similarly the rates of groundwater flow through the cross-section of unit width and whole thickness of the aquifer, under unit hydraulic gradient is defined as transmissivity of the aquifer. It is the product of aquifer thickness and permeability. The unit of transmissivity is metre square per day.

The volume of water released from or stored into the aquifer is defined by the storage coefficient (storativity) or specific yield. It is the amount of water released or stored into the change in the head of aquifer normal to the surface area of the aquifer. The storage coefficient is a non-dimensional quantity. The storage coefficient refers to the confined part of the aquifer whereas the specific yield refers to the unconfined part of the aquifer.

Types of Flows

As a well is pumped, the groundwater flows towards the pumping well. The fall of water level is measured in the pumped well as well as at a distant observation well. As long as the fall of water level in the well is measurable, the flow towards the pumping well is called unsteady state or nonequilibrium.

As pumping continues, the rate of fall of water level in the well decreases. After some time the fall of water level in the well becomes negligible. Under such conditions the flow towards the well is called steady state or equilibrium.

Pumping Test

Pumping test is a very simple technique. A well, which is screened in the aquifer, is pumped and the fall of water level (drawdown) is measured in the pumping well and/or in the nearby observation well at regular interval. As the pumping stops the water level in the well rises and this rise is also measured at suitable intervals. These measured water levels are used to determine the aquifer characteristics.

Objectives of the Test

Pumping tests are conducted on wells with different purposes. The determination of yield of the well could be one of the purposes. The pumping test data provide information about the yield and drawdown of the well. These data are used to determine the specific capacity or discharge-drawdown ratio of the well. Thus the production capacity of the well is determined which helps in selection of pump capacity etc.

Further the pumping test data provides valuable information about the aquifer characteristics. The two characteristics namely Transmissivity (T) and Storativity (S) are determined from such data. These tests are also referred as aquifer test, which gives information about aquifer behaviour. These informations are essential for regional groundwater flow studies.

Subsurface Conditions

It is very essential to have the knowledge of subsurface geological and hydrological conditions of the region where the tests are to be conducted. The informations such as aquifer thickness, the behaviour of top and the bottom layers of the aquifer and the extent of the aquifer etc., are essentially required to analyse the pumping test data. It is also essential to have the knowledge of nearby boundaries such as rivers or dykes. Such informations can be collected from the geological map, the lithologs and records.

Site Selection

In certain cases the selection of site may not be possible. For example when test is to be conducted on an existing well. However, in certain cases, for example, regional groundwater flow study, site selection becomes essential and in such cases the following points should be considered to select the suitable site for pumping test.

- The test site should be representative of a larger part of area and hydrogeological conditions should not change over a short distance.
- The test site should be far away from disturbances such as pumping well, rail or road (highway).
- Provision should be made to discharge the pumping water in such a way that it does not return to the aquifer.
- The gradient of water table or piezometric surface should be low around the test site.
- Constant supply of power should be assured.

Well Construction

After suitable site is selected, the pumping well is drilled. The well is screened in the aquifer zone and rest is cased. The well should penetrate the entire thickness of the aquifer.

In case there is more than one aquifer, separated by confining or semiconfining layers, it is suggested to test each zone separately by screening them individually. Similarly observation wells are drilled away from the pumping well.

The pumping well should be equipped with such pump, which can run for longer time without any problem. After the installation of pump, pumping at low rate should develop the well. The pumping should continue till the pumped water is free from sand. During the development of well, the response in the observation well is also observed to make sure that the discharge and duration are enough to produce measurable drawdown in the observation well.

TEST PERFORMANCE

Pre-pumping Measurements

Before the commencement of pumping test it is very essential to make measurements of water levels in the pumping as well as in the observation wells. The measurements are repeated after regular intervals (say half an hour). In case appreciable difference is noted, sufficient time is given to get water level stabilized. In some cases the measurement of water level prior to the test is carried out for a couple of days. This gives the trend of variation in the water level, which can be used for correction of drawdowns obtained during the test.

Measurements during Pumping Test

During the pumping test mainly two measurements are made. These are water level measurements in the wells and abstraction rate from the pumping well.

The water level measurements are made in all the wells during pumping test at regular interval. The fall of water level in the initial hours of the test is rapid. Therefore the water level is measured at short interval during the initial period. As the test continues the change in water level becomes less and less. Hence at the later stage the interval of measurement is increased. The most practical time intervals are as follows:

Time since pump started	*Time intervals*
0-5 minutes	½ minutes
5-60 minutes	5 minutes
60-120 minutes	20 minutes
more than 120 minutes	60 minutes

Similarly the water level measurements are carried out in the wells after the pumping stops. During the recovery period also the water level rises rapidly during the initial hours. Hence the measurements of water level in the initial hours should be carried out at short intervals. The measurement of the water level should be continued till near complete recovery occur.

The discharge in most of the pumping test is monitored. It is kept constant throughout the pump test. The discharge is controlled and maintained at a constant rate. The discharge is measured at frequent intervals and is noted on data sheet. A stopwatch and a container of known volume are used to measure the discharge. A meter (manometer) is connected to the discharging pipe, which indicates the variation in the discharge rate. The other methods of measuring discharge are by means of orifice weir, orifice bucket and jet stream method.

Pumping Duration

The rate of fall of water level decreases as the pumping continues. After some time the increase in drawdown is very small. In other words near steady state condition has been achieved. The time to achieve such condition depends on the type of aquifer and the pumping rate. Therefore, it is difficult to decide in forehand how long the test should continue. However during the test as the increase in drawdown reduces considerably the test can be stopped. In confined aquifer such condition is reached soon whereas in unconfined aquifer it takes little time.

DATA PROCESSING

The time drawdown and discharge data collected during the pumping test is processed before it is used for analysis.

The time data is converted into single set of unit. The time, which is recorded into different units, is converted into single unit (say in days). Similarly the water level measurements are converted into single unit (say metre). The water level measured during the pumping test is converted into the drawdowns. All the set of water level measured in different wells are converted into the drawdowns. Similarly the water level measured during the recovery period is also converted into the residual drawdown.

These data are then plotted on single or double logarithmic paper, which forms time drawdown curve.

The regional trend of water level prior to test is also analysed and, if necessary, suitable correction is applied to time drawdown data.

The time-drawdown curve plotted on log-log sheet indicates the nature of aquifer (as shown in Fig. 4). However the study of litholog and the behaviour of water level in piezometers at different depths is essentially required to recognize the aquifer type and its behaviours. For example if the aquifer response indicates semiconfined nature it is essential to observe the water level behaviour in the semiconfining layer.

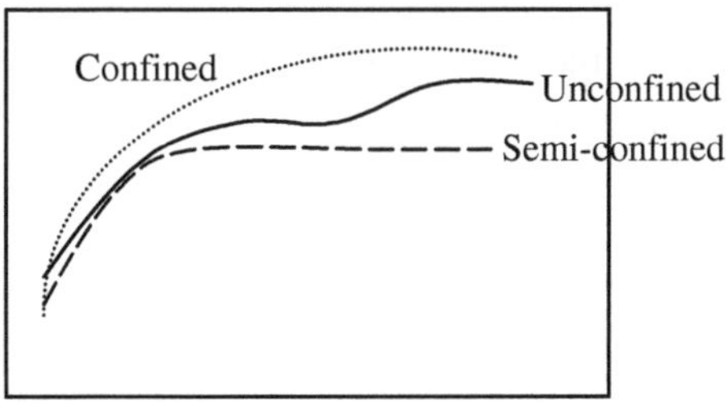

Figure 4. Characteristic curve for various aquifer conditions.

ANALYSIS OF PUMPING TEST DATA

Basic Assumptions

Darcy (1856) was first to establish experimentally that rate of flow across the porous media is proportional to the hydraulic gradient across it. He presented a formula as

$$q = \frac{Q}{A} = K\left(\frac{dh}{dl}\right) \tag{1}$$

where q is the rate of flow (Q) per unit cross sectional area (A being cross-sectional area), K is hydraulic conductivity; dh is difference in hydraulic head over a length of dl.

The above equation (1) has been considered as basis for the development of different formulas to calculate aquifer parameters under different conditions. Apart from the Darcy's law the following assumptions are made to develop solution of groundwater flow equations.

- The aquifer is extended to large area (infinite areal extent).
- The aquifer is homogeneous, isotropic and of uniform thickness.
- Prior to the pumping the piezometric surface or phreatic surface is (nearly) horizontal over the area influenced by pumping test.
- The pumping rate is constant.
- The pumping well penetrates the entire thickness of the aquifer.

As the well is pumped the groundwater flow towards the well becomes radially symmetrical which is shown in Fig. 5. The groundwater flow can be described as

$$\frac{\partial^2 s}{\partial r^2} + \frac{1}{r}\frac{\partial s}{\partial r} = \frac{S}{T}\frac{\partial s}{\partial t} \tag{2}$$

where s is drawdown at radial distance r and time t. The above equation is unsteady state groundwater flow equation in homogeneous, isotropic and confined aquifer. The boundary conditions can be described as follows:

- Initial drawdown in the well is zero,
$$s(0) = 0 \tag{3}$$
- Initial drawdown in the aquifer is zero,
$$s(r, 0) = 0 \tag{4}$$

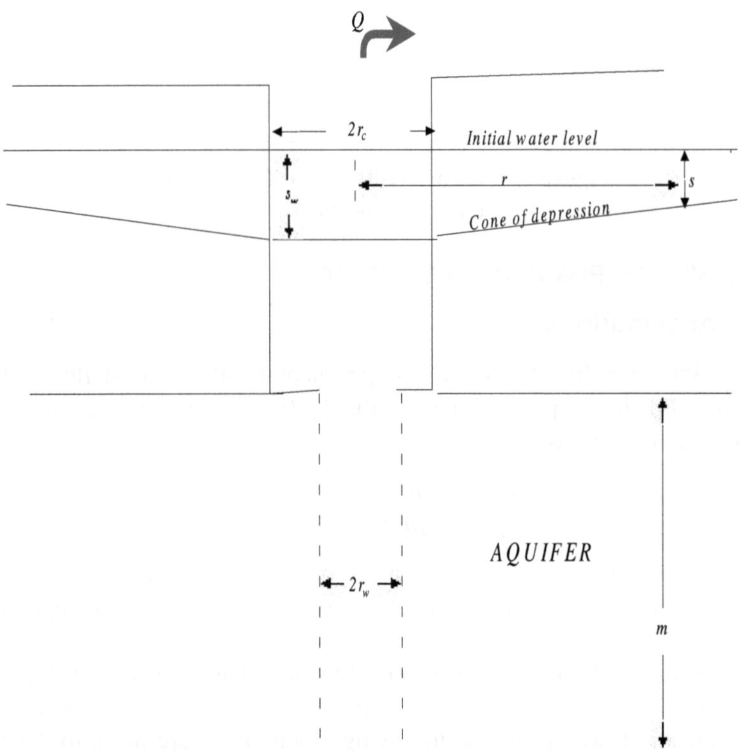

Figure 5. Flow towards the pumping well

- At any time the drawdown in the aquifer at the face of the well is equal to that in the well,

$$s(r_w, t) = s_w(t) \tag{5}$$

- At large distance the drawdown is zero at time t

$$s(\infty, t) = 0 \tag{6}$$

- The rate of discharge of the well is equal to the sum of the rate of flow of water into the well and the rate of decrease in the volume of water within well, which is not negligible in case of low permeable aquifer.

$$2\pi r_w T \frac{\partial s(r_w, t)}{\partial r} - r_c^2 \pi \frac{\partial s_w(t)}{\partial t} = -Q \qquad t > 0 \tag{7}$$

where s_w is drawdown in the well at time t, r_w is the effective radius of well screen, r_c is the radius of well casing, and Q is constant discharge during the test.

Using Laplace transform with respect to time and using the above conditions the solution to equation (2) can be written as

$$s = \frac{Q}{4\pi T} F(u, \alpha, \beta)$$

where $F(\ )$ is well function.

$$u = \frac{r^2 S}{4Tt}$$

$$\alpha = \frac{r_w^2 S}{r_c^2}$$

$$\beta = \frac{r}{r_w}$$

Pumping Test in Confined Aquifer

The pumping test in confined aquifer can be carried out under two different situations: (a) steady state and (b) non-steady state.

Steady State Flow in Confined Aquifer

The pumping test in confined aquifer, which shows negligible increase in drawdown during the test is called steady state test. Theim (1907) derived formula to estimate the aquifer parameter under such condition. The assumption as described in the previous section should satisfy the hydrogeological condition. In addition the flow towards the pumping well should be under steady state i.e. the negligible increase in drawdown. The formula to determine the transmissivity is given as

$$T = \frac{Q \ln (r_2 / r_1)}{2\pi(s_1 - s_2)} \tag{8}$$

where Q is the well discharge in m^3/d, T is the transmissivity of the aquifer in m^2/d, and s_2 and s_1 are steady state drawdowns in observation wells at distances r_2 and r_1 respectively.

At least two observation wells are required to use the above formula.

Unsteady State Flow in Confined Aquifer

The equation (8) can be used for the pump test in confined aquifer under steady state condition. However the steady state condition require long time to achieve. In some cases it is almost difficult to achieve the steady state condition. Theis (1935) developed methods to determine the aquifer parameters under non-steady state condition. The basic assumptions are same as mentioned in previous section. Additional assumptions are:

- The low towards the well is under nonsteady state.
- The water removed from the storage is discharged instantaneously with the decline of head.
- The diameter of the well is small enough so that the storage in the well is negligible.

The expression to determine the aquifer parameters are given as:

$$T = \frac{Q}{4\pi s} W(u) \tag{9}$$

and $$S = 4Ttu/r.r \tag{10}$$

where *W(u)* is well function, and s is the drawdown at distance r and time t.

Theis's Type Curve Method

Theis presented type curve which is the plotted values of *W(u)* and $1/u$ on log-log sheet. The field values of drawdown and time are plotted on similar log-log sheet. The field curve is matched with the type curve keeping the axes of the curves parallel (Fig. 6). In the matched position a match point is selected. The values of *W(u)* and $1/u$ for this match point is read from the type curve. Similarly the values of s and t are read from the field curve. Substituting these values of *W(u)*, $1/u$, s and t in equations (9) and (10) give the values of T and S.

Jacob's Semilog (slope) Method

Cooper and Jacob (1946) modified the Theis method and suggested an easy method to determine aquifer parameters. The method does not require curve matching. The time-drawdown data is plotted on semi-log paper (time on

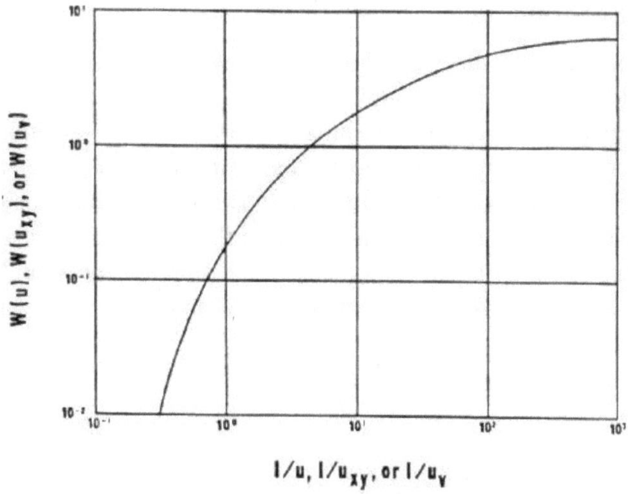

Figure 6. Theis type curve.

log axis and drawdown on linear axis). The additional condition required to use this method is that the test should be conducted for enough time so that the value of u should be less than 0.01. A straight line is drawn through the data points and the slope (Δs) of the straight line is determined. The time (t_0) is also noted where the straight line intercepts the time axis (i.e. at s = 0). The following formulas are used to calculate the aquifer parameters.

$$T = 2.3 \frac{Q}{(4\pi\Delta s)} \tag{11}$$

$$S = 2.25 \frac{Tt_0}{r \cdot r} \tag{12}$$

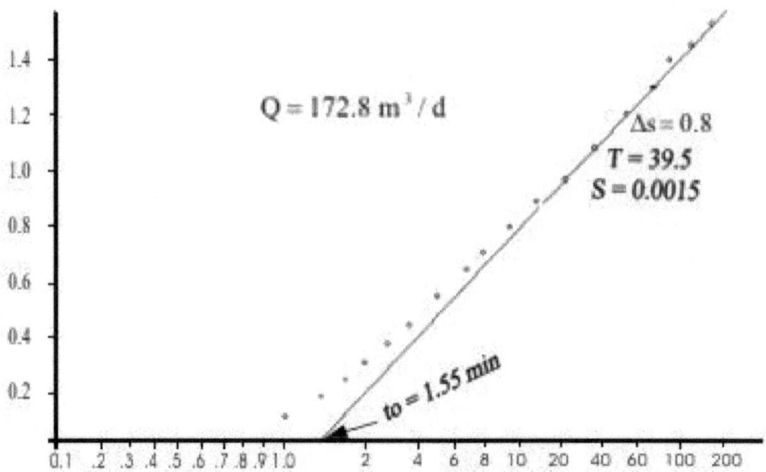

Figure 7. Jacob's semi-log plot.

Theis Recovery Formula

As the pumping stops the water level in the well starts rising. The water level data is collected in the similar way as during the pumping period. Residual drawdown is calculated similar to drawdown by subtracting the water level measurement prior to pumping, from the water level measurement during recovery period. The residual drawdown is plotted on semi-log paper against t/t' where t is total time and t' is the time since pump stopped. A straight line is fitted to the data plot and the slope of the straight line is determined as $\Delta s'$, the formula to calculate the aquifer is given as

$$T = 2.3 \frac{Q}{(4\pi\Delta s')} \qquad (13)$$

Semi-Confined Aquifer

Steady State Flow in Semi-confined (or leaky) Aquifer

The occurrence of confined aquifer in nature is not very common. The confining layers are found to be leaky in many cases such as alluvial aquifers. Under such situation the water is not only drawn from aquifer but also from the semiconfining beds. This is mainly due to pressure difference between the aquifer and semiconfined layers. Under such conditions the application of formula suggested in the previous section will give misleading results.

Hantush-Jacob's Method

Hantush and Jacob (1955) suggested semi-log method (similar to Jacob's method for confined aquifer), to find out aquifer parameters. The additional assumptions are made such as:

- Aquifer is semi-confined.
- Flow to the well is steady state.
- The water level in the semi-confining layers remains constant so that the leakage is proportional to the drawdown in the aquifer below it.

Further it is assumed that the value of $r/L < 0.05$ where L is the leakage factor defined as square root of (Tm'/k'); m' being the thickness of semi-confining layer and k' its permeability.

The values of maximum drawdown in the abstraction and observation wells are plotted on semi-log paper against their distances from the pumping well. The slope (Δs) of the straight line fitted to the data points is determined and substituted in the following formula to get aquifer parameter.

$$T = 2.3 \frac{Q}{2\pi\Delta s} \qquad (14)$$

Unsteady State Flow in Semi-confined Aquifer

The method to analyse the pumping test data from unsteady condition is similar to Theis method. The assumptions are same as described for steady state. The only additional assumption made is that the flow is under nonsteady state and water removed from the storage is instantaneously discharged with the declining of head. Walton (1962) described a method similar to Theis. He presented a family of type curves (Fig. 8). For each value of r/L there is a type curve. The field data is plotted on log log paper similar to the type curve. The match point is selected after suitable curve fitting is made. The values of well function $W(u, r/L)$, $1/u$, s, t and r/L are noted. The aquifer parameters are determined from the following formulas

$$T = \frac{Q}{4\pi s} W(u, r/L) \tag{15}$$

and

$$S = \frac{4Ttu}{r.r} \tag{16}$$

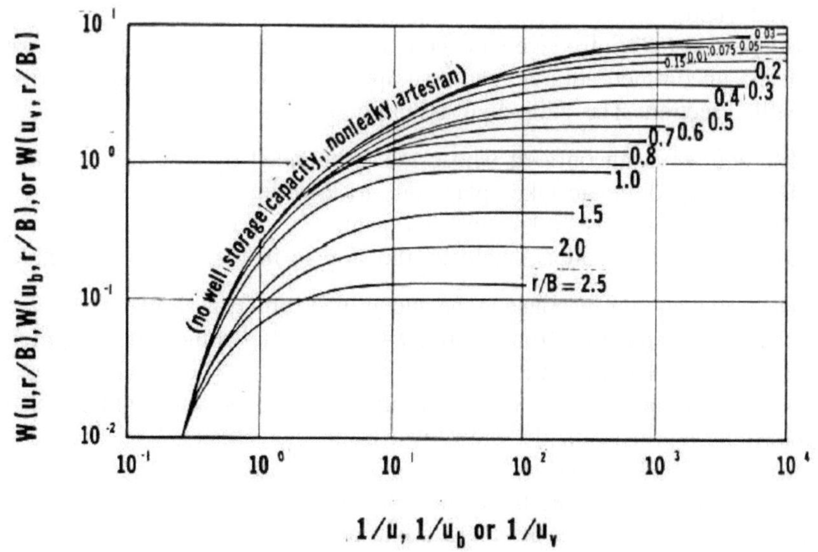

Figure 8. Leaky type curves.

Also the parameter c which is defined as hydraulic resistance of the semipervious layer is calculated as $c = L.L/T$, where the value of L can be calculated from the value of r/L.

Unconfined Aquifer

Steady State Flow in Unconfined Aquifer

In the pumping test of unconfined aquifer where the increase in the drawdown is negligible during pumping test, the Theim's eqn. (8) of confined aquifer can be used to calculate the transmissivity with following modification in the drawdown computation as:

$$s = s' - \left(\frac{s'^2}{2D}\right)$$

where D is the saturated depth of unconfined aquifer, s is the computed drawdown and s' is the drawdown observed. However we ignore many facts such as non-validity of Dupuit's assumption, seepage face, vertical flow etc near the well face.

Unsteady State Flow in Unconfined Aquifer

In the unconfined aquifer the water discharged by the pumped well is derived from the storage similar to confined aquifer. However additional drainage takes place i.e. the gravity drainage because of lowering of water table. The gravity drainage is not an instantaneous phenomenon. It takes place after a lag of time (delayed). Therefore an addition term is introduced to the flow equation, which defines the time lag release of water (i.e. delayed yield).

Boulton (1963) introduced method of analyzing the pumping test data from the unconfined aquifer. The assumptions involved are those described in earlier section. The additional assumptions are:

- The aquifer is unconfined whose top is bounded by free water table.
- The gravity drainage takes place due to lowering of water table.
- Flow to the well is under unsteady state.
- Diameter of the well is small i.e. the storage in the well is negligible.

The initial aquifer response in the unconfined aquifer is similar to confined i.e. the water is released from confined storage. After some time the gravity drainage starts and the rate of drawdown decreases. The gravity drainage does not continue for long, depending upon the fall of water table, it starts diminishing. As a result the drawdown again increases.

Boulton (1963) suggested family of type curves (Fig. 9) for different values of r/B to determine the aquifer parameters from the curve matching. The family of type curves are divided into two (i) initial A-type and (ii) later Y-type. The initial part of the field curve is matched with the A-type curves and the values of well function $W(u_a, r/B)$, $1/ua$, r/B, s and t are noted for the match point. The aquifer parameters are determined from the following formulas.

$$T = \frac{Q}{4\pi s} W(u_a, r/B) \tag{17}$$

and $$S = 4\ T.t.u_a/(r.r) \tag{18}$$

The storage coefficient value (S_a) thus calculated represents instantaneous release of water from the aquifer. Similar match is carried out with the later part of the field curve and type Y of the type curve having the same value

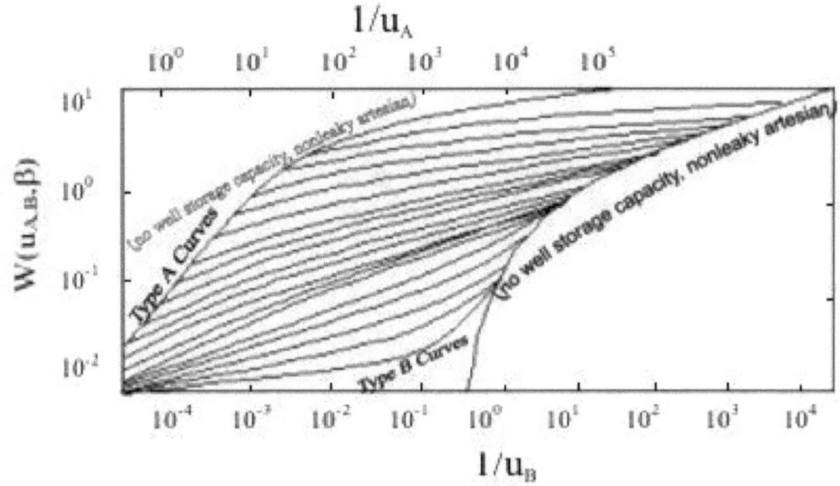

Figure 9. Boulton's Type curves.

of r/B as selected for early part match. The values of well function $W(u_y,r/B)$, $1/u_y$, s and t are noted for the match point. The aquifer parameters are again calculated using the following equations

$$T = \frac{Q}{4\pi s} W(u_y, r/B) \tag{19}$$

and

$$S = 4\,T.t.u_y/(r.r) \tag{20}$$

The equation (20) gives the value of specific yield of the material through which generally the drainage takes place.

The effective storage coefficient is the sum of S_a and S_y.

Numerical Method

During the pumping test in case of higher permeability of the aquifer the contribution from the aquifer becomes significant and well storage effect becomes insignificant. In such cases the method described by Theis (1935) can as well be used to estimate aquifer parameter. However, in the hard rock aquifers or in shallow thin aquifer due to poor permeability and high discharge rate, for most of the duration of the pumping phase, the well storage significantly affects the total discharge rate and the aquifer discharge to the well varies with time during pumping phase as found in a case study (Singh, 2000) which is shown in Fig. 10.

Therefore, the variable discharge rate test cannot be interpreted using Theis (1935) method, which gives ambiguous results. Further, in hard rock terrain where aquifers are of poor permeability, the aquifer response during the pumping period is almost negligible and hence it was suggested to include the recovery phase data to evaluate the aquifer parameters (Singh

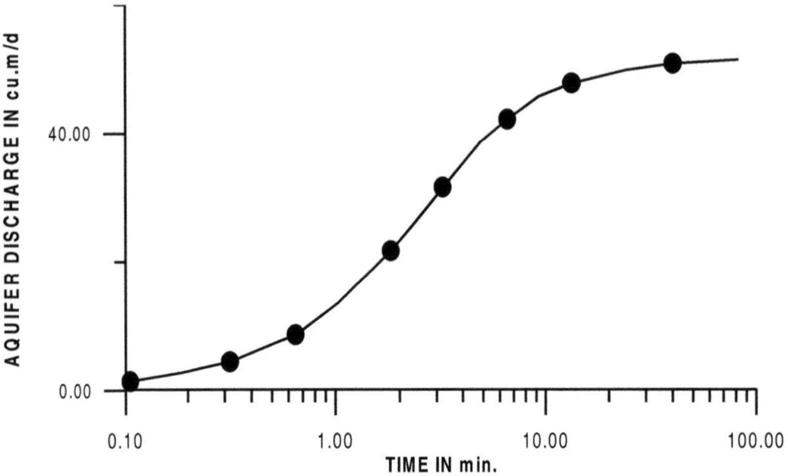

Figure 10. Variation in aquifer discharge rate.

and Gupta, 1986). Also, in the hard rock most of the shallow aquifers are of small-saturated thickness and during the pumping test there may be significant variation in the saturated thickness of the aquifer, particularly in the vicinity of the pumping well.

In order to consider the above-described boundary conditions, finite difference method (Rushton and Redshaw, 1979) has been considered to interpret the pumping test data.

The method involves solving the groundwater flow equation (1) using finite difference method. The method can also be employed to take into account variety of other boundary conditions, which are common in the field. The method requires discritization of the aquifer and the test duration. The radial distance from the centre of the pumping well is divided into increasingly discrete intervals (Δa = log r) as shown in Fig. 11.

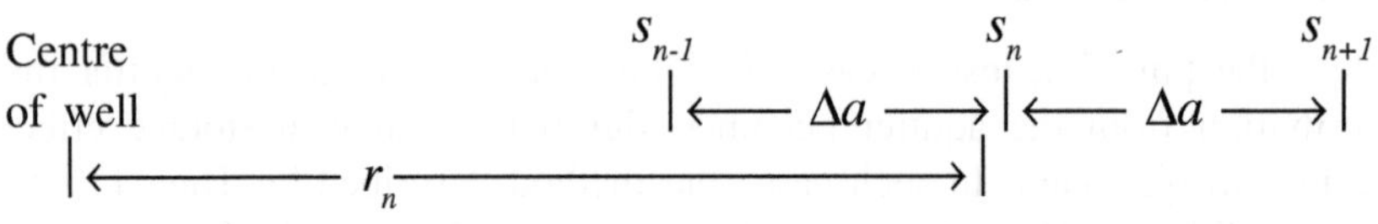

Figure 11. Radial finite difference mesh, logarithmic increase in mesh.

Similarly the duration of test is discritized as shown in Fig. 12.

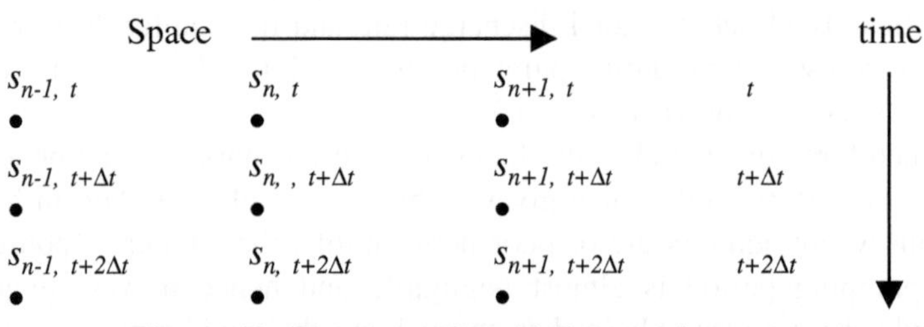

Figure 12. Representation of discrete time.

The boundary condition at the well (the discharge) and at the boundary is also prescribed in the similar terms as is expressed by Eqns (2) to (6). Thus finite difference expression is written as

$$\frac{mk_{\mathrm{r}}}{\Delta a^2}(s_{\mathrm{n+1}} - 2s_{\mathrm{n}} + s_{\mathrm{n+1}})_{t+\Delta t} = S\frac{r_{\mathrm{n}}^2}{\Delta t}(s_{\mathrm{n,t+\Delta t}} - s_{\mathrm{n,t}}) + Q_{\mathrm{t+\frac{1\Delta t}{2}}} r^2 \tag{21}$$

where s_{n} is the drawdown at the *n*th node of radial distance r and time t, k_{r} is the hydraulic conductivity and m is the saturated thickness of the aquifer.

The above equation (21) when written at various nodes of the finite difference model forms simultaneous equations, may be used to solve for drawdown.

The well storage is considered by assuming that aquifer extends into the region of the well. The properties of this region are considered differently so that it represents free water into the well in this model. The horizontal hydraulic resistance $\left(\frac{\Delta a^2}{mk_{\mathrm{r}}}\right)$ and time resistance $\left(\frac{\Delta t}{Sr_{\mathrm{n}}^2}\right)$ at the note representing well area are suitably modified to represent free water in the well. Initial guess values of aquifer parameters are used to calculate the time drawdown/recovery and matched with the observed time drawdown/recovery. The aquifer parameters are then varied to get a close match between observed and calculated time drawdown/recovery. The best fit of these curves gives representative aquifer parameters.

REFERENCES

Boulton, N.S. (1963). Analysis of data from non-equilibrium pumping tests allowing for delayed yield from storage. *Proc. Inst. Civ. Eng.*, **26:** 469-482.

Cooper, H.H. and Jacob, C.E. (1946). A generalized graphical method for evaluating formation constants and summarizing well field history. *Am. Geophys. Union Trans.*, **27:** 526-534.

Darcy, H. (1856). Les fountains publiques de la ville de Dijon. Paris, Victor Dalmont, p. 646.

Hantush, M.S. and Jacob, C.E. (1955). Nonsteady radial flow in an inifinite leaky aquifer. *Am. Geophy. Union Trans.*, **36:** 95-100.

Rushton, K.R. and Redshaw, S.C. (1979). Seepage and Groundwater Flow, John Wiley Publication, pp. 339.

Singh, V.S. (2000). Well storage effect during pumping test in an aquifer of low permeability. *Hydrological Sciences Journal*, **45(4):** 589-594.

Singh, V.S. and Gupta, C.P. (1986). Hydrogeological parameter estimation from pumping test on large diameter well. *Journal of Hydrology*, **87:** 223-232.

Theim, G. (1906). Hydrologische methoden. Leipzig, 56.

Theis, C.V. (1935). The relation between the lowering of the piezometeric surface and the rate and duration of discharge of a well using groundwater storage. *Am. Geoph. Union Trans.*, **16:** 519-524.

Walton, W.C. (1962). Selected analytical methods for well and aquifer evaluation, Ill. State Water Survey Hull., **49:** 81.

4

Application of Geostatistics in Hydrosciences

Shakeel Ahmed

Scientist, Indo-French Centre for Groundwater Research
National Geophysical Research Institute, Hyderabad-500007, India

PREAMBLE

Geostatistics based on the theory of regionalized variables has found applications in almost all domain of hydrogeology from parameter estimation to predictive modeling for groundwater management e.g., designing an optimal groundwater monitoring network, estimating parameters at unmeasured locations, groundwater model fabrication (optimal discretization), unbiased model calibration using estimation errors and in deciding the best models for prediction (Ahmed, 2001a).

Kriging was introduced as a least square estimator that improved on methods such as Squared distance, Inverse squared distance weighting or polynomial interpolation for which weighting was a determinant (Delhomme, 1978). In addition kriging yields both estimated values and estimated variances. Kriging, an estimator of the regionalized variable, is a potential tool in increasing the accuracy of the model calibration as it provides the best estimates of the system parameters. An advantage of kriging is that in the absence of measurement error it is an exact interpolator at measurement points (Delhomme, 1974). In hard rock areas where we encounter with complex geological setup, kriging of various parameters should be carefully done incorporating the available geological inputs.

Kriging, or best linear unbiased estimation (BLUE), has found many applications in mining, geology and hydrology. It shares with other variants of BLUE techniques the following features:

- The estimator is a weighted linear function of the known values with weights calculated according to the condition of unbiasedness and minimum variances. Unbiasedness means that on the average the error of estimation is zero. Minimum variance means that, again on the average, the squared estimation error is as small as possible.

- The weights are determined by solving a system of linear equations with coefficients that depend only on the variogram describing the structure of a family of functions (Kitanidis, 1997).

Kriging is an advanced interpolation procedure that generates an estimated surface from an *x-y* scattered set of points with *z* values (concentration or any other physico-chemical parameter for example). It is a weighted averaging method of interpolation derived from the theory of regionalized variable, which assumes that the spatial variation of a property, known as a 'regionalized variable', is statistically homogeneous throughout the surface and persists a definite structure. Kriging derive weights from the semi-variogram that measures the degree of spatial correlation among data points in a study area as a function of distance and direction between data points. The semi-variogram controls the way that kriging weights are assigned to data points during interpolation and consequently controls the quality of the results. Ordinary kriging is a well-known type of kriging interpolation technique of the kriging family that uses only the sampled primary variable to make estimates at un-sampled locations.

Matthew et al. (2002) carried out work using groundwater levels in the vicinity of pumping wells and these water levels are kriged using a regional linear and point logarithmic drift, the latter derived from the approximation to the Theis equation for drawdown in response to a pumping well. Residuals arising from using the most common (linear) drift to krige water levels in the vicinity of extraction wells often indicate large local departures from the linear drift, which correlate with areas of drawdown. The purpose of his work is to introduce a simple, single step kriging routine that improves on existing methods used with two-dimensional water level data. The linear-log kriging approach was developed as a practical solution to improve the level of interpretation possible from measured groundwater level data.

A large number of applications of geostatistics on groundwater hydrology include the first application by Delhomme (1974). Thereafter, a number of studies have been carried out in the field of hydrology. For example, Delhomme (1978), Gambolati and Volpi (1979), Mizell (1980), Darricau-Beucher (1981), Aboufirassi and Marino (1983, 1984), Neuman (1984), Hoeksema and Kitanidis (1984), Marsily and Ahmed (1987), Ahmed and de Marsily (1988), Thangarajan and Ahmed (1989), Dong et al. (1990), Ahmed and Murali (1992), Roth (1995), and Roth et al. (1996) amongst several others have demonstrated the use of geostatistics in groundwater hydrology. As a result of these studies it has become clear that multivariate and non-stationary geostatistics are comparatively more important in groundwater hydrology than some of the other geostatistical techniques. Further, some of the conventional techniques have to be suitably modified, and special procedures developed, for meaningful application of geostatistics in the field of hydrogeology. Some of the techniques used include kriging with linear regression and kriging in the presence of faults, which have been developed by Delhomme (1976). Conditional simulation has also been applied in aquifer modeling (Delhomme, 1979). Galli and Meunier (1987) and Ahmed (1987) have worked on kriging with an external drift extremely useful in practice for geohydrological parameters.

Ahmed and Marsily (1987) have compared a number of multivariate geostatistical methods for estimating transmissivity using data on transmissivity and specific capacity. Ahmed (1987) developed a special antisymmetric and anisotropic cross-covariance between residuals of hydraulic head and transmissivity based on the work of Mizell (1980) and used coherent nature of various covariances to cokrige transmissivity and hydraulic head for solving the inverse problem (Ahmed and de Marsily, 1987, 1989, 1993; Ahmed et al., 1990). Bardossy et al. (1986), Ahmed et al. (1988) and Kupfersberger and Bloschl (1995) have combined electrical and hydraulic parameters in geostatistical analysis.

THEORY OF REGIONALIZED VARIABLES

The most sophisticated technique for making predictions in groundwater hydrology, is numerical modeling of aquifer system either by solving the flow or the solute transport equations, wherein one simulates the discretized aquifer system utilizing various spatial parameters. Thus these parameters are required at a large number of locations depending upon the model scale or the discretization density. It is often very difficult to determine these parameters accurately in the field, and they are always measured at only a few places. Therefore, some tools are necessary to link the two situations, and to supply the parameter values for all the discretized volumes from the available observed data.

The deterministic approaches are not much applicable due to great variability of the hydrogeological parameters in space and time. The field heterogeneity of groundwater basins is often inextricable and very difficult to analyse with deterministic methods (e.g., Bakr et al., 1978; Delhomme, 1979; Ganoulis and Morel-Seytoux, 1985). In-situ measurements at the basin scale have shown that physical properties of the hydrological variables are highly irregular. However, this spatial variability is not, in general, purely random and the variables exhibit some kind of correlation in their spatial distribution. Matheron (1965, 1970, 1971) has given the name of "regionalized variables" to such parameters. According to him, "when a phenomenon spreads in space and exhibits certain spatial structure, we shall say that it is a regionalized variable. From the mathematical point of view, a regionalized variable is then simply a function of space but generally a very irregular function". Thus all the parameters used in groundwater hydrology, viz. transmissivity, piezometric heads, prediction, vertical recharge, etc., can be called regionalized variables.

Although the characterization and estimation of a Regionalized Variable (Re.V.) can be made on a purely deterministic basis, it is more convenient and common to introduce geostatistics in a probabilistic framework, bearing in mind that this artefact is only a tool for performing an estimation procedure. One, therefore, considers the aquifer studied as a realization of a random process. More precisely, let z be a property of the aquifer, say the transmissivity: we define Z as a Random Function (RF) $Z\ (x, \zeta)$. Here, x represents the coordinate in 1, 2, or 3 dimensional space, meaning that Z varies in space; ζ is called the "state variable" for a fixed ζ_1 value, $z\ (x, \zeta_1)$ is just an ordinary function of space, called a "realization" of $Z\ (x, \zeta)$. But ζ is assumed to be able to take an infinite number of values, and thus one can have an infinite number of "realizations" of the function Z. Each realization will, of course, be different, i.e. at any location x_o, two realizations ζ_1 and ζ_2 of Z will have different values:

$$z\ (x_o, \zeta_1) \neq z\ (x_o, \zeta_2)$$

In fact, for an infinite number of realizations, $z\ (x_o, \zeta)$ is a random variable, which can be characterized, for instance, by its probability distribution function (PDF), or by some of its first moments. The RF Z are assumed to have a "structure" that is to say that for any two locations in space x_1 and x_2, the random variables $z\ (x_1, \zeta)$ and $z\ (x_2, \zeta)$ are not independent: they are assumed to be correlated, and in general this correlation is a function of the separation between x_1 and x_2 in space.

Later on, we will use extensively the notion of the expected value of the RF Z: $E[z\ (x_o, \zeta)]$. This is sometimes also called the "ensemble average" of z at location x_o. This simply means that we take the average or expected value of the infinite number of realizations of the RF z at location $x_{o.}$ In other words, $E[z\ (x_o, \zeta)]$ is the first moment of the PDF of the Random Variable $z\ (x_o, \zeta)$, let us call it m_o. We can also define the second moment or variance σ_o^2, as

$$\sigma_o^2 = E[z\ (x_o, \zeta) - m_o^2] \qquad (1)$$

and the covariance between two points x_1 and x_2, as:

$$C\ (x_1, x_2) = E[(z\ (x_1, \zeta) - m_1)\ (z\ (x_2, \zeta) - m_2)] \qquad (2)$$

It is very important to realize that any averaging represented by the symbol E is taken over the ensemble of realizations, and not over space.

We can, of course, conceptually always define a RF $Z\ (x, \zeta)$, but this concept will be useful only if we can further assume that Z has two very important properties:

- Z displays some kind of stationarity, meaning that the PDF of $Z\ (x, \zeta)$ does not vary with x. For instance, 2nd order stationarity requires only that the first two moments of Z are not functions of x. The intrinsic hypothesis, defined later, is simply a 2nd order stationarity on the first increments of Z.

- Z satisfies the ergodic hypothesis. This assumption, which it is almost impossible to check in practice, says that Z displays, in space, the same behaviour as in the ensemble of realizations. In other words, by looking at one single realization, it is possible to infer the PDF of Z in the ensemble of realizations, or at least its first moments.

It is important to understand that these two rather theoretical hypotheses are just working hypotheses to enable us to develop a "tool" to make estimation. When we have this tool, we will check, by a special procedure, that the data are in agreement with the working hypothesis. If not, the "model" will be changed. Thus we do not claim that the property e.g. transmissivity is "stationary" or "ergodic" in nature. We simply propose a tool to estimate it, and will check that this tool is not inconsistent with the data.

In the remainder of this text, we will drop the notation for the state variable, and only define the RF $Z(x)$, bearing in mind that it can have an infinite number of realizations. The one available realization will be noted as $z(x)$.

It has been found, however, that in practice the hypothesis of 2nd order stationarity is often too restrictive, and not verified by some of the hydrogeological parameters, e.g. transmissivities. It often happens that the first moment m is indeed a constant, independent of x, but that the second moment σ^2 is apparently a function of the size of the domain of interest: the experimental variance (from the sample) does not appear to be finite when the size of the domain (and size of the sample) is increased. A weaker assumption, which can then be used is called the "intrinsic hypothesis". The increments of z, $[z(x+h) - z(x)]$ are in this case assumed to be stationary.

Constant mean:

$$E\,[z\,(x + h) - z\,(x)] = 0 \qquad \forall x,\ \forall h \tag{3}$$

Finite variance:

$$E\,[(z\,(x + h) - z\,(x))^2\,] = 2\gamma\,(h) \quad \forall x \tag{4}$$

$\gamma(h)$ is called the variogram, or sometimes the semi-variogram, and only depends on h.

When both the covariance and the variogram of a random function exists, i.e. in the stationary case, they are related, as can be demonstrated by:

$$\gamma\,(h) = C\,(0) - C\,(h) \tag{5}$$

or

$$\gamma\,(h) = \sigma^2 - C\,(h) \tag{6}$$

As h increases, the covariance decreases, but the variogram increases, and at a very large h the covariance becomes zero and the variogram takes the value of the field variance, i.e., after this separation, the variogram is constant, showing that there is no correlation among the data after that particular separation. This limiting value of the variogram is called its "sill" and the distance h at which it is reached is called the "range". However, in a truly intrinsic case, where the covariance cannot be defined, the variogram can increase indefinitely with the separation distance.

VARIOGRAPHY OF A HYDROGEOLOGIC PARAMETER

The theoretical aspects of geostatistics has been amply dealt with by earlier workers such as Matheron (1971), Journel and Huijbregts (1978), de Marsily (1986), Isaaks and Srivastava (1989), Samper and Carrera (1990), Deutch and Journel (1992), and Wackernagel (1995) amongst others.

Most of the hydrogeological parameters are defined and measured at points in a two-dimensional space. Therefore, all the derivations and examples are given here for two-dimensional space and point estimation is used.

The main steps involved in geostatistical techniques applied to hydrogeological parameters are: transformation of original variables if required, variography (i.e., structure analysis), cross-validation, estimation and backward transformation (if any). Variography that consists of modeling the spatial variability of a parameter is an important step and quality of the estimation depends on it. The procedure involves calculation of experimental variogram; its modelling by fitting valid variogram models, and its validation by estimation on points where the original field data are available.

Calculation of Experimental Variogram from Scattered Data

Unlike in mining or in many geophysical surveys, hydrogeological parameters are measured on scattered locations due to the fact that most of the parameters are collected from the wells, which exist in the vicinity of a village and not on a grid pattern. The low cost of the groundwater projects also often restricts systematic and extensive data collection. A generalised formula to calculate the experimental variogram from a set of scattered data can be written as follows (Ahmed, 1995):

$$\gamma(\underline{d},\underline{\theta}) = \frac{1}{2N_{\mathrm{d}}} \Sigma_{i=1}^{N_d} [z(x_{\mathrm{i}} + \hat{d}, \hat{\theta}) - z(x_{\mathrm{i}}, \hat{\theta})]^2 \tag{7}$$

where
$$d - \Delta d \le \hat{d} \le d + \Delta d, \theta - \Delta\theta \le \hat{\theta} \le \theta + \Delta\theta \tag{8}$$

with
$$\underline{d} = \frac{1}{N_{\mathrm{d}}} \Sigma_{i=1}^{N_d} \hat{d}_{\mathrm{i}}, \underline{\theta} = \frac{1}{N_{\mathrm{d}}} \Sigma_{i=1}^{N_d} \hat{\theta}_{\mathrm{i}} \tag{9}$$

where d and θ are the initially chosen lag and direction of the variogram with Δd and $\Delta\theta$ as tolerance on lag and direction respectively. $\underline{d}$ and $\underline{\theta}$ are actual lag and direction for the corresponding calculated variogram. N_d is the number of pairs for a particular lag and direction. The additional eqn. (9) avoids the rounding-off error of pre-decided lags and directions (only multiples of the initial lag and fixed values of θ only are taken in conventional cases). It is very important to account for every term carefully while calculating variograms. If the data are collected on a regular grid, Δd and $\Delta\theta$ can be taken as zero and $\underline{d}$ and $\underline{\theta}$ becomes d and θ respectively. Often, geohydrological parameters exhibit anisotropy and hence variograms should be calculated at least in 2 to 4 directions to ensure existence or absence of anisotropy. Of course, a large number of samples are required in that case.

Modeling of the Experimental Variogram

Calculation of the experimental variogram is subject to many approximations and its plot is often irregular. The experimental variogram is an estimate of the underlying variogram of the random function being studied. A theoretical variogram model is fitted to the experimental variogram. This procedure of fitting a permissible curve to an experimental variogram is called modeling. A theoretical variogram may be defined by a number of parameters e.g., sill, range, nugget effect and model type. The parameters of a variogram are often initially estimated by visual approximation and a smooth curve is fitted to the experimental variogram by this approximation technique. Sometimes, an experimental variogram is not satisfactorily fitted by any of the commonly used theoretical variograms. Linear combinations of several theoretical models are then often used and the resultant variogram is said to possess a nested structure. It is important to note here that variogram models must be positive definite functions so as to ensure unambiguous solution of the kriging system e.g., a negative variance of the estimation error. Nested models may be given as

$$\gamma^R(h) = \sum_{i=1}^{k} a_i \gamma_i(h) \tag{10}$$

where a_i are coefficient and $\gamma_i(h)$ are individual variograms. Fitting is usually done visually but often-automatic fitting using least squares and other techniques are also used. A measure of difference between theoretical and experimental variograms is always calculated to decide the best of several fits.

CROSS-VALIDATION TEST

Variograms obtained by modeling an experimental variogram are often not unique. It is, therefore, necessary to validate it. Cross-validation is performed by estimating the random function at the points where realizations are available (i.e. at data points) and comparing the estimate with data (Ahmed and Gupta, 1989). In this exercise, measured values are removed from the data set one by one and the same is repeated for the entire data set. Thus, all the measurement points, the measured value (z), the estimated value (z^*) and the variance of the estimation error (σ^2) become available. This leads to computing following statistics:

$$\frac{1}{N}\sum_{i=1}^{N}(z_i^{\text{obs}} - z_i^*) \approx 0.0 \tag{11}$$

$$\frac{1}{N}\sum_{i=1}^{N}(z_i^{\text{obs}} - z_i^*)^2 \approx \min \tag{12}$$

$$\frac{1}{N}\sum_{i=1}^{N}(z_i^{\text{obs}} - z_i^*)^2 \Big/ \sigma_i^2 = 1 \tag{13}$$

$$\left|z_i^{\text{obs}} - z_i^*\right| \Big/ \sigma_i \leq 2.0 \qquad \forall i \tag{14}$$

Various parameters of the variogram model are gradually modified to obtain satisfactory values of eqns. (11) to (14). During the cross-validation many important tasks are accomplished, such as

- Testing the validity of the structural model.
- Deciding the optimum neighbourhood for estimation.
- Selecting suitable combinations of additional information particularly in case of multivariate estimation.
- Sorting out unreliable data.

SPECIALIZED APPLICATIONS OF GEOSTATISTICAL VARIOGRAPHY

Use of Geological Soft Data in Variography

Often, qualitative information such as the geological and structural features of the area is not incorporated although they usually have much control on the variability of the parameters. Ahmed et al. (1995) have carried out variographic analysis of transmissivity, hydraulic head and depth of the aquifer basement of a weathered rock aquifer in Vaiper river basin of South India by dividing the aquifer into different zones. They found that in the zones free from geological structural features, as evidenced from a lineament map, the spatial characters of the regionalized variables are very different from the zones with the lineaments. They have concluded that in such areas zone-wise estimation may be preferred over simply using an anisotropic model.

Use of Data with Large Sampling and Analytical Errors

In practice a few special procedures have to be followed. Ahmed and de Marsily (1987) have shown that erroneous data may give an extra nugget effect equal to the variance of the error. De Smedt et al. (1985) provided a case study where the nugget effect was not modeled on assuming it to be the result of erroneous data. Although geohydrological data may have sampling errors, the nugget effect could also be the result of high variability at a scale smaller than the average minimum sampling interval. In case there are erratic values, the nugget effect can be reduced and its optimized value can be obtained by eliminating the erratic values during the procedure of cross-validation.

Procedures for a Fast Cross-Validation

Ahmed and Gupta (1989) have carried out an efficient way of performing cross-validation. One should first vary the range to arrive at an optimal value by checking equation (12). Equation (12) does not vary when the variogram model is multiplied by a constant. Therefore, the sill may be varied to arrive at a satisfactory value of eqn. (13). A suitable neighbourhood can then be chosen in the next step. If errors of estimation are still unacceptably high, a new model may have to be chosen.

ESTIMATION OF A REGIONALIZED VARIABLE (KRIGING)

Brief Description of the Theory of Estimation

The kriging technique was originally developed by Matheron (1965, 1971) to estimate regionalized variable such as the grade of an ore body at known location in space, given a set of observed data. Its application to groundwater hydrology has been described by a number of authors, viz. Delhomme (1976, 1978, 1979), Delfiner and Delhomme (1973), Marsily et al. (1984), Marsily (1986), Aboufirassi and Marino (1983, 1984), Gambolti and Volpi (1979, 1979a) to name a few.

If $Z(x)$ represents any random function (for instance, the transmissivity field in an aquifer) with values measured at n locations in space $z(x_i)$, $i = 1 \ldots n$ and if the value of the function Z has to be estimated at the point x_0, which has not been measured, the kriging estimate is defined as:

$$Z^*(x_o) \sum_{i=1}^{n} = \lambda_i \; z(x_i) \tag{15}$$

where $Z^*(x_0)$ is the estimation of function $Z(x)$ at the points x_0 and λ_i are the weighting factors. Now we impose two conditions to eqn (15), i.e. the unbiased condition and the condition of optimality.

The unbiased condition (also called universality condition) means that the expected value of estimation error or the difference between the estimated $z^*(x_0)$ and the true (unknown) $z(x_0)$ value should in the average be zero:

$$E\,[z^*(x_o) - z(x_o)] = 0 \tag{16}$$

The condition of optimality means the variance of the estimation error should be minimum.

$$\sigma^2 = \text{var } [z^*(x_o) - z(x_o) \,] \text{ minimum} \tag{17}$$

Note that the variance of estimation error σ^2_k is not a priori variance σ^2 of the random variable z. It is instead a measure of the uncertainty in the estimation of Z at an unmeasured location. By definition σ^2_k at a measured location is zero.

Using the expected value $E[Z(x)] = m$ and rewriting eqn (16) with the help of eqn. (15) we get:

$$\sum_{i=1}^{n} \lambda_i = 1 \tag{18}$$

Developing eqn. (11) we get:

$$\sigma_k^2 = \sum_{i=1}^{n} \lambda_i \sum_{j=1}^{n} \lambda_j E[Z(x_i)\, Z(x_j)] + E\,[Z(x_o)^2] - 2\sum_{i=1}^{n} \lambda_i\, E\,[Z(x_i)\, Z(x_o)] \tag{19}$$

Introducing a coefficient called Lagrangian multiplier μ and adding the term $-2\mu\, (\sum_{i=1}^{n} \lambda_i - 1)$ to the above equation we get:

$$Q = \sigma_k^2 - 2\mu \left(\sum_{i=1}^{n} \lambda_i - 1 \right)$$

$$= \sum_{i=1}^{n} \lambda_i \sum_{j=1}^{n} \lambda_j\, C(x_i\,,\, x_j) + C(0) - 2\, C\,(x_i\,,\, x_o) - 2\mu \sum_{i=1}^{n} \lambda_i + 2\mu \tag{20}$$

To minimize the above expression which is quadratic in λ and μ we equate $\partial Q/\partial \lambda_i$, $i = 1......n$ and $\partial Q/\partial \mu$ to zero and obtain the following kriging equations:

$$\sum_{j=1}^{n} \lambda_j\, C(x_i, x_j) - \mu = C(x_i, x_o) \qquad i = 1.....n \tag{21}$$

$$\sum_{j=1}^{n} \lambda_j = 1 \tag{22}$$

where $C(x_i, x_j)$ is the value of covariance between two points x_i and x_j. Now to express the value σ^2_k we multiply both sides of eqn (21) by $\sum_{i=1}^{n}\lambda_i$ and put the value of $\sum_{i=1}^{n}\lambda_i \sum_{j=1}^{n}\lambda_j \; C(x_i, x_j)$ into eqn (20) to get:

$$\sigma_k^2 = C(0) - \sum_{i=1}^{n}\lambda_i \, C(x_i, x_o) + \mu \tag{23}$$

which is the variance of the estimation error. Equations (21) and (22) are a set of $(n + 1)$ linear simultaneous equations with $(n + 1)$ unknowns and on solving them we get the values of λ_i, $i = 1 \ldots n$ which are later used to calculate the estimated value by eqn. (15) and the variance of the estimation error by eqn (23). Therefore, the square root of the expression given by eqn. (23) gives the standard deviation σ_k. That means true value will be within $z^* (x_0) \pm 2\sigma_k$ with the 95% confidence.

When we deal with intrinsic case i.e. working with variogram, which is common in hydrogeology the kriging eqns. (21) to (23) are simply modified as follows:

$$C\,(x_i, x_j) = C\,(0) - \gamma\,(x_i, x_j) \tag{24}$$

$$C\,(x_i, x_o) = C\,(0) - \gamma\,(x_i, x_o) \tag{25}$$

Equations (24) and (25) hold good only when both the covariance and the variogram exists, i.e. in the case of stationary variables. In case the covariance cannot be defined we can derive the following kriging equations:

$$\sum_{j=1}^{n}\lambda_j \, \gamma\,(x_i, x_j) + \mu = \gamma\,(x_i, x_o) \tag{26}$$

$$i = 1\ldots\ldots n$$

$$\sum_{j=1}^{n}\lambda_j = 1 \tag{27}$$

and the variance of the estimation error becomes:

$$\sigma_k^2 = \sum_{i=1}^{n}\lambda_i \, \gamma\,(x_i, x_o) + \mu \tag{28}$$

Expression in a matrix form:

$$\begin{bmatrix} \gamma_{11}\ \gamma_{12}\ \gamma_{13}\cdots\cdots\cdots\gamma_{1n}\ 1 \\ \gamma_{21}\ \gamma_{22}\ \gamma_{23}\cdots\cdots\cdots\gamma_{2n}\ 1 \\ \gamma_{31}\ \gamma_{32}\ \gamma_{33}\cdots\cdots\cdots\gamma_{3n}\ 1 \\ \\ \\ \gamma_{n1}\ \gamma_{n2}\ \gamma_{n3}\cdots\cdots\cdots\gamma_{nn}\ 1 \\ \\ 1\ldots\ldots 1\ldots\ldots 1\ldots\ldots\ldots\ldots\ldots 1\ldots 0 \end{bmatrix} \begin{bmatrix} \lambda_1 \\ \lambda_2 \\ \lambda_3 \\ \\ \\ \lambda_{n\lambda} \\ \\ \mu \end{bmatrix} = \begin{bmatrix} \gamma_{10} \\ \gamma_{20} \\ \gamma_{30} \\ \\ \\ \gamma_{n0} \\ 1 \end{bmatrix} \tag{29}$$

where γ_{nm} means $\gamma(x_n, x_m)$. In the short form it is written as:

$$[A]\,[\gamma] = [B]$$

or

$$[\gamma] = [A]^{-1}\,[B] \tag{30}$$

We have seen previously that after a certain distance, called the range, which also shows the zone of influence, the variogram generally takes an asymptotic value. Thus for estimating any variable at a particular point, it is not very useful to consider the data points beyond the zone of influence, called the neighbourhood of the estimated point x_0. Therefore, if we consider the points, which lie inside the zone of influence for estimating the variable at any given point, a smaller matrix will be used. This procedure is called the 'moving neighbourhood'. However, when we used the moving neighbourhood for each point of estimation, a different matrix has to be inverted separately for calculating the situation and this becomes very time consuming. If we work with the so-called 'unique neighbourhood', where all the available data points are considered simultaneously, it is sufficient to calculate the inverse of matrix $[A]$ only once. A simple multiplication is needed to get the kriged estimate to any location.

It is also clear that the value of $z^*(x_0)$ is equal to $[\lambda]^T\,[Z]$ where $[Z]$ is the column matrix of the measured values at the measurement points; therefore we can easily write:

$$z^*(x_o) = [\lambda]^T\,[Z] = \{[A]^{-1}\,[B]\}^T\,[Z]$$
$$= [B]^T\,[A]^{-1}\,[Z] \tag{31}$$

Since $[A]$ is a square, symmetric matrix it is unchanged when transposed

also $$z^*(x_0) = [B]^T[C] \quad \text{where} \quad [C] = [A]^{-1}\,[Z]$$

So in the case of the unique neighbourhood, matrix $[C]$ is calculated once and used for the estimation of all the points.

Similarly the matrix equation for the variance of the estimation error can be written as follows:

$$\sigma_k^2 = [\lambda]^T\,[B] = [B]^T\,[A]^{-1}\,[B] \tag{32}$$

UNIVERSAL KRIGING MOSTLY APPLIED TO WATER LEVEL ESTIMATION

Water levels generally being a non-stationary variable, Ordinary Kriging technique is not applicable and the technique of Universal Kriging is thus applied. The water level estimate at any point p (say a well) using Universal Kriging is written as follows:

$$h_p^* = D_p^* + \sum_{i=1}^{N} \lambda_i\,(h_i - D_i) \tag{33}$$

where h_p^* and D_p^* are the estimated water level and the drift respectively at any unmeasured well p; h_i and D_i are the values of water level and drift at well i and λ_i are kriging weights. The water levels h are split into a smoothly varying deterministic part D and the residual $(h–D)$. A bounded variogram could be obtained for the residual, as this is a stationary random function.

Modeling the drift correctly is difficult as it is not possible to estimate parameters of the drift and that of the variogram of residuals from a single data set. Drift depends on the nature of water level variation, and could be linear, quadratic or of any higher order. Usually a drift is approximated by polynomials of the space coordinates. For example, a linear drift can be written as:

$$D_i = a + bx_i + cy_i \quad \forall i \tag{34}$$

A quadratic drift can be written as:

$$D_i = a + bx_i + cy_i + dx_i^2 + ey_i^2 + fx_iy_i \tag{35}$$

where a, b, c, d, e and f are the drift coefficients and are constants. It is possible to estimate a drift by ordinary kriging but it has got the indeterminacy problem, as that requires the knowledge of the variogram of residuals.

The kriging weights and the drift coefficients can be determined as follows:

$$\sum_{i=1}^{N} \lambda_i \gamma_{ij} + a + bx_j + cy_j = \gamma_{jp} \qquad j = 1 \text{ ------------- } N \tag{36}$$

$$\sum_{i=1}^{N} \lambda_i = 1 \tag{37}$$

$$\sum_{i=1}^{N} \lambda_i x_i = x_p \tag{38}$$

$$\sum_{i=1}^{N} \lambda_i y_i = y_p \tag{39}$$

and the variance of the estimation error is given by:

$$\sigma_p^2 = \sum_{i=1}^{N} \lambda_i \gamma_{ip} + a + bx_p + cy_p \tag{40}$$

where γ are the variograms of residuals determined by the variographic analysis and i, j, p denote different points.

VARIOGRAPHIC ANALYSIS OF WATER LEVEL IN MAHESHWARAM WATERSHED

The water level data from a small watershed in granitic terrain (Fig. 1) with a fair density of measurement are analyzed geostatistically. The statistics of water level are shown in Table 1. The reduced level of groundwater varies from 591 m to 646 m in the area within the year. Although the mean is more or less constant, there is some difference of the water levels measured at different times of the year. The lowest and the highest variance months have very close mean. This shows that water levels have gone down in some areas and they have risen also at the same time in other areas. The variance increases during post-monsoon periods as the aquifer is hydro-geologically disturbed and the variability of rainfall recharge adds to the variability of the water levels while the water level during pre-monsoon periods are less disturbed due to absence of recharge component and pumping is more or less uniform resulting in a decrease of the variance of water level.

Table 1. Statistics of water level (AMSL)

Month	*Number of Wells monitored*	*Minimum*	*Maximum*	*Mean* (m)	*Variance* (σ^2)	*Standard Deviation* (σ)	σ/m
January, 2000	26	595	639	616	134	11.6	0.01883
April, 2000	27	593	636	613	124	11.2	0.01827
June, 2000	18	592	637	615	158	12.6	0.02048
July, 2000	28	591	639	614	130	11.4	0.01856
August, 2000	24	593	642	616	141	11.9	0.01931
September, 2000	28	596	646	618	147	12.1	0.01957
October, 2000	29	595	646	617	140	11.8	0.01912
November, 2000	29	593	642	616	136	11.6	0.01883
December, 2000	29	593	640	615	130	11.4	0.01853
January, 2001	29	593	640	615	129	11.3	0.01837
Average				615.5	136.9	11.69	0.01898

All the above statistical parameters in metres (m) while variance in m^2

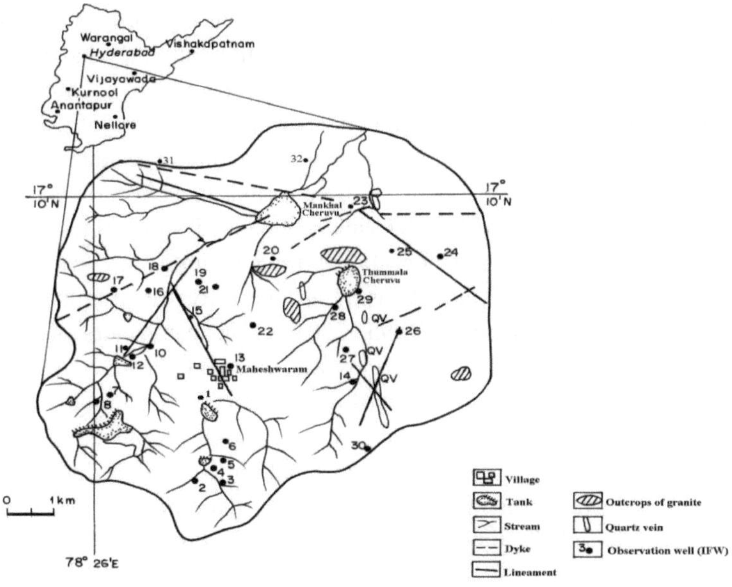

Figure 1. Location map of studied area with drainage system.

ESTIMATION AND MODELING OF EXPERIMENTAL VARIOGRAM OF WATER LEVEL

Calculation of experimental variograms and modeling with a theoretical variogram of the parameter is the first and primary step of the geostatistical estimation. Initially in the study area monthly water level data from 32 observational wells all from the existing private wells are collected for the study. Theory of 'regionalized variables' has been applied to analyze the water level variability. Geostatistical variographic analysis of the monthly water level was carried out starting from January 2000. During this process the variograms were calculated using individual monthly water level data with variable lags and corresponding tolerances to get the acceptable experimental modeled variogram. A theoretical variogram is fitted graphically by visual trial and error method. However, only stationary random function can have bounded variograms that becomes possible to model with theoretical variogram. The generalized formula used to calculate experimental variogram from a set of scattered data as given in eqn. (7) has been used for calculation of the variogram. The presence of drift in a parameter produces an unbounded variogram and it is not possible to model it with a finite sill and range. Therefore, drift has to be removed directly or indirectly to obtain a bounded variogram.

During the geostatistical analyses of water level data, initially unbounded variograms were obtained due to presence of a clear drift in water levels (Fig. 2). Since a typical water level map shown in Fig. 1 depicts a single direction of major flow, it was possible to calculate directional variograms. The experimental variograms calculated in the direction of flow and perpendicular to the direction of flow is shown in Fig. 3. It is clear that the major drift terms are present in the flow direction, and they have practically no contribution in the perpendicular direction. Thus calculation of experimental variogram in the perpendicular direction of major flow provides a bounded variogram that can be taken as variogram of residual. Calculation of experimental variograms for all the time periods in a direction perpendicular to the main flow direction (N-S) yielded bounded variograms.

Experimental variograms for all the time periods were calculated and modeled with theoretical models using an interactive programme "PLAYG" developed at NGRI (Ahmed, 1995, 2001a). The parameters of theoretical variograms for different time periods are given in Table 2. As shown in Table 2 all the variograms have zero nugget effect confirming continuity of the parameter as well as low variability at the origin. A higher sill and lower range are found for the variograms calculated for the periods during monsoon showing high variability of the phenomena. Experimental and theoretical variograms for all the time period are shown in Fig. 4.

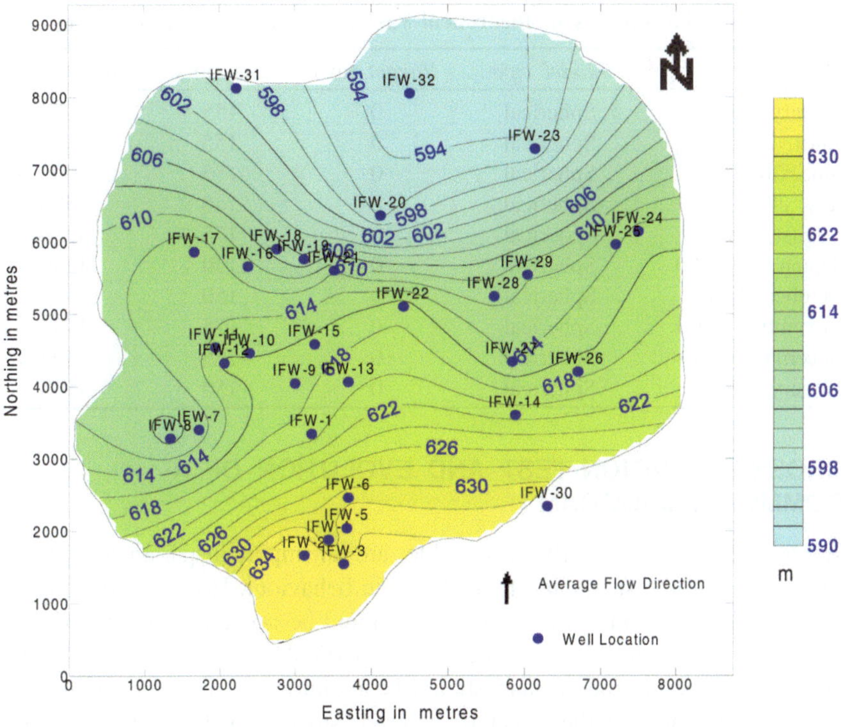

Figure 2. Water level Contour map, Jan. 2000 (Surfer without kriging). (Colour reproduction on Plate 4)

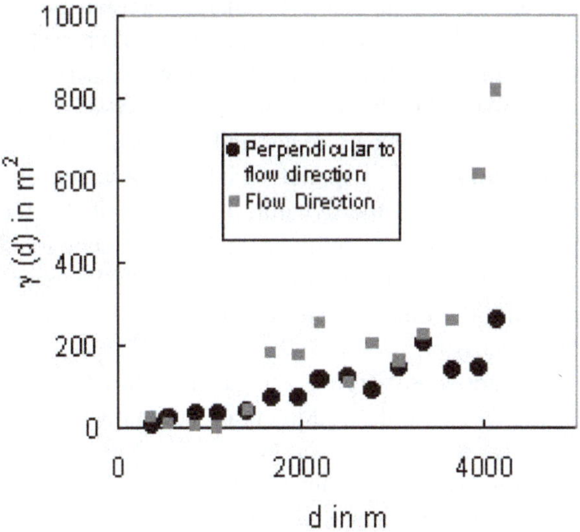

Figure 3. Experimental variogram of water level in the aquifer in two directions.

Table 2. Parameters of theoretical variogram after graphical fitting

Month	*Model type*	*Nugget* (m^2)	*Sill* (m^2)	*Range* (m)
January, 2000	Spherical	0	155	4000
April, 2000	Spherical	0	155	5000
June, 2000	Spherical	0	175	4000
July, 2000	Spherical	0	140	4500
August, 2000	Spherical	0	175	3750
September, 2000	Spherical	0	190	4200
October, 2000	Spherical	0	150	3500
November, 2000	Spherical	0	175	3500
December, 2000	Spherical	0	165	3000
January, 2001	Spherical	0	150	3500

CROSS-VALIDATION TEST AND EVOLUTION OF A COMMON VARIOGRAM

Water levels, being directly measurable, are an important parameter for the study of aquifer systems and their dynamic behaviour. However, the spatial and temporal variability of water levels should be accounted for in order to understand the connectivity of the aquifer particularly in hard rock regions. Water levels being time variant should be monitored at an adequately high frequency to indicate the effect of external stresses such as pumping and recharge.

Although water level in an unconfined aquifer is a continuous variable showing spatial variability, variations in rate of change arise due to the heterogeneity of the aquifer and the rainfall recharge. Geostatistical techniques using the theory of regionalized variables (Matheron, 1965) have been applied to study water levels in aquifers by a number of workers (Gambolati and Volpi, 1979; Mizell, 1980; Marsily et al., 1984; Dagan, 1985; Ahmed, 1985; Marsily, 1986; Dong et al., 1990; LaVenue and Pickens, 1992; Ahmed and Marsily, 1993 etc.). Both time and spatial variability of water levels should be studied jointly to understand the dynamic behaviour of aquifer systems. Rouhani (1990) has attempted a multivariate geostatistical model jointly using the time and space variability of water levels. However, practical application of analyzing the time and space variability of water level jointly has not been very successful.

Since water levels are time varying and are monitored using the same network of observation wells at desired intervals, estimating them for all the time periods following all the steps of geostatistical estimation becomes quite cumbersome. However, to account for the temporal variation in water levels it is possible to group water levels for certain time periods having similar behaviour and analyze them geostatistically for its spatial variability. Ahmed (1985) carried out such an exercise and showed that there could be

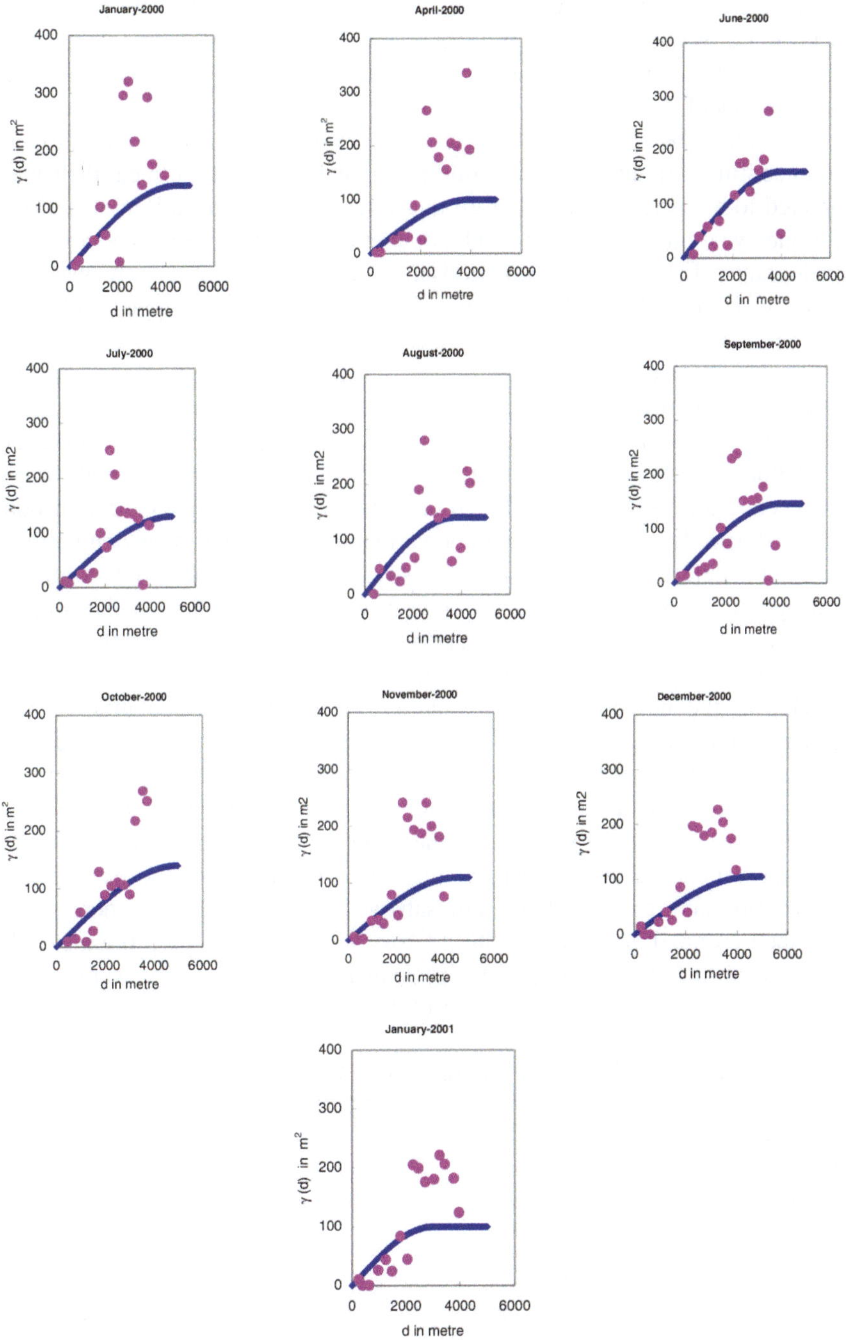

Figure 4. Experimental (dots) and Theoretical (curves) variograms for all periods.

a single variogram representing spatial variability for all the time periods in a year. La Venue and Pickens (1992) have used transmissivity and water levels jointly through groundwater flow equation and kriging to calibrate the aquifer model.

Generally the water table aquifer exists in the area under semi-confined or unconfined conditions. Universal kriging technique with a linear drift was applied to analyze the available groundwater levels during different periods for one cycle during the year 2000. An attempt was made to evolve common variogram(s) for different time periods in a year so that the frequently observed water levels could be estimated on the grids of an aquifer model and used for calibration.

As the major rainfall in India occurs during the monsoon season (July to October) and most of the water related activities, either natural such as recharge, or man made, such as agriculture with groundwater extraction for irrigation, are closely related to this period. It was therefore, hypothesized that a common variogram for the groundwater levels in monsoon or monsoon affected periods and another common variogram for the non-monsoon periods could be determined or evolved. Also an adequately derived common variogram for all the months of a year was used to analyze the spatial variability of water levels from various months in the year and the results of using the corresponding variograms were compared.

Due to the number of approximations made during calculation of the experimental variograms, the fitted models are bound to have ambiguities. Therefore, it was necessary to validate the variogram. Following eqns from (11) to (14) were used to cross-validate the variogram. Following norms decided by these equations are calculated on the average basis to decide the best theoretical variogram representing a true variability of the parameters. The variograms are modified until satisfactory values of these norms are obtained (Ahmed and Gupta, 1989). Cross validation was performed for all months starting from January 2000 to January 2001 and revised variograms for each time period were obtained (Table 3).

Table 3. Variogram obtained after Cross-Validation Test

Month	*Model type*	*Nugget* (m^2)	*Sill* (m^2)	*Range* (m)
January, 2000	Spherical	0	140	4500
April, 2000	Spherical	0	100	4000
June, 2000	Spherical	0	160	4000
July, 2000	Spherical	0	130	5000
August, 2000	Spherical	0	140	3750
September, 2000	Spherical	0	147	4200
October, 2000	Spherical	0	140	5000
November, 2000	Spherical	0	110	4500
December, 2000	Spherical	0	105	4500
January, 2001	Spherical	0	100	3000

To avoid calculation of experimental and theoretical variograms for each time period, common variograms for the entire period and/or groups of periods were worked out. Four different types of common variograms were determined and were cross-validated using observed values for all the months and the results were compared.

(i) A common variogram estimated by taking resulting nugget effect, sill and range as the mean of the individual variograms called a mean variogram.
(ii) A variogram with a high sill value (the highest sill obtained) and a low range (lowest range obtained) to represent the type of aquifer present in a hard-rock area such as the one studied.
(iii) A common variogram by averaging the nugget, sill and range for the monsoon periods was evolved (called monsoon variogram).
(iv) A common variogram was evolved in a similar way for non-monsoon period (called non-monsoon variogram).

All four common variograms (Fig. 5) were used for performing a cross-validation test with the observed value from the representative periods/months, and the results are compared in Table 4. In addition to the cross-validation test, values of water levels for the month of January 2000 were kriged using all the four considered variograms simultaneously. It is well established that kriging estimates are mostly affected by the particular field values and are less sensitive to the variograms; the kriged estimates for the four cases as depicted in Fig. 6a-d provide interesting features. Except a few lows that could be attributed to the local pumping, the water level contours are different in the northern and southern parts of the area.

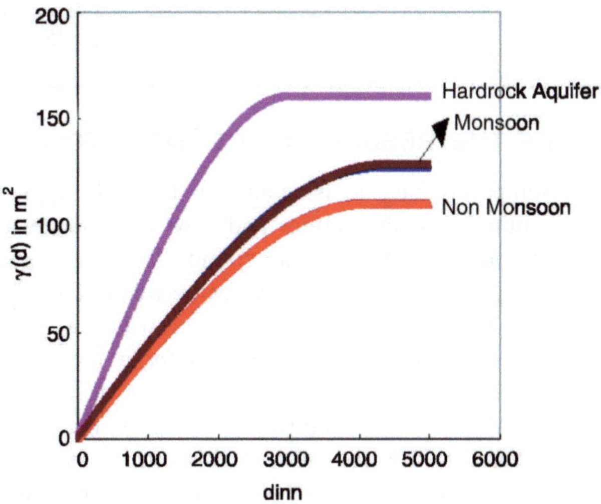

Figure 5. Various types of common variograms studied.

Table 4. Comparison of Cross-Validation with various common variogram

Month	*Month wise*		*Yearly Mean Variogram* $\gamma(d) = 0.0 + 127.2$ sph(4245)		*Considering hard rock variability* $\gamma(d) = 0.0 + 160$ sph (3000)		*Non-monsoon* $\gamma(d) = 0.0 + 110$ sph(4100)		*Monsoon* $\gamma(d) = 0.0 + 128.4$ sph(4390)	
	Eqn. 12	Eqn. 13	Eqn. 12	Eqn. 13	Eqn. 12	Eqn. 13	Eqn. 12	Eqn. 13	Eqn. 12	Eqn. 13
January 2000	38.45	1.03	38.10	1.07	43.28	0.65	38.37	1.20		
April 2000	24.78	0.69	24.62	0.57	29.97	0.36	24.99	0.65		
June 2000	63.49	1.02	65.06	1.38	60.58	0.77	64.74	1.54		
July 2000	35.75	0.97	35.92	0.83	40.86	0.50	36.13	0.93		
August 2000	44.67	0.72	47.34	0.95	45.16	0.51			47.93	0.99
Septem-ber 2000	36.44	0.80	36.29	0.93	41.51	0.56			36.36	0.96
October 2000	29.62	0.84	29.73	0.78	33.49	0.47			29.91	0.80
Novem-ber 2000	29.32	0.83	29.28	0.67	34.45	0.41			29.27	0.69
Decem-ber 2000	27.91	0.86	27.82	0.66	32.84	0.41			27.80	0.68
January 2001	33.50	0.66	28.52	0.67	33.50	0.41	28.87	0.75		

DISCUSSION OF THE RESULTS AND CONCLUSION

A number of assumptions have to be made for the field application to a technique, e.g., continuity of the aquifer and flow of groundwater. In this area not less than four sets of joints are found. The most prominent set strikes N-S on average basis. The strike varies between N10°E to N10°W. The joints are either vertical or dip very steeply towards west with angles 70° to 75°. All these joints and its system help in the percolation of groundwater from the surface. They also help to act as conduits for the transmission of groundwater. There are two sets of prominent dykes and the strike of the first set is E-S70°E and the second set strikes N50°E to N60°E. There are joints and fractures in the dykes also but the surfaces are curved.

These fracture also act as conduits for the transmission of surface water to the ground. The area is also traversed by lineaments, identified through

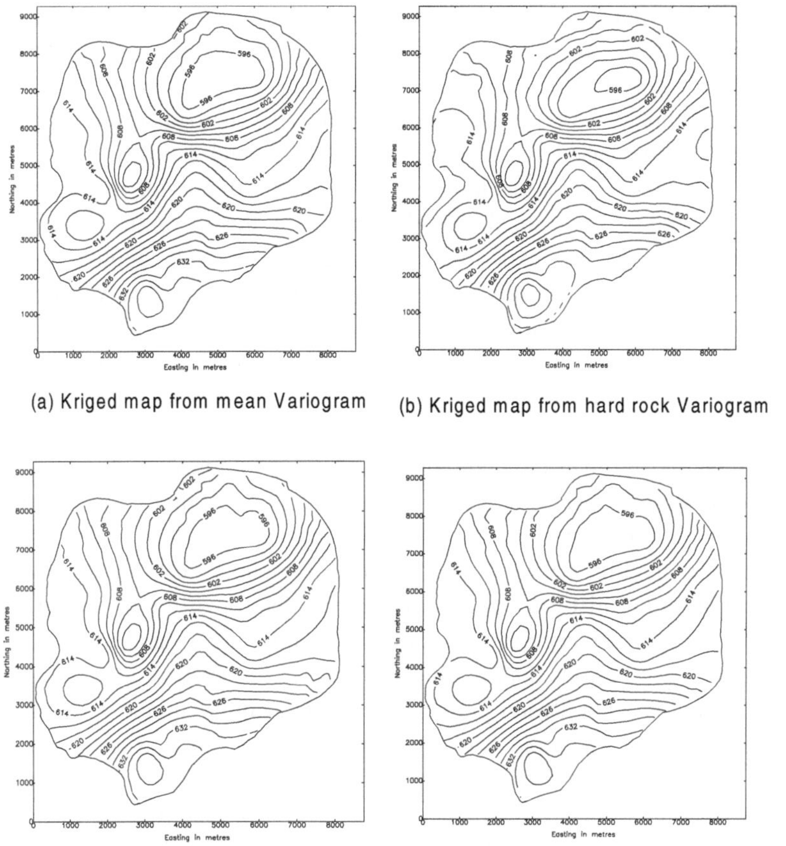

(a) Kriged map from mean Variogram (b) Kriged map from hard rock Variogram

(c) Kriged map from non-monsoon Variogram (d) Kriged map from monsoon Variogram

Figure 6. Water-levels contour resulting from kriged values using different variogram for the month of January 2000.

the study of aerial photographs and land-sat imageries. The high yielding wells aligned in particular direction on the ground coincide with those lineaments. In Fig. 6, only the dykes and lineaments are shown but not the joints as the joints are too small to be represented on the plan. The above fact ensures the continuity of the aquifer and allows us to calculate variogram using water level data from the entire area.

The calculation of variogram of the residuals of water levels has been possible as the groundwater flow conditions are simple and linear drift exists. In case, the water levels are more disturbed and it exhibits quadratic or higher order drifts, it becomes difficult to estimate the variogram of the residuals using directional variogram.

The averaging of the variograms is also allowed, as this is very similar to the case of nested variogram or nested covariance (Journel and Huijbregts, 1978; Wackernagel, 1995) where a variogram can be a weighted average of several independent variograms. However, according to Marsily and Ahmed (1987), the cross-validation tests are quite strong tools and if a variogram is capable of reproducing the observed data to a satisfactory level, it is not very important to worry as to how one has arrived to that variogram. We have extensively used this property of the cross-validation in our present study. The cross validation tests are much more important in this case as obviously seen from Fig. 3 that the fitting of the theoretical variograms through the respective experimental variograms are very approximate and that situation could not be improved due to the limited observed data and the constraint of calculating the variogram in a particular direction.

The results are further compared in Fig. 5, which shows dominance of the observed data over the variograms, and also that estimates using the hard rock variogram are not acceptable.

The number of measurement points for hydrogeological studies are always less resulting in a sparse data set as compared to other fields such as mining etc. Application of the theory of regionalized variable thus requires special effort in analyzing the variability of the parameter. This study shows that performing geostatistical analysis of a parameter for different time periods, it is not absolutely necessary to carry out variographic analysis separately for all the time periods that, at times, are quite cumbersome and ambiguous.

Although it is not always possible to evolve a single unique variogram for all time periods, the study with the help of cross-validation test has shown that if the year cycle is divided into two parts, monsoon and non-monsoon periods, it is possible to evolve a common variogram for each part. The common variograms during the validation test could satisfactorily reproduce the measured values of water level without losing much on the outcome (square mean estimation error and square mean reduced error) as compared to cross validation test using individual variograms. This was possible for water levels as both the input (rainfall recharge) and the output (groundwater withdrawal for irrigation) more or less follow a cyclic pattern each year. Even in the case of low or high rainfall years, the spatial variability of water levels could be assumed to remain same.

The evolution of common variogram helps in analyzing water levels for all the time periods that could be used, for e.g. calibration of an aquifer model etc. The study also helps in estimating water levels at any time period when all the wells could not be monitored due to any reason. This would have, otherwise, been extremely difficult because the variographic analysis in the absence of sufficient measurement becomes much more ambiguous.

ESTIMATION OF WATER LEVEL TIME SERIES

Water level is the only parameter that is measured from the aquifer directly and also depicts the dynamics of the aquifer. It is measured normally in static condition but in spite of all precautions taken while measuring the water level there are some minor variations due to its time varying nature. In aquifer modeling, particularly in the calibration of the model in transient condition a sufficiently long time series of water level is required on all the wells that are used. So to deal with two different situations that is to have a sufficiently long time series of water level at any well, geostatistics was applied to fill the data gap, at the time when the water level could not be measured. With this we obtained a mixed time series of water level at any well containing measured water levels without any error term and the estimated water level with estimation errors at uniform frequency. Results of a few wells have been reported here.

ONE DIMENSIONAL VARIOGRAM OF WATER LEVEL IN TIME

In the present study using the water-level data from 25 experimental observation wells, the 1-D variogram analysis using eqns (7) to (9) was carried out individually for each well to obtain the water level variability in time. The variogram was calculated and fitted with the theoretical variogram at all the monitoring wells (IFP's).

WATER LEVEL ESTIMATION IN TIME

Ordinary kriging of the geostatistics was used for estimation as given the monsoon climate in the area and if there is no heavy pumping the water level time series could be assumed stationary. The estimation were made at the month(s) for each well where the water level was not monitored due to some unavoidable cause e.g., due to the pumping in the nearby well. Thus the water level was estimated using the existing observed data and the corresponding variogram's model parameters. The estimation was done using the block estimation technique by ordinary kriging using eqns (44) to (47), which gives the estimated water level along with the variance of the estimation error.

CONCLUSIONS

This new procedure helps in reconstructing the time series of water levels in all the control wells with uniform data. These estimated values are plotted along with the variance of the estimation error (in the form of 2σ values) and with the observed water level in the series as hydrographs. This study also helps in an unbiased model calibration as the simulated head could be verified against some value that was absent otherwise. But the plot of the error permits to check that the simulated heads are within the permissible limits. Figures 7 (a-d) show such series graphically for four wells in the area.

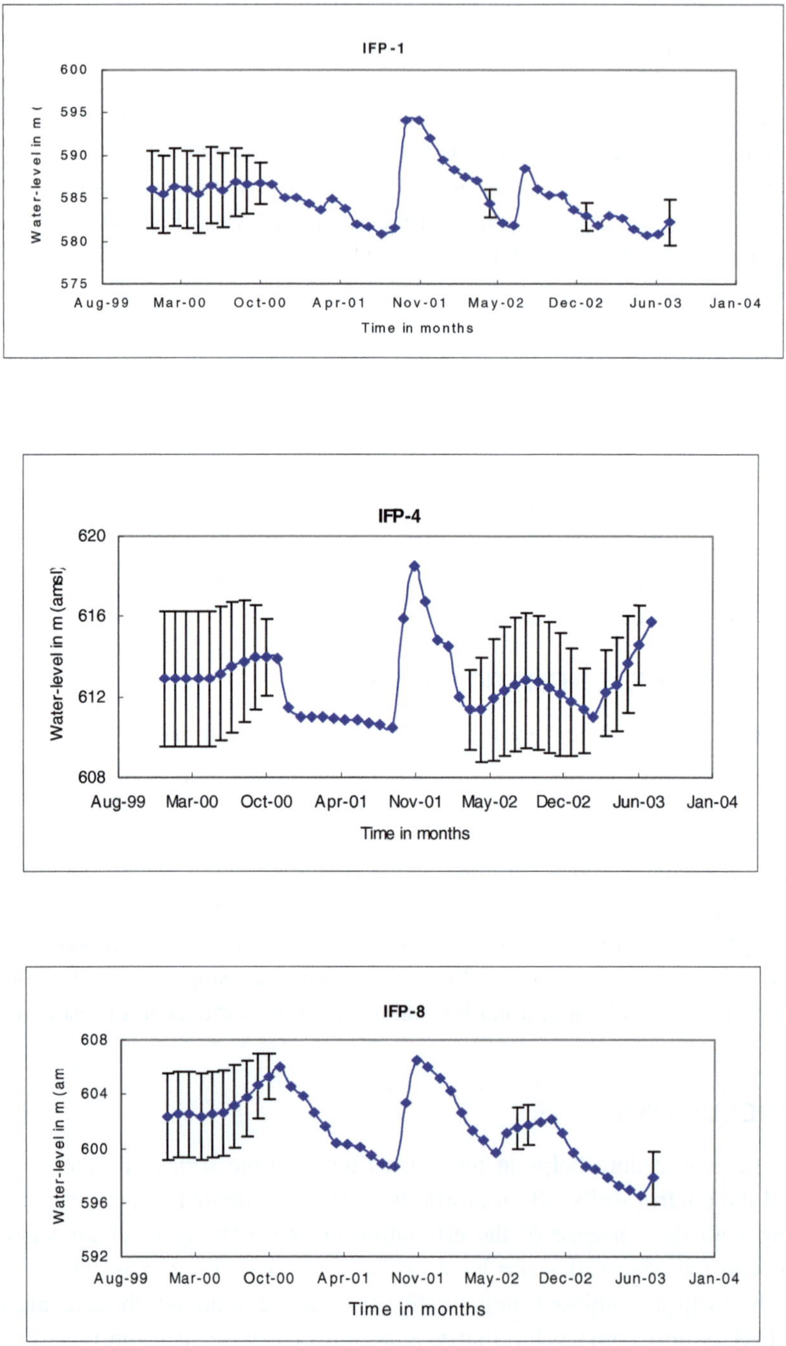

Figure 7a-c. Observed and estimated water level time series along with its standard deviation: a-Well IFP-1; b-IFP-4; and c-Well IFP-8.

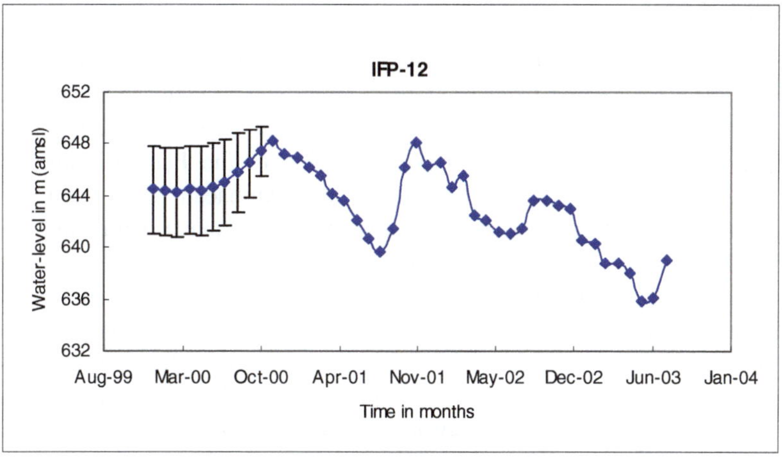

Figure 7d. Observed and estimated water level time series along with its standard deviation: Well IFP-12.

ANALYSES OF FLUORIDE CONTENT IN GROUNDWATER

Fluoride is often found in the aquifers belonging to hard rock terrain due to the presence of fluorine-bearing minerals. Since fluoride pollution is a class of natural pollution, it adds two major constraints to the problem of tackling the groundwater pollution, in general. The source of pollution is unidentified as well as the highest or the initial concentration is also unknown. Most of the data are obtained by analyzing the water samples randomly collected from the wells. In the absence of the knowledge about its source, constraints of feasibility of data collection and random sampling makes the monitoring network, generally, inadequate. The geostatistical technique based on the theory of regionalized variables (Matheron, 1971), which provides two unique advantages viz., (i) the estimates with the variance of the estimation error and (ii) independence of the calculation of revised variance of the estimation error on the measured values, can help in the following two ways:

- Regionalize the information obtained from the random sampling to delineate the maximum and minimum concentrations with their error bounds and thus identifying the possible location of source(s) of Fluoride Pollution, and
- Decide to add new monitoring wells and/or avoid redundant sampling to design an optimal network for the monitoring of this natural pollutant.

A case study from an aquifer in Anantapur district of Andhra Pradesh is presented for the application of the above method. The technique of Ordinary Kriging was used which is well documented in the literature e.g., Journel and Huijbregts (1978), Marsily (1986), Isaak and Srivastava (1989), Delhomme (1978) etc. and so they are not described here.

STUDY AREA

The area studied is situated in Anantapur district of Andhra Pradesh, India (Fig. 8). The geological formations in the area can broadly be divided into an older group of metamorphic rocks belonging to the Archaean and a younger group of sedimentary rocks belonging to the Petrozoic age. The latter covers the northeastern part of the area. Archaean rocks occupy the remaining part, which consists of schists, gneisses, quartz veins and basic dykes. The granite rocks show pronounced changes in colour, mineralogy, chemical composition, structure and texture within a short distance. The principal fluorine bearing minerals such as fluorite and fluoropatite, are responsible for high concentration of fluoride in groundwater, as they release fluoride under normal pressure and temperature. The factors governing distribution of fluoride in groundwater depend upon the amount of fluorine in the source rock, duration of contact of water with rocks, temperature, rainfall and vegetation. The dissolved solids including fluoride in water increase considerably due to high evapotranspiration under the hot, semi-arid climate conditions prevalent in the region. The other factor controlling the amount of fluoride present in water is the amount of weathering of the granite rocks of the area.

AVAILABLE DATA AND THEIR GEOSTATISTICAL ANALYSIS

Water samples from 146 dug-wells have indicated the presence of fluoride almost in all the wells (Ahmed and Murali, 1992). The values ranged from 0.4 to 6.0 ppm (with a mean of 1.98 ppm and a variance 0.813 ppm^2) with 99 wells showing the F content above the permissible limit of 1.5 ppm. This is quite obvious from the value of mean of the data.

Variographic analysis by calculating a mean experimental variogram and cross-validation tests were performed as usually carried out in the geostatistical analysis. The parameter was taken as isotropic as already verified by Ahmed and Murali (1992). The finally taken variogram is as follows:

$$\gamma(h) = 0.038 + 0.016 \text{ sph } (7.0)$$

Then using the validated variogram model, F was estimated at a uniform grid of 1 km side. The maps of F estimates and the variance of the estimation error together were used to produce a new map (Fig. 9) showing the following:

(i) Low fluoride zone (F<1.5 ppm) with high estimation reliability.
(ii) High fluoride zone (F>1.5 ppm) with high estimation reliability.
(iii) Low fluoride zone with low estimation reliability and
(iv) High fluoride zone with low estimation reliability.

The estimated F values range from 0.6 ppm to 4.05 ppm with the value of standard deviation of the estimation error (σ_k) ranging from 0.1 to 1.152. However, the objective has been to design a network of monitoring wells so that on estimating F content anywhere in the area, the standard deviation of

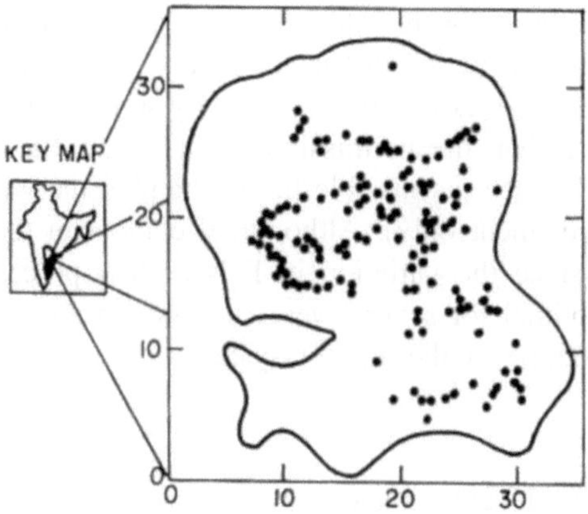

Figure 8. Location of wells in the existing monitoring network.

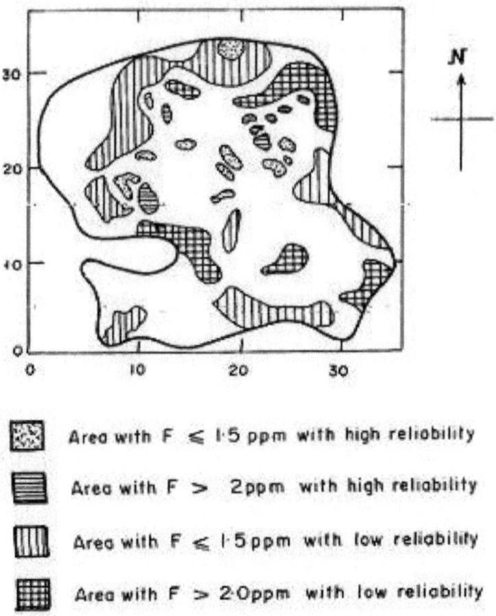

Figure 9. Estimated fluoride in the entire district with their estimation reliability.

the estimation error should not exceed a pre-set value σ_c= 0.5. Therefore, the monitoring wells were removed from the area where σ_k were less than σ_c and at the same time additional wells were selected in the area where σ_k values were more than the σ_c values (Rouhani, 1985; Gao et al., 1996). After several trials, a revised network was arrived having only 119 wells (Fig. 10). Using only 119 wells selected for monitoring, the calculated σ_k ranged between 0.113 and 0.585. Of course, a few points where σ_k was slightly greater than the $\sigma_{c,}$ were ignored. They were mainly points at the corners or falling just out of the district boundary. This exercise helped in reducing the above four zones into two zones only so that one can demarcate the areas with F less than the permissible limit (allow the users to drink water) and F more than the permissible limit (abandon the wells in the area or use water after treatment only). Although F content in the groundwater is time variant but since the variation in F is a slow process, an optimal network could be optimal for several years unless some major remediation modifies the flow regime of the aquifer system.

CONCLUSIONS

A procedure has been developed and applied to design an optimal network of monitoring Fluoride pollution with a desired precision of its estimation elsewhere. The random monitoring wells were re-selected and a revised network of 119 wells (which is less in number compared to the existing network of 146 wells) was prepared. The revised network ensures that F estimates made anywhere in the area will provide an estimated value with a desired error of estimation. Fluoride being time variant pollutant requires frequent monitoring and an optimal monitoring network will also add cost-effectiveness to the data collection.

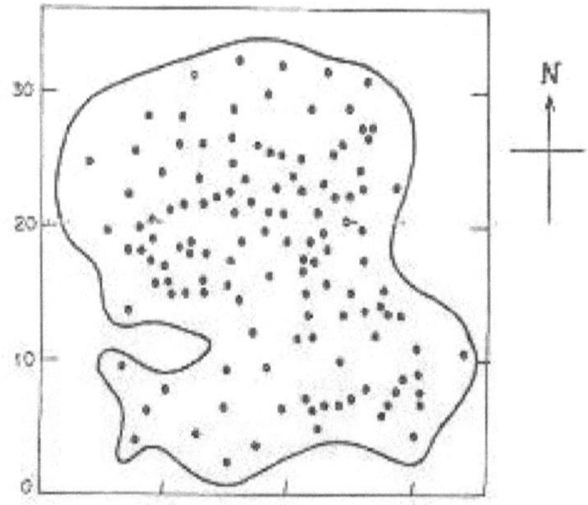

Figure 10. Location of wells in the revised optimal network.

GENERAL CONCLUSIONS

The deterministic approaches are not much applicable due to great variability of the hydrogeological parameters in space and time. The field heterogeneity of groundwater basins is often inextricable and very difficult to analyse with deterministic methods. Another advantage of using stochastic approach is that it provides the variance of the estimation error together with the estimated values. Of course, there are many advantages of these methods particularly in groundwater modeling:

(i) The closer the values of the aquifer parameters to reality, the faster will be the model calibration. Better estimated values (with lower estimation variance) are initially assigned to the nodes of aquifer model using geostatistical estimation.

(ii) An assumption is made in Aquifer Modeling that a single value of system parameter represents the entire mesh (Of course, very small). Averaging over a block in two or three dimension can be obtained through block estimation.

(iii) An optimal mesh size and number of nodes in discretizing aquifer system can be obtained and best location of new control points can be predicted.

(iv) A confidence interval given by the standard deviation of the estimation error provides a useful guide to T modification at each mesh and to check that the calculated heads fall inside the confidence interval of the observed heads.

(v) A performance analysis of the calibrated model can be achieved to decide the best-calibrated model using variance of the estimation error, which can be used for prediction.

The geostatistical estimation (particularly the block estimates) has the advantage of providing uniform value over the mesh of the aquifer and also over the isolated and variable size of the meshes. The most important contribution of the geostatistical estimates is with the fact that it provides the values of the estimation variance and with this magnitude the aquifer model could be calibrated at all the meshes in an unbiased way. The following criterion could be used:

$$\left|\frac{h_i^m - h_i^*}{\sigma_i}\right| < 2.0 \qquad \forall i$$

where h_i^m, h_i^* and σ_i are the model head, estimated head and standard deviation of the estimation error at the ith mesh.

Using these constraints, the model is calibrated at all the meshes and would be free from the biasness of the aquifer modeler. The model parameters should then be changed on the meshes where the above criteria are not satisfied. A faster and unbiased model calibration thus could be achieved.

Evaluation of a common variogram is risky, in general, but the advantages of skipping the variographic analysis which is much more cumbersome for the sparse data, provide extra reward. Water levels thus could be estimated for a large number of time periods without much effort. This will also help in keeping a smaller time interval for calibration of aquifer model in the transient condition.

The analyses of Fluoride analyses in groundwater using geostatistics has provided two fold applications viz., (1) estimation of F at all the locations in the area and (2) providing zones of potable and unsafe water as well as indicating places where more monitoring are required.

Acknowledgements: Most of the case studies reported come from a project funded by the IFCPAR, New Delhi and author is thankful to the Director, NGRI for providing facilities to carry out the research and prepare this manuscript.

REFERENCES

Aboufirassi, M. and Marino, M.A. (1983). Kriging of water levels in Souss aquifer, Morocco. *J. Math. Geol.*, **15(4):** 537-551.

Aboufirassi, M. and Marino, M.A. (1984). Cokriging of aquifer transmissivities from field measurements of transmissivity and specific capacity. *J. Math. Geol.*, **16(1):** 19-35.

Ahmed, S. (1985). Analyse geostatistique de quelques variables hydrogeologiques. Report C.F.S.G. S-184, CGMM, Ecole des Mines, Fontainebleau, France.

Ahmed, S. (1987). Estimation des transmissivites des aquiferes par methods geostatistique multivariables et resolution du probleme inverse. Doctoral thesis. Ecole Nationale Superieure des Mines, Paris, France, 148 pp.

Ahmed, S. (1995). An interactive software for computing and modeling a variogram. *In:* Mousavi and Karamooz (eds.) Proc. of a conference on "Water Resources Management (WRM'95)", August 28-30. Isfahan University of Technology, Iran, 797-808.

Ahmed, S. (2001). Rationalization of Aquifer Parameters for Aquifer Modelling Including Monitoring Network Design, Modelling in Hydrogeology. UNESCO International Hydrological Programme. Allied Publishers Limited, 39-57.

Ahmed, S. and Gupta, C.P. (1989). Stochastic spatial prediction of hydrogeologic parameters: role of cross-validation in krigings. International Workshop on "Appropriate Methodologies for Development and Management of Groundwater Resources in Developing Countries", Hyderabad, India. February 28-March 4, 1989. Oxford & IBH, IGW, Vol. III, 77-90.

Ahmed, S. and Marsily, G. de (1993). Cokriged Estimation of Aquifer Transmissivity as an Indirect Solution of the Inverse Problem: A practical Approach. *Water Resour. Res.*, **29(2):** 521-530.

Ahmed, S. and Marsily, G. de (1987). Comparison of geostatistical methods for estimating transmissivity using data on transmissivity and specific capacity. *Water Resour. Res.*, **23(9):** 1717-1737.

Ahmed, S. and Marsily, G. de (1988). Some application of multivariate kriging in groundwater hydrology. *Science de la Terre, Serie Informatique*, **28:** 1-25.

Ahmed, S. and Marsily, G. de (1989). Cokriged estimates of transmissivity using jointly water levels data. *In:* M. Armstrong (ed.), Geostatistics, pp. 615-628. Kluwer Academic Publishers.

Ahmed, S. and Murali, G. (1992). Regionalization of fluoride content in an aquifer. *Jour. of Environmental Hydrology*, **1(1):** 35-39.

Ahmed, S., Marsily, G. de and Gupta, C.P. (1990). Coherent structural models in cokriging aquifer parameters—Transmissivity and water levels. *In:* Proc. of

International Conf. on Water Resources in Mountainous Regions Lausanne, Switzerland, Aug. 27-Sept. 1. Internat. Asso. of Hydrogeol. Pub., Vol. XXII, Part 1, 173-183.

Ahmed, S., Marsily, G. de and Talbot, A. (1988). Combined use of hydraulic and electrical properties of an aquifer in a geostatistical estimation of transmissivity, *Groundwater*, **26(1):** 78-86.

Ahmed, S., Sankaran, S. and Gupta, C.P. (1995). Variographic analysis of some hydrogeological parameters: Use of geological soft data. *Jour. of Environ. Hydrology*, **3(2):** 28-35.

Bakr, A.A., Gelhar, L.W., Gutjahr, A.L. and McMillan, J.R. (1978). Stochasitc analysis of spatial variability in subsurface flow. Part 1: Comparison of one and three dimension flows, *Water Resour. Res.*, **14(2):** 263-271.

Bardossy, A., Bogardi, I. and Kelly, W.E. (1986). Geostatistical analysis of geoelectric estimates for specific capacity. *J. Hydrol.* **84:** 81-95.

Dagan, G. (1985). Stochastic modeling of groundwater flow by unconditional and conditional probabilities. *Water Resour. Res.*, **21(1):** 65-72.

Darrican-Beucher, H. (1981). Approche geostatistique du passage des donnees de terrain aux parameters des modeles en hydrogeology. Doctoral thesis., Ecole Nationale Superieure des Mines, Paris, France, 119 pp.

De Smedt, F., Belhadi, M. and Nurul, I.K. (1985). Study of spatial variability of hydraulic conductivity with different measurement techniques. *In:* G. de Marsily (ed.), Proc. of Conference on the Stochastic Approach to Sub-surface Flow, Motvillargen, France, June 3-6.

Delfiner, P. and Delhomme, J.P. (1973). Optimum Interpolation by kriging. *In:* "Display and Analysis of Spatial Data". J.C. Davis and M.J. McCullough (eds.), 96-114, Wiley and Sons, London.

Delhomme, J.P. (1976). Application de la theorie des variables regionalisees dans les sciences de l'eau. Doctoral Thesis. Ecole Des Mines De Paris Fontainebleau, France.

Delhomme, J.P. (1978). Kriging in the hydrosciences, Advances in Water Resources, 1 no. 5: 252-266.

Delhomme, J.P. (1979). Spatial variability and uncertainty in groundwater flow parameters: a geostatistical approach. *Water Resour. Res.*, **15(2):** 269-280.

Delhomme, J.P. (1974). La cartographie d'une grandeur physique a partir des donnees de differentes qualities *In:* Proc. of IAH Congress. Montpelier. France. Tome X, Part-1, 185-194.

Deutsch, C.V. and Journel, A.G. (1992). GSLIB, Geostatistical Software Library and User's guide. Oxford Univ. Press, New York, 340 pp.

Dong, A., Ahmed, S. and Marsily, G. de (1990). Development of geostatistical methods dealing with boundary condition problems encountered in fluid mechanics. *In:* Proceedings, 2nd European Conference on the Mathematics of Oil Recovery, Arles, France, September 11-14 (1990). edited by Guerillot and Guillon, 21-30, Technip, Paris.

Galli, A. and Meunier, G. (1987). Study of a gas reservoir using the external drift method. *In:* G. Matheron and M. Armstrong (eds.), Geostatistical Case Studies, 105-120. D. Reidel Publ. Co., Dordrecht.

Gambolati, G. and Volpi, G. (1979a). Groundwater contour mapping in Venice by Stochastic Interpolators, 1. Theory. *Water Resour. Res.*, **15(2):** 281-290.

Gambolati, G. and Volpi, G. (1979). A conceptual deterministic analysis of the kriging technique in hydrology. *Water Resour. Res.*, **15(3):** 625-629.

Ganoulis, J. and Morel-Seytoux, H. (1985). Application of stochastic methods to the study of aquifer systems. Technical Documents in Hydrology. UNESCO, Paris.

Gao, H., Wang, J. and Zhao, P. (1996). The updated Kriging Variance and Optimal sample design. *Math.Geol.*, **28(3):** 295-313.

Hoeksema, R.J. and Kitanidis, P.K. (1984). An application of the geostatistical approach to the inverse problem in two-dimensional groundwater modeling. *Water Resour. Res.*, **20(7):** 1003-1020.

Isaak, M. and Srivastava, R.M. (1989). An introduction to applied geostatistics. Oxford Univ. Press, New York, 561 pp.

Journel, A.G. and Huijbregts, Ch J. (1978). Mining Geostatistics. Academic Press, London, 26-95.

Kitanidis, P.K. (1997). Introduction to Geostatistics: Application to Hydrogeology. Cambridge University Press, pp. 249.

La Venue, A.M. and Pickens, J.F. (1992). Application of a coupled adjoint sensitivity and kriging approach to calibrate a groundwater flow model. *Water Resour. Res.*, **28(6):** 1543-1569.

Marsily, G. de (1986). Quantitative Hydrogeology: Groundwater hydrology for Engineers. Academic Press, 203-206.

Marsily, G. de., Lavedan, G., Boucher, M. and Fasanini, G. (1984). Interpretation of interference tests in a well field using geostatistical techniques to fit the permeability distribution in a reservoir model. *In:* Geostatistics for Natural Resources Characterization, part 2, edited by G. Verly et al., 831-849. D. Reidel, Hingham, Mass.

Marsily, G. de and Ahmed, S. (1987). Application of kriging techniques in groundwater hydrology. *Jr. Geol. Soc. of India*, **29(1):** 47-69.

Matheron, G. (1963). Traité de Géostatistique appliquée. Mémoires du BRGM, No. 14, Edition Technip.

Matheron, G. (1965). "Les variables regionalisèes et leur estimation." Masson, Paris.

Matheron, G. (1970). LeTheorie des variables regionalisees et ses applications. Tech. Rep. N 5, CGMM, Ecole des Mines, Fontainebleau, France.

Matheron, G. (1971). The theory of regionalized variables and its application. Paris School of Mines, Cah. Cent. Morphologie Math., 5. Fontainebleau.

Matthew J. Tonkin and Steven P. Larson (2002). Kriging Water Levels with a Regional-Linear and Point-Logarithmic Drift. *Groundwater*, **40(2):** 185-193.

Mizell, S.A. (1980). Stochastic analysis of spatial variability in two-dimensional groundwater flow with implications for observation-well-network design. Ph.D. thesis. N.M. Inst. of Min. and Technol., Socorro.

Neuman, S.P. (1984). Role of geostatistics in subsurface hydrology. *In:* G. Verly, M. David, A.G. Journel and Marechal, A. (eds.), Geostatistics for Natural Resources Characterization, pp. 787-816. Proc. NATO-ASI, D. Reidel Publ. Co., Dordrecht.

Roth, C. (1995). Contribution de la geostatistique a la resolution du probleme inverse en hydrogeologie. Doctoral thesis. Ecole Nationale Superieure des Mines, Paris, France, 195 pp.

Roth, C., Chiles, J.P. and de Fouquet, C. (1996). Adapting geostatistical transmissivity simulations to finite difference flow simulators. *Water Resour. Res.*, **32(10):** 3237-3242.

Rouhani, S. (1985). Variance reduction analysis. *Water Resources Research*, **21(6):** 837-846.

Rouhani, S. (1990). Multivariate geostatistical approach to space-time data analysis. *Water Resour. Res.*, **26(4):** 585-591.

Samper, F.J. and Carrera, J. (1990). Geoestadistica—Aplicaciones la Hidrologia Subterranea. Barcelona Univ., 484 pp.

Thangarajan, M. and Ahmed, S. (1989). Kriged estimates of water levels from the sparse measurements in a hard rock aquifer. *In:* C.P. Gupta et al. (eds.), Proc. of Internat. Groundwater Workshop (IGW-89), Hyderabad, India, Feb 28 to March 4. Oxford and IBH, New Delhi, **1:** 287-302.

Wackernagel, H. (1995). Multivariate Geostatistics: An introduction with applications. Springer Publication, 256 pp.

5

Augmentation of Groundwater Resources through Aquifer Storage and Recovery (ASR) Method

C. Barber

Center for Groundwater Studies
c/o Flinders University, PO Box 2100
Adelaide, South Australia 5001

PREAMBLE

Augmentation of groundwater has been practiced for centuries, to allow more sustainable use of groundwater particularly for irrigation supplies. In general, recharge is achieved through surface infiltration into shallow, unconfined aquifers. More recently, aquifer storage and recovery (ASR) has been developed to allow direct injection into deeper aquifer systems, including those which contain poor quality groundwater (brackish to hyper saline) thereby creating freshwater storage where none existed previously. The main constraint to recharge by both infiltration and injection is pore clogging from particulates, from bio-film formation and chemical precipitation. The latter is brought about by reactions between infiltrating/injectant water and resident groundwater, and between these mixtures and mineral phases, which make up the aquifer matrix. These reactions can have adverse impacts not only on pore clogging, but also on degradation of groundwater quality, for example from solubilisation of iron, or formation of dissolved sulphides. Examples of approaches used to assess possible impact on groundwater quality are presented for an arid region in NW China where an assessment was made of storage of treated sewage effluent in an intermontane aquifer system prior to re-use. This involved on-site pilot studies and modeling. Approaches involving geo-chemical modeling and assessment of likely impacts of injection of potable water in confined aquifers in less arid southern Australia will also be presented, highlighting the difficulties in predicting reactions where little is known of the reactivity of injectant and groundwater with aquifer mineralogy. Pilot studies will be reported from these sites and the approaches to assess impacts on groundwater quality will be compared. An approach using relatively simple hydro-geochemical modeling for prediction and interpretation of water quality trends has shown the greatest promise in these studies.

INTRODUCTION

Augmentation of groundwater has been practiced for centuries, particularly in arid and desert areas where natural recharge is intermittent (Pyne, 1995). The artificial augmentation of groundwater is defined as a process by which excess water is directed into the ground, either by spreading on the surface, using recharge wells or by altering natural conditions to increase infiltration to replenish an aquifer. It is a way of storing water underground in times of water surplus to meet demands in times of shortage.

Pavelic and Dillon (1997) have reviewed over thirty international case studies of artificial recharge using water and wastewater. Generally it was reported that most studies used surface infiltration to recharge the groundwater, although injection wells and ASR were becoming more common recently, and where recharge of confined or semi-confined systems was being attempted, injection and recovery wells were the only possible option.

With ASR, the injection well is used both for injection and recovery, thus overcoming what has been a major drawback with recharge using wells—that is for regular well development to remove clogging solids and bio-films on well screens and adjacent aquifer media (Pyne, 1995). Single well recharge and recovery also increases the likelihood of recovering injected water, which is important where there are significant differences between injected water and groundwater (eg. where the latter is saline, brackish or polluted). Injection wells can be developed regularly during recovery phases, and these become self-cleaning, maintaining well capacity for both injection and recovery. Additionally, the use of recharge wells overcomes the need for large areas of land needed for surface infiltration techniques of artificial recharge. An additional major advantage of using well injection is that this opens up the possibilities for confined and semi-confined aquifers to be used for storage of injected water. The combined use of wells for both injection and withdrawal of water has greatly improved the economics of ASR.

Clogging of injection wells with solids and bio-films is still a significant factor in ASR, which needs to be overcome. This can be minimized by pre-treatment of effluent prior to injection. Thus partial removal of suspended solids (SS) to levels of around 10 mg/l has been suggested, although storm water injection with SS levels averaging 30 mg/l in Australia has been shown to be sustainable over several years in carbonate aquifers.

Surface spreading of water, particularly treated effluent to augment groundwater recharge, on the other hand, takes advantage of natural water treatment as the recharge infiltrates through the vadose (unsaturated) zone before recharging groundwater. This process has been called Soil-Aquifer Treatment (SAT) or geo-purification. Often, improvement in the quality of infiltrating water is the objective of a SAT system. Water or treated wastewater extracted following SAT treatment and aquifer storage is often used without further treatment for a range of non-potable uses. The latter include recreational uses (e.g. irrigation of golf courses, wetland and stream recharge) as well as for landscape irrigation where direct contact of water with humans is negligible. Water, which is to be used for potable purposes, may require some additional treatment (Pyne, 1995).

SAT using effluent with periodic infiltration and drying to avoid clogging and to promote nitrification, de-nitrification and other pollutant removal or retardation should be sustainable indefinitely for a wide range of non-potable uses. Primary treated sewage effluent (i.e., settled effluent) has been successfully used for recharging groundwater at land-spreading sites. Kanarek and Michail (1996) report on a SAT system for treated wastewater in Israel, which has operated successfully for nearly 20 years, treating flows of 270,000 m^3/d of secondary-treated sewage effluent from a population of around 1.3 million. The system has successfully achieved filtration, disinfection and nitrogen removal through nitrification and de-nitrification during cyclic infiltration into a sand aquifer, and provided treated water for reuse in irrigation.

The desired role of SAT is stated to be removal of those constituents, which pose the greatest threat to public health. Unfortunately, some processes, which remove pollutants, are not efficient in a natural setting and this is the case even in customized treatment plants. Not all pollutants present in sewage are removed or degraded at the same rates or to the same extents. Natural conditions, which favour attenuation of some contaminants, also favour persistence of others.

The chemical and biochemical reactions, which can take place in both ASR and SAT applications, are many and varied, and these are difficult to predict and assess. There is a need to develop an approach to try and evaluate likely reactions, which may affect the efficiency or sustainability of artificial recharge operations, as well as adversely affecting groundwater quality or some intended reuse. This paper summarizes and evaluates two approaches, which have been used in pilot studies of ASR in Australia and of SAT in northwest China.

EVALUATION OF IMPACTS OF AR/ASR ON GROUNDWATER QUALITY

The objectives of predicting geochemical interactions during artificial recharge are varied, and can be summarized as follows:

- Prediction of contaminant transport and "breakthrough"
- Reactions between mixing aqueous phases
- Reactions between aqueous phases and aquifer matrix minerals.

There are a variety of solute transport models which can be used to address contaminant transport, and given appropriate and realistic coefficients for defining partition between phases (sorption, reaction, decay, gas/liquid exchange) then contaminant behaviour during artificial recharge can be assessed using modeling, if flow processes are also reasonably well understood. The main problem with this is in calibrating models with realistic coefficients describing contaminant attenuation.

The use of physical laboratory or test-cell models of aquifer systems (column studies, sand tanks etc) have been popular for assessing contaminant behaviour, and these also have been used for assessing likely reaction between infiltrating or injected water and aquifer minerals (eg see Rinck-Pfeifer et al., 1998). Although this approach has the advantage of control over physical and chemical conditions, it is debatable whether these reflect likely conditions during artificial recharge. By nature, physical models cannot be representative of aquifer conditions at the scale of even a pilot study of artificial recharge because of (often unknown) heterogeneity of aquifer media and groundwater flow. Physical models are perhaps best suited to obtain data under controlled conditions on reaction kinetics. These also have the advantage of visibility and accessibility, which are major drawbacks in subsurface field investigations.

Comparison of results of laboratory column studies, plume modeling and field-scale tracer testing of degradation and attenuation of mono-aromatic (BTEX) hydrocarbons (Thierrin et al., 1993) suggest that problems of scaling up and representativity are significant. Results from the testing carried out during this research (as compound half-lives determined from the tracer test, from plume modeling, and from large [100 mm diameter, 2 m long] columns) are shown in Table 1. It is clear that determinations made in the field at and close to plume scale are significantly different for toluene from the much smaller scale testing using the laboratory sand columns infiltrated with groundwater from the test site spiked with BTEX compounds. Other data for benzene, ethyl benzene, and xylene isomers is not precise enough from column testing where no degradation was observed over the duration of the testing (40 days). Clearly, some degradation should have been observed over this period, if degradation rates were similar to those observed at field scale. These results, although not relating to artificial recharge, highlight the difficulties in reproducing field-scale conditions at small (laboratory column) scale, particularly where conditions are dominated by micro-biological activity—essentially the conditions observed in AR/ASR and SAT systems.

Table 1. Comparison of results of tests to assess degradation of BTEX hydrocarbons in anoxic groundwater at different scales (plume scale through modeling, from in-plume injection of deuterated BTEX tracers, and in laboratory column experiments) (from Thierrin et al., 1993)

BTEX compound	*Compound half-life in days*		
	Within plume tracer test	*Plume modeling*	*Laboratory Columns*
Benzene	>800	>800	>42
Toluene	100 ± 40	120 ± 25	0.3 ± 0.1
Ethyl benzene		230 ± 30	>42
p-xylene	225 ± 75		>42
m- and p-xylene		170 ± 10	
o-xylene		125 ± 10	>42

Field-scale reactions between mixing aqueous phases, minerals, organic carbon and ion exchange phases can be assessed relatively simply using equilibrium mixing-cell models such as PHREEQC (Parkhurst et al., 1999). Thus this type of modeling approach can be used to provide a wide range of assessment of redox and pH-dependent reactions, which could impact on the efficiency of an artificial recharge system. These assessments can also be used to provide quantitative evaluation of monitoring data of artificial recharge schemes.

In AR/ASR, modeling is important for assessment of mineral dissolution reactions (eg carbonate minerals), which could locally increase hydraulic conductivity but also have an adverse impact on aquifer structural integrity by dissolution of aquifer matrix and reducing the extent of mineral cementation. Likewise, mineral precipitation would decrease hydraulic conductivity and contribute towards aquifer clogging. Redox reactions in particular could give rise to both beneficial and undesirable impacts, through infiltration or injection of often-oxygenated surface water into anoxic aquifers. Thus reactions between dissolved oxygen and sulphidic minerals (eg pyrite) are likely to rapidly reduce DO, lower pH and induce pH dependent reactions such as carbonate dissolution, although the low solubility of oxygen in water limits the impacts of these reactions. Progressive lowering of redox potentials, where infiltrating water contains assimilable organic carbon, provides conditions for de-nitrification, iron and manganese reduction and solubilisation, and sulphate reduction. These reactions are well known sequences described by Stumm and Morgan (1996) and their modeling is relatively straightforward. This type of mixing cell model thus provides a potentially useful tool for preliminary appraisal of aqueous-mineral reactivities during artificial recharge, of use in preliminary planning as well as providing a sound basis for evaluation of monitoring data.

The approach has drawbacks through reliance on equilibrium being attained between aqueous and mineral phases. This may not be problematic, as models such as PHREEQC now include reaction kinetics, which can be used to take account of non-equilibrium conditions. Aqueous and mineral phase heterogeneity is probably still the most significant impediment to utility of this type of model.

Traditionally, little preliminary prediction of the effects of mixing and reaction between recharged water and natural groundwater is carried out, and in general these effects are only investigated by monitoring during preliminary or pilot studies. This is perhaps not surprising considering the variability in composition of water being recharged, and the natural variability in composition of groundwater, but mostly because of natural heterogeneity in mineralogical composition of aquifer media, and of flow processes (inter granular or fracture/fissure flow where the latter bypasses inter granular pore fluids and interaction is mainly through diffusive interchange between pore water and fractures). The challenge is to find a sensible conceptual model of aquifer behaviour during artificial recharge, which provides realistic appraisal of system behaviour.

A possible approach is developed here by reference to two case studies. The first of these is a feasibility study for SAT in China, which used conventional laboratory and field studies to assess likely system performance. Second, a preliminary field study of potable water ASR is considered, where the model PHREEQC was used to assess system performance using field monitoring data during a pilot phase injection trial. The latter approach was based on previous experience in preliminary modeling of possible ASR schemes in Australia.

CASE STUDY—SOIL AQUIFER TREATMENT IN PR CHINA

China has one of the lowest per capita quantities of water resources, being ranked 88th in the world. Of 479 cities investigated in China in recent years, more than 300 were facing water shortages and 40 more cities had severe shortages. Recycling of water generally is very low, and often wastewater is discharged without any treatment or only partial treatment into surface water, or onto land and thence into groundwater. Untreated wastewaters are also used directly for irrigation, which has caused soil, crop and water pollution. The latter has given rise to even greater shortages of potable water. National water management departments and environmental agencies in China have formally indicated that wastewater treatment and reuse schemes need to be developed urgently, to help in reducing the water supply problems (Barber and Zhu, 1999).

Currently in China, the wastewater discharge rate is more than 100 million m^3/d. More than 80% of this is disposed directly to the environment without any treatment, giving rise to serious pollution. As a developing country, China is unable to fund construction of a large number of municipal drainage systems and wastewater treatment plants. Therefore, natural treatment systems are being developed and implemented, especially wastewater treatment in-situ followed by reuse. If this can be implemented in China, the water supply and wastewater pollution situation in Northern China, especially in the Northwest District, will be significantly eased.

Given the reliance on groundwater and the presence of extensive aquifer systems in Gansu, it was considered that wastewater treatment and reuse could usefully be combined with aquifer storage of treated wastewater, for example in winter when there was no demand for reuse of treated water for irrigation. A project was consequently set up to investigate the feasibility of using SAT at the city of Jiayuguan in northern Gansu.

The overall treatment and reuse scheme proposed for the demonstration site involved the following components:

- Pre-treatment of effluent to reduce suspended solids and provide a clear effluent from which pathogenic bacteria and viruses could be removed using UV disinfection to allow immediate effluent reuse or aquifer storage depending on demand;
- Infiltration of primary-treated effluent through shallow trenches or spreading basins which are subject to cyclic infiltration and drying to avoid bio-film formation and to promote nitrogen removal;
- Soil Aquifer Treatment (SAT) during percolation through a vadose zone to filter and disinfect effluent, precipitate and adsorb nutrients and contaminants, nitrify ammonia during drying cycles and to denitrify effluent during infiltration cycles;
- Storage of treated effluent in the saturated zone of the aquifer system prior to recovery through pumping wells; and
- Reuse of treated effluent for irrigation.

For the scheme to be sustainable and environmentally acceptable, the scheme would have the following treatment targets or objectives:

- Treated wastewater entering the saturated zone after SAT would conform to irrigation water standards to protect human health and for environmental sustainability.
- Effluent within the saturated zone of the aquifer and irrigation return flows would not impact adversely on drinking water supplies.

Given the difficulties in predicting quantitatively the extent of soil-aquifer treatment, and in designing infiltration and injection systems, it is always recommended that pilot schemes be operated prior to designing full-scale aquifer storage/recovery and reuse schemes. This was carried out for process optimization at the Jiayuguan site, and biochemical reactions of the proposed scheme were evaluated through laboratory column experiments and field infiltration studies, as well as through some preliminary modeling. This is perhaps a more standardized approach, which is widely applied to

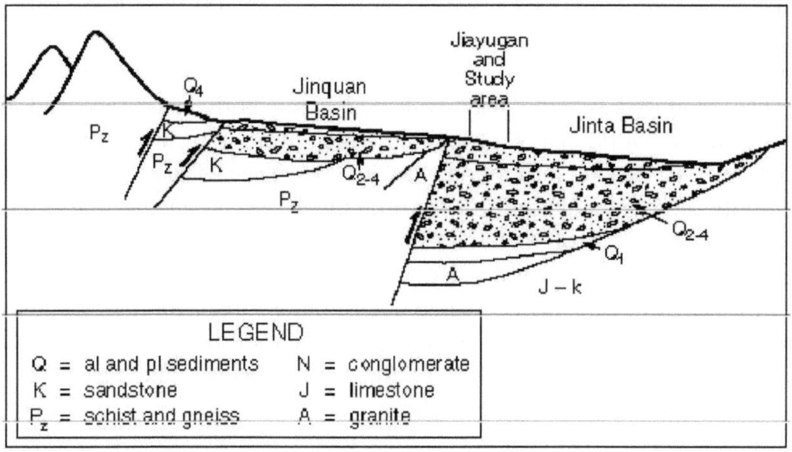

Figure 1. Schematic NE-SW geological cross-section of western and eastern basins at the Jiayuguan SAT test site, Gansu Province, PR China, showing the study area and the position of the city in the eastern basin. The vertical scale is exaggerated.

provide preliminary data on likely system operation, but suffers from problems of heterogeneity and scale. At the Jiayuguan test site, groundwater occurs in Quaternary and older mixed alluvial and colluvial sediments which have accumulated within a graben structure. The sediments form two basins in the Jiayuguan region. The western basin (Jinquan Basin) is separated from the eastern basin (Jinta Basin) by a fault zone and basement high on the western side of the city separating the two basins (Fig. 1). In the region of the fault, the water table is close to the surface (10 to 25 m) with the thickness of the aquifer only 40 to 60 m. The aquifer is much deeper east of the fault within the eastern basin, and the hydraulic head may be found up to 100 m below ground surface. Groundwater for the industrial and domestic needs of the city is abstracted from the west basin from bore fields in the shallowest water level zones near the fault. The aquifer is recharged by snowmelt and from the Beida River, which itself is fed by snowmelt.

Groundwater is extracted at the eastern parts of the Jinta Basin for irrigation and domestic use. Salinities of groundwater are reported to be around 500 mg/l TDS or less.

The groundwater flow is generally towards the fault zone. Hydraulic conductivities within the gravel deposits are very high in the western basin, with values over 200 to 350 m/d in some places. Maximum yields of individual wells can be greater than 10,000 m^3/d.

The city of Jiayuguan lies within the eastern basin at its western end close to the fault (see Fig. 1). The proposed site for infiltration and treatment of wastewater is fully within the eastern basin to the southeast of the city. The piezometric head of groundwater beneath the site is around 100 m below surface, and the salinity of groundwater is less than 500 mg/l TDS. Very few wells have been drilled in this part of the basin, only four being within a 6 km radius of the treatment site, the nearest being about 3 km away.

The proposed scheme was evaluated using both laboratory and field experimental work, using sediments which were typical of those at the proposed site where infiltration was to take place. These materials were naturally quite variable and it was difficult to assess whether these indeed were representative of materials through which effluent would percolate. However, it was considered these tests would provide preliminary data on infiltration rates and hence required infiltration areas for the volumes of effluent (20,000 m^3/d), and on the possibility of obtaining the required biochemical reactions inherent in the SAT process. The results of the infiltration trials are shown in Table 2.

Table 2. Experimental results for 1.5 m long laboratory columns of soils from the Jiayuguan demonstration site infiltrated with sewage effluent, simulating the SAT process

	Influent (mg/l)	*Effluent* (mg/l)	*Removal* (%)
COD	73	27	62
BOD_5	41-115	16-20	62-83
Ammoniacal-N	11-21	8-10	33-47
Nitrate-N	26-44	19-32	21-27

These generally indicate that a degree of treatment can be achieved through simple cyclic infiltration through natural soils at the site followed by drying cycles. However, given the expected likely infiltration area of around 15 ha and naturally large variability in the glacial sediments across the proposed site, the data could only provide at best a first approximation to what might happen in reality. It was, however, recognized that this work was only a part of a feasibility study. Similarly, the field and laboratory experimental work was of necessity at small scale, and given time constraints could not be extended over periods of more than a few infiltration and drying cycles. The likely representativity of those tests could also be brought into question, although it was clear that there was scope for removal of contaminants from the effluent.

Probably the main reason for abandonment of the scheme was the much greater uncertainty in predicting infiltration and storage of effluent in groundwater beneath the site. Initial information from very limited regional surveys suggested a very different hydrogeological setting to that found by preliminary drilling. Thus the sediments beneath the site were an extensive layered sequence of gravels, sands, silts and extensive clay lenses. Groundwater was possibly semi-confined, and any infiltration of effluent would have been extremely complex. With hindsight, it is clear that (at least at this site) the first appraisal should have been drilling to investigate the local conditions beneath the site. Given a favourable outcome from this, then laboratory testing and modeling and particularly a more detailed pilot investigation, at a scale more appropriate for this large-scale infiltration, would have been needed to provide data for process optimization.

CASE STUDY—EVALUATION OF GEOCHEMICAL CHANGES DURING POTABLE WATER ASR, JANDAKOT, WESTERN AUSTRALIA

Conjunctive use of groundwater and surface water has been a feature of the water supply system in Perth for many years, and there are extensive unconfined (surficial aeolian sand aquifers) and confined/semi-confined Mesozoic aquifers in sedimentary rocks in the Perth Basin which are exploited for water supplies. The city is becoming increasingly reliant on groundwater and, given below-average rainfall over the last decade, is actively investigating artificial recharge options. In general, storm water is used extensively to recharge surficial unconfined aquifers through drainage basins, to augment storage and use of shallow groundwater for irrigation of parks and gardens (Scatena and Williamson, 1999). There are perhaps limited opportunities for additional storage of potable water in the surficial sediments, as these are directly connected with wetlands and other groundwater-fed ecosystems on the coastal plain, which would be adversely impacted if groundwater levels rose significantly. However, there are deeper, confined sedimentary aquifers where potable water ASR would be feasible (Scatena and Williamson, 1999). One such scheme, and evaluation of geochemical impacts on the groundwater system and injected water, is described here.

Table 3. Simulation details for PHREEQC modeling

Injectant "storage" phase	Simulation 1	Equilibrium with phases siderite, pyrite and ion exchange (CEC 0.03)
	Simulation 2	Equilibrium with phases siderite, pyrite, calcite and ion exchange (CEC 0.03)
	Simulation 3	Equilibrium with phases siderite, pyrite and ion exchange (CEC 0.03), and limited reaction with calcite (dissolution limited)
	Mix	No mineral phase reaction (dissolution or precipitation)

PHREEQC was used to assess through geochemical modeling possible chemical reactions on mixing of groundwater and injectant (potable water from a reservoir) and reaction and exchange with possible aquifer mineral matrices during a trial potable water ASR injection into the Cretaceous/ Jurassic Leederville Formation at Jandakot, near Perth in Western Australia. The study was carried out between July 2000 and January 2001, with injection of over 40,000 kl of water over 10 days into a mainly sandstone and interbedded shale aquifer. Storage of groundwater was over a 16-week period, and recovery of injected water took place over a 7-week period.

The aquifer was around 250 m thick at the site, the top of the aquifer was approximately 100 m below surface, and shale units and groundwater salinity generally increased with depth. Groundwater in the upper part of the aquifer had TDS of around 660 mg/l, whilst deeper in the aquifer, these increased to over 1900 mg/l. Both mixing and reactions during aquifer "storage" (up to week 16) and during "recovery" (post week 16) have been considered.

Conceptual models of potential mineral reactions (Table 3) were defined through consideration of mineralogical analyses of limited amounts of drill core, from mineral-saturation relationships with natural groundwater from the site using PHREEQC, from analysis of trial injection and from over 30 initial simulations of possible water-mineral reactions. Some limited analysis of Fe(II) and Fe(III) species, sulphur species and sulphur stable isotopes (S^{34}/S^{32}) have also been used to assess speciation and reactions identified by modeling.

Reactions between mineral phases siderite, pyrite, calcite (or ankerite), and ion exchange were considered in more detailed modeling of injectant/groundwater mixes with low concentrations of organic carbon. Model output and monitoring data were compared (eg see Figs. 2-4) by estimating the proportion of background groundwater to injectant based on chloride concentration, assuming this anion behaves conservatively: the fraction of groundwater (f) was given by

$$f = \frac{Cl_{sample} - Cl_{injectant}}{Cl_{groundwater} - Cl_{injectant}}$$

Simulations were also compared with simple mixing of injectant and groundwater assuming no reaction. Sample results for dissolution, precipitation and ion exchange by mineral phases is shown in Table 4, and actual and simulated concentrations for bicarbonate, iron and sulphate for various aqueous mixtures of injectant and groundwater are given in Figs. 2-4.

Iron geochemistry and its behaviour during ASR are particularly important. Analysis and modeling indicates that Fe(II) is predominant in the aquifer, and also with injectant when equilibrated with mineral phases in the aquifer. This was confirmed by Fe(II)/Fe(III) species analysis.

The conceptual model, which shows the closest fit to actual data for iron and other constituents, involves dissolution of pyrite and partial dissolution of calcite (or possible ankerite) and precipitation of siderite, along with exchange of divalent cations (Ca) for monovalent cations (Na). The models also indicate that the small amounts of dissolved oxygen in injectant are utilized for oxidation of sulphide derived from pyrite dissolution, and not for oxidation of iron. These results suggest that geochemical conditions are not favourable for activity of iron oxidizing bacteria.

Table 4. Dissolution (–ve) and precipitation (+ve) and ion exchange for mineral phases in simulations where *f* = 0.1 (90% injectant) during the storage phase. Values in Moles/L

	Siderite	*Pyrite*	*Calcite*	*Exchange phases*		
	$FeCO_3$	FeS_2	$CaCO_3$	CaX_2	NaX	FeX_2
Groundwater	0.03	11.89	-0.23	5.65	8.65	0.017
Simulation 1	-0.292	-0.056	-	5.83	7.08	0.356
Simulation 2	0.063	-0.056	-1.53	7.17	5.38	0.013
Simulation 3	0.041	-0.056	-1	6.74	6.04	0.034

There are some significant differences between actual and predicted concentrations of bicarbonate, iron and sulphate (Figs. 2-4). Modeling indicates that sulphate reduction, which would provide additional removal of iron as iron sulphide, should be minimal due to the low concentrations of dissolved organic carbon in injectant and groundwater, and both iron and sulphate ought to increase as the proportion of groundwater increases. In fact, both decrease during the storage phase, and additionally bicarbonate increases substantially above that expected from calcite dissolution. The latter suggests that sulphate reduction, and possibly iron sulphide precipitation, is occurring. This sequence of reactions is difficult to reconcile with expected thermodynamic behaviour, which is modeled using PHREEQC, unless organic carbon becomes available during ASR to promote sulphate reduction following initial pyrite oxidation. Most presumed sulphate reduction takes place during the storage phase when injectant makes up the major part of the recovered water (eg 80-90%), which suggests that reduction takes place

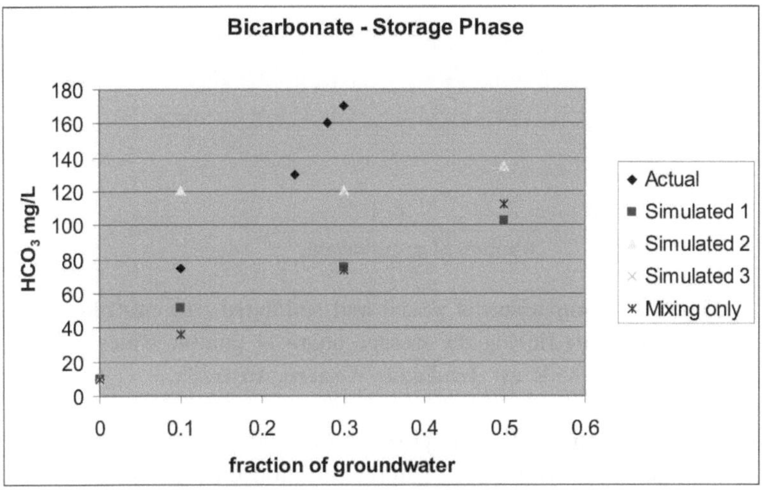

Figure 2. Comparison of actual and simulated concentrations of bicarbonate during the storage phase of potable water ASR at Jandakot, Western Australia.

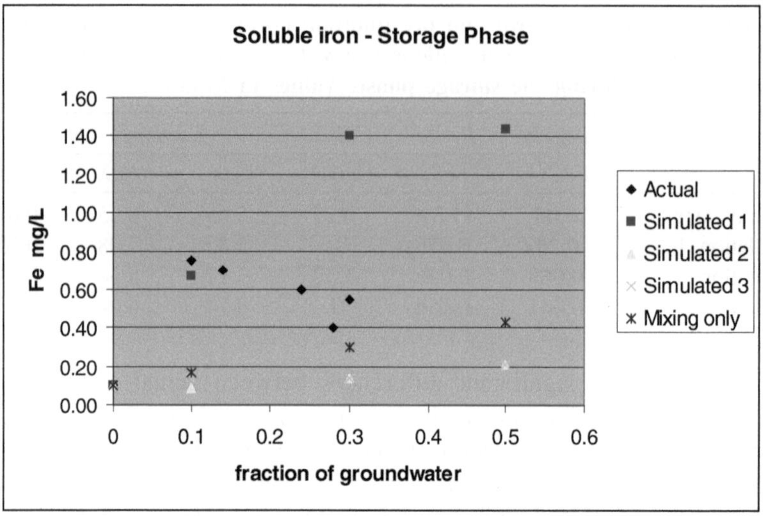

Figure 3. Comparison of actual and simulated concentrations of soluble iron during the storage phase of potable water ASR at Jandakot, Western Australia.

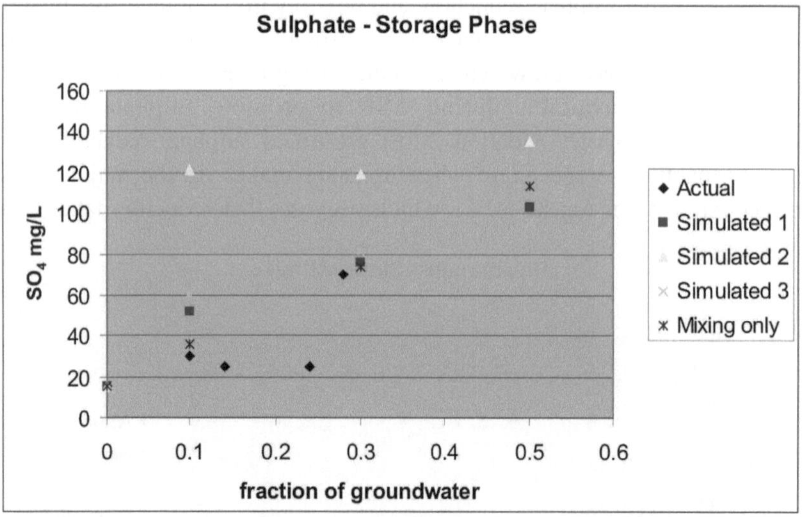

Figure 4. Comparison of actual and simulated concentrations of sulphate during the storage phase of potable water ASR at Jandakot, Western Australia.

close to the injection well. It is possible that this zone may be enriched in organic matter possibly through natural filtration during injection, although this is conjectural and requires further investigation. It is concluded that the low soluble iron concentrations are at least partially due to sulphide precipitation.

Sulphate shows the greatest variability in concentration during the storage phase, and the trends over time during storage are quite complex. Modeling predicts small increases in sulphate arising from dissolution of pyrite and oxidation of dissolved sulphide by dissolved oxygen, as indicated above. However, sulphur isotope fractionation from the Jandakot site (Rattray, 2001) suggests processes of initial pyrite oxidation followed by sulphate reduction occurring during ASR, a pattern noted previously during ASR in South Australia (Rattray, 1998).

During the recovery phase, groundwater and injectant were being pumped mainly from the upper part of the Formation, and modeling suggests lower equilibration between mineral phases and injectant and mixes of this with groundwater. Here, concentrations follow more closely the predicted trends for mixing between injectant and groundwater, with limited ion exchange, limited dissolution of pyrite and siderite and minimal dissolution of calcite. There is little evidence from modeling or S-isotope ratios that sulphate reduction takes place during the recovery phase when groundwater and injectant outside of the immediate location of the well are being recovered (Rattray, 2001). This supports the above conclusion that sulphate-reducing conditions have developed very close to the injection well during the injection and storage phases.

CONCLUSIONS

It is concluded from the case study on potable water ASR that the use of PHREEQC geochemical modeling initially in batch-reaction/mixing mode provides a useful methodology for identifying a conceptual model of water mixing and mineral reactions which impact on chemical changes of water stored in ASR schemes. However, considering the dynamic conditions close to the injection well, without sulphur stable isotope analysis the differences between predicted and actual concentrations of iron, sulphate and bicarbonate would have been difficult to interpret. It is thus clear that modeling by itself cannot be relied on to provide realistic results, but should be used as simply another tool to assist in evaluating system behaviour during artificial recharge, or as in this case during ASR.

It seems premature to suggest that deterministic modeling of water quality impacts during artificial recharge can be realistic and accurate with current understanding, largely because it is difficult to establish a conceptual model of mineral/water reactions, and because of apparent transient changes during storage very close to the ASR well. It seems likely that modeling of SAT would have similar, possibly increased, difficulties compared to those experienced with the ASR case study. Similarly, physical models of SAT or ASR, even larger scale models in laboratory or field, are too small in scale to be able to provide realistic indications of system behaviour. However, these could be useful for determination of possible reaction rates and exchange capacities under controlled conditions. Even so, these may still be of questionable value, and great care is needed with use of such data in predictive modeling.

From the studies so far, the most useful approach to determination of water quality impacts during artificial recharge would involve the following:

- Identification of aqueous phase end-members and variability in space and time.
- Identification of likely mineral assemblages which have importance during aquifer storage, from mineral/groundwater saturation indices for groundwater(s) in the near vicinity of the AR/ASR scheme.
- Prediction of possible water quality impacts using most probable aqueous phase compositions and most likely mineral assemblages, using reaction/mixing cell models such as PHREEQC possibly in tandem with laboratory studies to assess aqueous/mineral reaction kinetics.
- Use model output as simply another tool for interpretation of groundwater monitoring data, and combine these with more detailed monitoring such as determination of stable isotope ratios for sulphur, carbon, and possibly nitrogen, developing a conceptual understanding of processes and their impacts on groundwater quality in time and space.

Finally, given a good conceptual understanding of system behaviour, it may be possible to develop a deterministic model, which can then be used to optimize an AR scheme, or at least to minimize adverse impacts on water quality and improve overall system sustainability.

REFERENCES

Barber, C. and Zhu, K. (1999). Augmentation of water supplies for irrigation: A feasibility study of the use of soil aquifer treatment, aquifer storage, recovery and reuse of wastewater in an intermontane basin, Gansu province, PR China. Proc. Intl Conference on Water Resource Management in Intermontane Basins, Chiang Mai, Thailand, 2-6 February, 1999. Water Research Centre, Chiang Mai University.

Kanarek, A. and Michail, M. (1996). Groundwater recharge with municipal effluent: Dan region reclamation project, Israel. *Water Science and Technology*, **34:** 227-233.

Parkhurst, D.L. and Appelo, C.A.J. (1999). User's guide to PHREEQC (Version 2)—A computer program for speciation, batch-reaction, one-dimensional transport, and inverse geochemical calculations. U.S. Geological Survey Water-Resources Investigations Report 99-4259, 312 pp.

Pavelic, P. and Dillon, P.J. (1997). Review of international experience in injecting natural and reclaimed waters into aquifers for storage and reuse. Centre for Groundwater Studies report No 74, May 1997.

Pyne, R.D.G. (1995). Groundwater recharge and wells. A guide to aquifer storage and recovery. Lewis Publishers, Florida, USA.

Rattray, K.J. (1998). Geochemical reactions induced in carbonate bearing aquifers through artificial recharge. Unpublished MSc Thesis. Flinders University of South Australia.

Rattray, K.J., Martin, M. and Xu, C. (2001). Aquifer Storage and Recovery at Jandakot, Perth, Western Australia. Unpublished report. Water Corporation of Western Australia.

Rinck-Pfeifer, S.M., Ragusa, S.R. and Vandevelde, T. (1998). Column experiments to evaluate clogging and biogeochemical reactions in the vicinity of an effluent injection well. Proc. 3rd Intl Symposium on Artificial Recharge of Groundwater—TISAR 98. 21-25 Sept. 1998, Amsterdam, Netherlands. J.H. Peters et al. (eds).

Scatena, M.C. and Williamson, D.R. (1999). A potential role for artificial recharge in the Perth region: A pre-feasibility study. Centre for Groundwater Studies Report No. 84, August 1999.

Stumm, W. and Morgan, J.J. (1996). Aquatic Chemistry, 3rd Edition. Wiley Interscience, New York.

Thierrin, J., Davis, G.B., Barber, C., Patterson, B.M., Pribac, F., Power, T.R. and Lambert, M. (1993). Natural degradation of BTEX compounds and naphthalene in a sulphate reducing groundwater environment. *Hydrological Sciences*, **38(4):** 309-322.

6

Environmental Impact Assessment, Remediation and Evolution of Fluoride and Arsenic Contamination Process in Groundwater

N. Madhavan and V. Subramanian[1]

Research Associate, School of Environmental Sciences
Jawaharlal Nehru University, New Delhi 110067
[1]Professor, School of Environmental Sciences
Jawaharlal Nehru University, New Delhi 110067

PREAMBLE

Natural geologic environments in India affect human health in a variety of ways through interactions between geochemical, hydrologic, and biologic processes and human activities. Numerous national-scale cases serve as examples, fluorosis in Rajasthan and arsenicism diseases in West Bengal regions. The areas of fluorine-related endemic ailments, which occur in many parts of Indian states that include Tamil Nadu, Andhra Pradesh and Karnataka, have a distribution that tends to match that of regions with high-fluorine rocks and aquifers. Environmental factors such as climate, along with the human activities and cultural customs, can also enhance health impacts in areas with high natural background concentrations of hazardous geochemical compounds. The serious arsenicism in parts of West Bengal state is related not only to the high arsenic content (as high as 28 μg/g) in recent Ganges alluvium deposits but also to the semi-arid conditions on the desert in the sulphide mining Rajasthan of the western part of India.

INTRODUCTION

Fluoride is the most electronegative of all chemical elements and is, therefore, never encountered in nature in the elemental form. It is seventeenth in the order of frequency of occurrence of the elements, and represents about 0.06 to 0.09% of the earth's crust (Wedephol, 1974). Fluoride is an essential ion for all living beings from the health point of view. It helps in the normal mineralisation of bones and formation of dental enamel. Fluoride when consumed in inadequate quantities (less than 0.5 ppm) causes health problems like dental caries, lack of formation of dental enamel and deficiency of mineralization of bones, especially among the children (WHO, 1996). On the contrary, if fluoride is consumed or used up in excess (more than 1.0 ppm), it can cause different kinds of health problems, which equally affect both young and old (WHO, 1996). Higher fluoride concentration exerts a negative effect on the course of metabolic processes and an individual may suffer from skeletal fluorosis, dental fluorosis, nonskeletal manifestation or a combination of the above (Susheela and Kharb, 1990; Susheela and Kumar, 1991). The incidence and severity of fluorosis is related to the fluoride content in various components of environment, viz. air, soil and water. Out of these, groundwater is the major contributor to the problem. Fluoride is a typical lithophile element under terrestrial conditions. The bulk of the element is found in the constituents of silicate rocks, where the complex fluorophosphate apatite, $Ca_{10}(PO_4)_6F_2$, seems to be one of the major fluoride mineral (Subramanian, 1980, Madhavan and Subramanian, 2001). Next with regard to fixation of the bulk of fluoride come some complex hydroxy-silicates and hydroxy-alumino-silicates, in which the hydroxyl ions (OH) may be largely replaced by fluoride (Omueti and Jones, 1977), as is the case in amphiboles and minerals of the mica family (biotite and muscovite). In many rocks, especially in those of late magmatic stages of evolution, the fluoride in such silicates may even greatly exceed the amount fixed in apatite. Not uncommonly one encounters CaF_2 as a constitution of magmatic rocks; much more rarely one finds villiaumite (NaF), in a few nephaline syenites (Deshmukh et al., 1995). In recent times the interest in fluoride has greatly increased, owing to its importance in the precipitation of fixation of phosphate in minerals like fuorapatite, and to the recognition of pathological conditions in man and animals, described as fluorosis (Agrawal et al., 1997).

DISTRIBUTION OF FLUORIDE IN SOIL

Fluoride occurs mainly in the silicate minerals of the earth crust at a concentration of about 650 ppm, and ranks 13th among the elements (Wedephol, 1974). Fluoride is ubiquitous in the environment and is always present in plants, soils, and phosphatic fertilizers. The fluoride concentrations in these materials are on the order of 3×10^0, 3×10^2, and 3×10^4 ppm for plants, soils, and phosphatic fertilizers, respectively (Sparks, 1995). The following mean values, in ppm of F, are reported for various geological materials: basalts, 360; andesites, 210; rhyolites 480; phonolites, 930; gabbros and diabases, 420; granites and granodiorites 810; alkalic rocks, 1,000; limestones, 220; dolomites, 260; sandstone and graywackes, 180; shales, 800; and oceanic sediments, 730 (Fleischer and Robinson, 1963). It is an essential element in the following minerals: fluorite, apatite, cryolite, topaz, phlogophite, lepidolite, and other less important minerals.

Very common soil minerals, such as biotite, muscovite, and hornblende may contain as much as several percent of fluoride and, therefore, would seem to be the main source of fluoride in soils. However, Steinkoenig (1919) believed that micas, apatite, and tourmaline in the parent materials were the original source of fluoride in soils. It appears, therefore, that the fluoride content of soils is largely dependent on the mineralogical composition of the soils inorganic fraction.

Total fluoride concentrations in United States soils range from <100 ppm to over 1,000 ppm (NAS, 1971). For world soils, the median value reported ranged from 200 to 300 ppm (Bowen, 1979 and Rose et al., 1979), whereas an average of 430 ppm was reported for United States surface soils (Shacklette and Boerngen, 1984). The fluoride contents 30 soil profiles (n = 137) of representative soils of varied texture, parent material, and geographic distribution varied from trace amounts to 7070 ppm for an unusual Tennessee soil containing phosphate rocks (Robinson and Edgington, 1946). Thc average for surface samples of these soils was 292 ppm. In general, fluoride was concentrated in colloidal material (Adriano, 1986).

Adriano (1986) observed that the natural fluoride content of soil increases with increasing depth, and that only 5% to 10% of the total fluoride content of soil is water-soluble. Under natural conditions, the concentrations of fluoride in soil solutions are usually below 1 ppm, but under severely polluted soils by fluoride they may reach levels of about 10 ppm (Polomski et al., 1982). In soils in the humid temperate climate, fluoride could be readily lost from minerals in the acid horizons (Omueti and Jones, 1977). A substantial amount of this fluoride is retained in subsoil horizons, where it is complexed with Al that is most likely associated with phyllosilicates (Huang and Jackson, 1965). The extent of fluoride retention depends on the amount of clay and pH; therefore, the pattern of fluoride distribution with depth follows clay pattern closely. Fluoride occurs only in trivial amounts in the organic matter fraction of these soils examined by Omueti and Jones (1977).

Several investigators observed that the solubility of fluoride in soils is highly variable and has the tendency to be higher at pH below 5 and above 6 (Larsen and Winddowson, 1971; Gilpin and Johnson, 1980). Fluoride solubility in soil is complex and may be controlled by solid phases even more insoluble than CaF_2. In addition, fluoride solubility may be related to the solubility of Al or other ionic species with which it forms complexes

(Arnesen, 1998; Rajwanshi et al., 1999). Soils having high pH and low levels of amorphous Al species, clay, and organic matter generally sorb little fluoride (Omueti and Jones, 1977). Thus, it appears that the predominant retention mechanism is that of fluoride exchange with the OH group of amorphous materials, such as Al-hydroxides (Flühler et al., 1982; Barrow and Ellis, 1986; Bond et al., 1995; Anderson et al., 1991). In this case, fluoride resulting in a simultaneous release of Al and Fe replaces the crystal lattice OH^- of clay minerals. Other fluoride retention mechanisms include the binding of fluoride to soil cations (e.g., Ca^{2+}, Al^{3+}), or fluoride precipitation as CaF_2, as in calcareous soils (Gupta et al., 1982; Slavek et al., 1984). Fluoride occurs in soils as the singly charged fluoride ion, or occasionally as a component of such complex anions as $(BF_4)^-$, $(AlF_6)^{3-}$, and $(SiF_6)^{2-}$ (Hopkins, 1977).

DISTRIBUTION OF FLUORIDE IN GROUNDWATER

Groundwater is an important resource for mankind existence and economic development. Due to the scarcity of surface water, Rajasthan has to depend on groundwater resources to a great extent. In arid and semi-arid areas of the state, groundwater is the only water resource for drinking and agriculture purposes. The total groundwater resources of the state are 11,715 million-hectare metres, which appear to be inadequate compared to the rapidly increasing demands for domestic and industrial uses. The depth of water varies widely throughout the state. To the east of Aravali, the depth is comparatively shallower than in the west. It generally varies between less than 10 and 25 m in the eastern part, whereas in the western part, it ranges between 20 and 80 m. The water level slopes towards the east and southeast on the eastern side, whereas to the west of the Aravali, it slopes towards west and the northwest (Handa, 1975).

The problem of high fluoride concentration in groundwater resources has now become one of the most important toxicological and geo-environmental issues in India. During the last three decades, the high fluoride concentration in water resources and the resultant disease 'fluorosis' is being highlighted considerably throughout the world. India is also confronting the same problem and about million people in villages are consuming water having high fluoride (Agrawal et al., 1997; Choubisa, 1998; Susheela, 1999).

Groundwater fluoride in high level is present in all the 31 districts and has become a serious health related issue in 23 districts of Rajasthan (Agrawal et al., 1997; Maithani et al., 1998; Datta et al., 1999). The studies carried out by the above earlier workers noticed that the total number of problem villages, which have high fluoride content in groundwater and a substantial proportion of the population are using high-fluoride water sources. In Ajmer district the widespread fluoride-rich water and the prevalence of fluorosis in certain areas are also indicative of features of regional and local geological significance. Some of the important places having fluoride-rich groundwater, along with the important geological features responsible for the fluoride incidence are described by Madhavan and Subramanian (2002). The fluoride concentration in groundwater varies from trace to maximum of 44 ppm (Table 1). The main source of fluoride is the naturally fluoridated drinking water derived from pyroxene or amphibole minerals, which commonly contain 3 to 5 ppm fluoride and sometimes thermal spring bring as high as 10 to 15 ppm fluoride.

Table 1. Nature of fluoride concentration in groundwater, India

Place	*Lithology*	*F, ppm*
Cuddapah (AP)	Granite	0.3-8.0
Guntur (AP)	charnokites	0.6-2.5
Medak (AP)	archean crystalline	0.4-6.9
Nalgonda (AP)	granite gneisses	0.4-20.0
Pennar river basin (AP)	granite schist and gneisses	<1.0-1.0
Visakhapatnam (AP)	khondalite, lepynite and charnokite	0.2-8.4
Vamsadhara river basin (AP)	granite gneisses	<0.5-3.4
Karbi Anglong—Assam	cretaceous sandstone	0.4-20.6
Bihar	schist and gneisses	0.1-2.5
Delhi	amphibolites and quartzites	0.2-32.5
Gujarat	basalt and quaternary sediments	0.1-40.0
Karnataka	granite and hornblende schist, gneisses	0.4-7.4
Kerala	charnockites and quaternary	0.1
Shivpuri (MP)	gneisses	0.2-6.4
Orissa	granite schist and gneisses	0.1-16.4
Punjab	recent alluvium	0.1-16.5
Ajmer (RJ)	calc schist and gneisses, biotite schist	0.1-12.0
Barmer (RJ)	sandstone and alluvial deposits	tr-13.2
Churu (RJ)	sandstone and alluvial deposits	tr-30.0
Dungarpur (RJ)	schist and gneisses, fluorapatite	0.1-10.0
Jaipur (RJ)	mica schist and quartzite	4.5-28.1
Jaisalmer (RJ)	sandstone and alluvial deposits	0.2-4.6
Nagaur (RJ)	limestone, biotite schist	tr-44.0
Pali (RJ)	limestone, biotite schist	5.6
Pushkar valley (RJ)	calc schist and gneisses, biotite schist	0.19-13.5
Sirohi (RJ)	schist and gneisses, fluorspar	<1.0-6.0
Southeastern Rajasthan	calc schist and gneisses, biotite schist	0.2-16.2
Udaipur (RJ)	amphibolites, gneisses and schist	0.1-11.7
Tamilnadu	charnokite, gneisses and schist	0.51-4.0
Uttar Pradesh	recent alluvium	0.1-17.5
West Bengal	recent alluvium	0.1-2.2
Indian rivers	various lithology	0.5

The problem of fluorosis in the Ajmer district has been known for a fairly long time. The first case of skeletal fluorosis was reported in 1959 from Jobner in Jaipur district (Agrawal et al., 1997). Later Bhargava et al. (1978), reported the prevalence of dental fluorosis in parts of Ajmer district. In the last few years many studies on the fluoride issue have been carried out and fluorosis was correlated with high concentration of fluoride in drinking water and a number of cases of fluorosis, both in human beings and domestic animals were reported from the state (Agrawal et al., 1997).

Fluoride incidence in groundwater is mainly a natural phenomenon, influenced basically by the local and regional geological setting and hydrogeological conditions. The chief source of fluoride in groundwater is the fluoride-bearing minerals in the rocks and soils.

The weathering and leaching processes, mainly by moving and percolating water, play an important role in the incidence of fluoride in groundwater. The various factors that govern the release of fluoride into water, by the fluoride-bearing minerals are: (i) the chemical composition of water, (ii) the presence and accessibility of fluoride minerals to water, and (iii) the contact between the source mineral and water (Keller, 1979).

DISTRIBUTION OF FLUORIDE IN PLANTS AND ANIMALS

Fluoride is not an essential plant element but is essential for animals. However, continuous ingestion by animals of excessive amounts of fluoride can lead to the disorder fluorosis, and sub-optimal levels in the diet can have an equally damaging effect. Therefore, plant content of fluoride is of interest to livestock producers (Kabata - Pedias, 1989; Keerthisinghe et al., 1991; Stevens et al., 1995). In endemic fluorosis areas, the main fluoride source is either the water or the dust-contaminated feedstuff (including pastures) or both (Underwood, 1977). The chronic fluorosis of sheep, horses, and cattle locally known as Darmous, that occurs in parts of North Africa is caused by contamination of water supplies and feedstuff with dusts originating from rock phosphate deposits and mines (Underwood, 1977).

Surface water used for drinking and cooking in most areas generally contains < 1 ppm fluoride, or even 0.1 ppm or less (Underwood, 1977). In regions, where water comes from amphibolite and mica schist, it contains high fluoride content. Fluorosis in human has occurred, as in China, Southern India, Srilanka and Africa, where concentrations of fluoride as high as 20 to 40 ppm have been reported (Jones et al., 1977; Dissanayake, 1991; Sangodoyin and Ogedengbe, 1991; Deshmukh et al., 1995; Latha et al., 1999).

Plants growing in areas free of fluoride air pollution usually contain only trace amounts of fluoride, usually <10 ppm in the foliage (Kumpulainen and Koivistoinen, 1977). Kumpulainen and Koivistoinen (1977) reported that cereals usually contain <1 ppm fluoride, where fluoride tends to accumulate in the outer layer of the grain and in the embryo. The fluoride contents of both leaf and root vegetables do not differ appreciably from those of cereals, with the exception of spinach, which is unusually enriched in fluoride. Potato peelings can contain as much as 75% of the total fluoride in the whole tuber. Tea is one of the most fluoride enriched drink, with about two thirds of the fluoride in leaves being soluble in the beverage (Fung et al., 1999).

Brewer (1965) reported that the availability of soil fluoride to plants is controlled by the following factors: pH, soil type and amount of clay, Ca and phosphate. Liming acid soils to around pH 6 to 6.5 would decrease the levels of soluble fluoride compounds present in or added to the soil. The use of phosphate fertilizers having low fluoride concentrations can also be beneficial, as it can reduce fluoride toxicity, presumably through ion competition, at low pH values as long as the fluoride level does not exceed 180 ppm.

Several anthropogenic sources of fluoride can enrich the soil with this element (Jinadasa et al., 1993; Sahu et al., 1998). Phosphatic fertilizers, especially the super phosphates, are perhaps the single-most important source of fluoride in agricultural lands. A repeated application of rock phosphates containing several percent of fluoride was shown to have significantly elevated the fluoride contents of soils (Omueti and Jones, 1977). Typical additions of phosphate fertilizers (50-100 kg P_2O_5 per ha/yr) could elevate soil fluoride content by 5 to 10 ppm/yr (Gilpin and Johnson, 1980). Other sources that may affect soil status of fluoride are precipitation, fallout of combustion products of coal and other industrially polluted air, and fluoride bearing insecticides (Adriano and Doner, 1982).

UPTAKE OF FLUORIDE BY THE SOIL (DEFLUORIDATION)

The maximum contaminant level for fluoride in drinking water is set at 1.5 ppm by WHO (1996). Because of the fluorosis risk involved, India has already lowered the maximum contamination level to 1.0 ppm; as against the US-EPA (1999) current allocation level for fluoride with a view of health consciousness increased into 4 ppm. Various treatment methods have been adapted to remove fluoride from drinking water. Fluoridation by definition, adjustment of fluoride in water supplies to an optimal concentration between 0.7 to 1.2 ppm (Alabdula'aly, 1997). If the fluoride concentration in the water source is beyond the permissible level consistently, it is essential to consider some remedial measures to prevent the incidence of fluorosis.

These include:

(a) Methods based upon the addition of chemicals to cause precipitation or co-precipitation during coagulation, aluminium sulphate and lime are mixed slowly for five minutes then left for settling. The method can be carried out in a bucket in a single household or in a large village scheme. This method is invented in India and used to a small extent in villages (Bulusu and Nawalakhe, 1990). The limitation of adding lime in this method, fluoride is removed by precipitation as calcium fluoride

(Yang et al., 1999). However, solubility of calcium fluoride is such that fluoride below 8 ppm remains in drinking water (Vora and Joshi, 1998). The use of bone char, which is carbonized to black or gray pieces at high temperature, is a method, which is cheap, safe and easy to handle. When water is filtered through crushed bone char in a column or a bucket the fluoride is taken up by the bone char (Mehrotra et al., 1999). Bone can also be added to fluoride water in a container and defluoridation by stirring (Killedar and Bhargava, 1993). Defluoridation can either way be done on a household scale and centrally in a village. But the limitation of use of bone char may not be acceptable for predominantly Hindu population for drinking water. Fluoride can also be precipitated with calcium and phosphate, but only when in contact with an appropriate surface such as bone char or activated alumina (Killedar and Bhargava, 1993; Mehrotra et al., 1999).

(b) In method based upon ion exchange, lanthanum impregnated silica gel absorbs fluoride ion at the normal pH (Wasay et al., 1996). Literature also shows that various hydrous oxides of rare earth elements exhibit high adsorption capacity for anions (Serrano et al., 2000).

(c) In methods based upon membrane separation, reverse osmosis technology is the membrane filtration based water purification system. This method removes all the dissolved solids including fluoride. Elimination of excess fluoride in potable water by electrolysis using an aluminium anode removes excess amount of fluoride from 4-5 ppm to 0.5-1.0 ppm, without significant changes in levels of other ions has been developed (Ming et al., 1987). The limitation of using this method is expensive, so it is not affordable by average income family in India.

(d) In method based upon adsorption, activated alumina is the best way to reduce fluoride levels down to below 1 ppm when fluoride levels are below 15-20 ppm in the drinking water. It has been reported that the treated water from Nalgonda technique contains residual aluminium in the range of 2.1 to 6.8 ppm under various operating conditions whereas, 0.3 ppm is in the treated water through activated alumina column. The present WHO (1996) limit for aluminium in drinking water is 0.2 ppm, which may undergo further revision due to its reported neuro toxicity (Rajwanshi et al., 1999). Activated carbon can be prepared from different materials such as wood, lignite, nutshells, rice husk, sawdust, coconut shell, fly ash etc (Arulanantham et al., 1992; Singh et al., 2000). Process involves heating to high temperature, with or without pre-treatment. Untreated powdered activated carbon showed significant fluoride uptake along with enormous cation at acidic pH, whereas little or no removal was observed at neutral pH (Madhavan and Subramanian, 1999).

(e) Studies have suggested a number of chemical and physical factors, which are responsible for the binding of fluoride in soils. These factors have been associated with the form and concentration of Ca, Fe, Al, clay, humus and pH (Omueti and Jones, 1977; Polomski et al., 1982; Slavek et al., 1984; Wang and Reardon, 2001).

(f) Defluoridation by stilbite, natrolite, serpentine, bentonite, kaolinite and apophyllite has been investigated both for laboratory samples of aqueous fluoride solutions and for fluoride rich potable water samples (Kau et al., 1998). Fluoride ion interaction with synthetically prepared goethite has been investigated (Jinadasa et al., 1993) over a range of pH values (4-9) and fluoride concentration (10^{-3}-10^{-5}M). The amount of fluoride retained by goethite suspensions was found to be a function of pH, media ionic strength, fluoride concentration, and goethite concentration (Jinadasa et al., 1993).

FLUORITE RESOURCES IN INDIA AND ITS USES

In India large known deposits of fluorite (fluorspar or calcium fluoride) are located in Amba-Dongar (21°59': 75°01') and Karipani (22°01': 74°08') and a small occurrence of fluorspar in Ajmer district are reported (GSI, 1963) in Khairot (25°53': 74°01') and Barla (26°29': 75°01'). Native phosphates, as apatite, or rock phosphates are highly valued now as artificial fertilisers of manures, either in the raw condition or after treatment with sulphuric acid to convert them into acid or super phosphates. The main occurrences of phosphatic deposits in India are located in Jhamar-Khtra, cretaceous of Trichinapally, Mussoorie, Udaipur, Jaisalmer (Rajasthan), Dalbhum in Singhbhum, Bihar and a low grade quano deposits of recent origin is found in some of the Laccadive group of coral islands (GSI, 1977; Wadia, 1994).

The largest use of calcium (or sodium) fluoride is in steel making, in which it prolongs the fluidity of the ingot, thus facilitating the escape of undesirable gaseous products. Fluorite is also used as a flux in the smelting of Ni, Cu, Au and Ag, as rodenticides, and as catalyst for certain organic reactions. Fluorapatites are used industrially for the preparation of phosphoric acid, fertilizers, and mineral feeds. Cryolite is used as an insecticide, and in molten form, as an electrolyte in the production of Al from alumina. Fluoride and its compounds are used in producing U (from the hexafluoride form, UF_6) for the nuclear industry, high-temperature plastics, air conditioning, and refrigeration. Hydrofluoric acid is extensively used for etching the glass of light bulbs, etc. The effect of F-bearing fertilizers on the fluoride content of plants is usually insignificant. However, plants growing in the vicinity of industries such as aluminium smelting can have substantially increased fluoride contents (Adriano, 1986). Fluoride anomalies could become very useful geochemical proximity indicators and also the geochemical behaviour of fluoride seem to be one of the most important indicator for ore forming processes (Pb-Zn deposits) in carbonate sediments (Schneider and Möller, 1977).

The atmospheric release of fluorine by industries is of ecological importance. The steel industry is the major source of atmospheric fluoride in the United States and the third largest in Canada (NRCC, 1977). The primary aluminium production industry is the third largest source of atmospheric fluoride in the United States and the largest source in Canada (Adriano, 1986). Earlier study shows fluoride in recent sediments act as a trap of various toxic metal contaminants in an estuarine environment (Krumgalz et al., 1990).

The fluorinated sulphur compound known as sulphur hexafluoride or SF_6 is an almost perfect dielectric (nonconductor). This invisible gas is inert, nontoxic, nonflammable and five times denser than air. In the specialised applications, it serves as an insulating gas in high-voltage electrical switchgear, including circuit breakers. SF_6 is also used as a cover gas in magnesium smelters and for plasma-etching tungsten in the electronics industry. In Germany, it was used until recently to fill sound-insulating windows and because of its adiabatic properties, by automakers to fill tyres for a firmer ride. SF_6 reportedly is still used in sealed stereo speaker cabinets to improve bass response, and until recently it was SF_6, not air, that gave Nike air sneakers their bounce. But SF_6 also has a dark side—one that clouds the outlook for some basic applications of the gas. Because of its efficient absorption of infrared energy at certain wavelengths and its extremely long lifetime in the upper atmosphere (estimated to be about 3200 years), SF_6 is considered the most potent of all known greenhouse gases. The atmospheric accumulation of these gases, many scientists believe, is causing a slow but discernible warming of the earth's climate. The Intergovernmental Panel on Climate Change (IPCC) estimates the global warming potential of SF_6 to be some 23,900 times greater, per molecule, than that of carbon dioxide, the greenhouse gas of greatest concentration in the atmosphere.

Although its atmospheric concentration is a hundred million times lower than that of CO_2 and its estimated contribution to total man-made global warming to date is only 0.1% that of CO_2, SF_6 was among six types of greenhouse gases targeted for emissions reduction at the 1997 Kyoto Summit. On an equivalent basis of global warming potential, SF_6 accounts for 14% of the reductions in greenhouse gas emissions that the United States and more than 150 other nations agreed to make by 2012.

ARSENIC IN ENVIRONMENT

Arsenic is one of the so-called heavy metals (metalloids), which is toxic in the environment at low concentrations (ppb levels). Compounds of arsenic were once dreaded (especially arsenic trioxide because of its tasteless and odourless characteristics) among the elites and aristocrats due to its uses in poisoning others. Saga of arsenic seems to have taken a full turn. Arsenic mobilized in ground water due to natural reaction has been responsible for causing deaths or adverse health impacts on millions of people especially residing in Bengal delta plain.

Arsenic contamination in ground water has almost become a calamity especially in Bangladesh and some parts of West Bengal in India. In

Bangladesh alone the ground water arsenic occurrences in 52 out of its 64 districts are estimated to be affecting 40 million people in an area of 118,012 km^2 (Karim, 2000). The ground water in seven districts of West Bengal, India, covering an area of 37,000 km^2 with the population of 34 million, has been contaminated with arsenic. In 830 villages/wards more than 1.5 million people, out of the total population, drink the arsenic contaminated water (Mandal et al., 1998). People from these countries have been reported to be suffering from diffuse melanosis, spotted melanosis, leucomelanosis, mucusmembrane melanosis, diffuse keratosis, spotted keratosis, hyperkeratosis, gangrene, squamous cell carcinoma, skin, kidney, lung, liver and bladder cancer. It is estimated that millions of people who usually drink water containing arsenic from 0.05 to 0.04 mg/l are at risk of health hazards (Anawar et al., 2002 a; Das et al., 1996).

Arsenic in nature is found in its elemental form as well as associated with some other elements. Some of the most common naturally occurring arsenic bearing minerals are realgar (AsS), orpiment (As_2S_3), arsenopyrite (FeAsS) etc. Thermodynamic studies reveal that arsenopyrite has similar Eh-pH stability range to pyrite, except under acid conditions (pH 4) where arsenopyrite should transform to realgar or orpiment (Craw et al., 2003).

Now a days, almost all the world's arsenic is obtained as a byproduct of the smelting of copper, lead, cobalt and gold ores. The quantity of arsenic generally associated with lead and copper ores may range from trace to 2-3%, whereas the gold ores contain up to 11% (Azcue and Nriagu, 1994).

The main uses of arsenic compounds in ancient times were pharmaceutical and medicinal while main modern uses of arsenic compounds lie in agriculture (as pesticides), livestock (feed additives), electronics (solar cells, semiconductor applications), industries (in glassware and ceramics) and in metallurgy (as alloys and battery plates).

Typical background concentrations of arsenic in fresh water ranges from 1 to 10 μg/l (Tamaki and Frankenberger, 1992; Azcue, 1995). After the arsenic calamity came to be exposed the WHO has lowered the upper limit for arsenic content in drinking water from 50 μg/l to 10 μg/l. However, financial considerations have made many countries to reduce the arsenic limit on hold. Many third world countries still continue with the 50 ppb of arsenic limit.

ARSENIC MOBILISATION IN GROUNDWATER

The controversy over mechanism of arsenic mobilization in groundwater in Bengal delta basin is a raging debate within scientific community. Any attempt to mitigate the arsenic problem must first resolve this debate. This debate is primarily centered around whether the principal causes of high arsenic concentrations in some surface water is the reductive dissolution of hydrous Fe oxides or the release of adsorbed/combined As or any digenetic changes in the minerals structure itself or a combination of all.

Three different mechanisms have been proposed to explain the arsenic mobilization in the groundwater of Bengal delta basin.

According to Mallick and Rajagopal (1996), Mandal et al. (1998), Roychoudhary et al. (1999) and Das et al. (1996), arsenic concentration is caused by oxidation of sulphide minerals as water tables are lowered by pumping. According to them, the reducing environment in the younger region of the Ganges delta may have resulted in the formation in which pyrite and arsenic is mobilized by the oxidative dissolution of these solids.

Acharya et al. (1999, 2000) have suggested that arsenic pollution is caused by ion exchange by phosphate from fertilizers (or any other source of phosphorous) and that arsenic toxicity in ground water in Bengal basin is caused by its natural settings, but appears to be triggered by the recent anthropogenic activities. According to them, there is a strong co-relation between arsenic affected area and geo morphology and Quaternary geology of Bengal basin and that arsenic toxicity in groundwater in the Ganges delta and some low lying area in the Bengal basin is confined to middle Holocene sediments. They suggest that arsenic appears to occur adsorbed on iron hydroxides coated sand grains and clay minerals and is transported in soluble form and co-precipitated with, or is scavenged by, Fe (III) and Mn (IV) in the sediments, which began preferentially entrapped in fine-grained and organic rich sediments during mid-Holocene sea level rises in deltaic and some low lying areas of the Bengal basin. It was liberated subsequently under reducing conditions and mediated further by microbial actions. Intensive extraction of groundwater for irrigation and application of phosphate fertilizers possibly triggered the recent arsenic release of arsenic to groundwater.

Bhattacharya et al. (1997), Nickson et al. (1998, 2000) and Mcarthur et al. (2001) have suggested that arsenic pollution in the Bengal basin was a natural process whereby arsenic released during reductive dissolution of iron oxyhydroxides, a process that also reduces sorbed arsenate to arsenite (Zobrist et al., 2000). Their hypotheses assign great importance to the role of organic matter; in particular, peat with sediments, in generating anoxia in groundwater and have concluded that peat is the redoxed driver for reduction of FeOOH.

ARSENIC SPECIES AND OTHER HEAVY METALS

Arsenic is perhaps unique among the heavy metalloids and oxyanion-forming elements (e.g. As, Se, Sb, Mo, V, Cr, U, Re) in its sensitivity to mobilization at the pH value typically found in groundwater (pH 6.5 to 8.5) and under both oxidizing and reducing conditions. Arsenic can occur in the environment in several oxidation states (–3, 0, +3 and +5) but in natural water mostly found in inorganic form as oxyanions of trivalent arsenite (As III) or pentavalent arsenic (As V) (Smedley and Kinniburgh, 2002). Redox potential (Eh) and pH are the most important factors controlling arsenic speciation. In groundwater of neutral pH, both arsenate and arsenite are present, but the predominant species is arsenate as $H_2AsO_4^-$. The distribution of arsenic between arsenate and arsenite is significant due to higher toxicity and mobility of arsenite (Tareq et al., 2003). Under oxidizing conditions, $H_2AsO_4^-$ is dominant at low pH (less than 6.9), while at higher pH $HAsO_4^{2-}$ becomes dominant. Under reducing conditions at pH less than about pH 9.2, the unchanged arsenite species $H_3AsO_3^0$ will predominate (Yan et al., 2000).

Organic As forms may be produced by biological activity, mostly in surface waters, but are rarely quantitatively important. Organic forms may, however, occur where waters are significantly impacted by industrial pollution. The dominant organic forms found are dimethyl arsinic acid [DMAA; $(CH_3)_2AsO(OH)$] and monomethyl arsonic acid [MMAA; $CH_3AsO(OH)_2$] where As is present in both cases in the pentavalent oxidation state. Proportion of these two species has been reported to increase in summer as a result of increased microbial activity (Hasegawa, 1997). The organic species may be more prevalent close to the sediment-water surface (Hasegawa et al., 1999).

Most toxic trace metals occur in solution as cations (e.g., Pb^{2+}, Cu^{2+}, Ni^{2+}, Cd^{2+}, Co^{2+}, Zn^{2+}), which generally become increasingly insoluble as the pH increases. At the near-neutral pH, typical of most ground water, the solubility of most trace metal cations is severely limited by precipitation as or co-precipitation with, an oxide, hydroxide, carbonate or phosphate mineral, or more likely by their strong adsorption to hydrous metal oxides, clays or organic matter. In contrast, most oxyanions including arsenate tend to become less strongly sorbed as the pH increases (Dzombak and Morel, 1990).

Other toxic trace elements like vanadium, strontium, barium, iron, zinc, copper and cromium, generally associated with arsenic and recognized as transitory or permanent natural contaminants in ground water have also been determined (Nicolli et al., 1989).

Arsenic enrichment does not follow the enrichment mechanism of Mn and other transition metals (Zn, Ni, Cr, Co) but its behaviour is quite similar to that of iron (Stummeyer et al., 2002) and is hosted by the insoluble residual phase of the suspended matter matrix.

One of the major sinks of arsenic in aquatic environment is the iron-rich sediments (Peterson and Carpenter, 1986) but adsorption of arsenic in mN-oxides has also been reported (Takamatsu et al., 1985; Saifullah et al., 1998). Once it is in sediment, arsenic diagenesis can be largely controlled by Fe, Mn, and organic matter redox reactions, hydrolysis and diffusive/advective transport (Brannon and Patrick, 1987; Belzile and Tessier, 1990). In some cases the relative role of Fe(II) and Mn(II) oxides as arsenic scavenger may rather be unclear (Peterson and Carpenter, 1986).

ARSENIC IN SOIL

Arsenic as natural element can be detected in almost all soils. The arsenic content of soils is related to the geologic substratum, and a rather wide range of arsenic levels have been found in soils around the world with an average of 5-10 mg/kg in uncontaminated soils (Ferguson, 1989) or 5 ppm (Vinogradev, 1959; Backer and Chasnin, 1975) or 6 ppm (Bowen, 1979).

There are differences in the arsenic content of soils derived from various igneous rock types. The level of arsenic in soils derived from granite is lower than in those derived from basalt (Yan-Chu, 1994). The parent materials of soil are usually sedimentary rocks. During the formation of these rocks, arsenic is carried down by precipitation of iron hydroxides and sulphides. Therefore, iron deposits and sedimentary iron are rich in arsenic. Naidu et al. (1998) have also reported that arsenic concentrations are significantly higher in calcareous meta-sedimentary rocks than in litho-chemical units containing largely igneous rocks, non-calcareous meta-sedimentary rocks and clastic sedimentary rocks.

Arsenic, however, may accumulate in soils due to human activities such as waste discharges of metal processing plants, burning of fossil fuels, mining of arsenic containing ores and use of arsenical pesticides (Ferguson, 1989). Mines, especially, become a point source in its vicinity as far as arsenic pollution is concerned. Arsenic concentrations gradually decrease with distance from the mine (Jung et al., 2002).

The principal factors influencing the concentrations of arsenic in soil are the parent rocks and human activities. The process of pedogenesis influences the behaviour of arsenic in different parent rocks. Factors such as climate, the organic and inorganic components of the soils and redox potential status also affect the level of arsenic in soils (Yan-Chu, 1994).

Arsenic occurs mainly as inorganic species in soils but also can be converted to organoarsenic compounds by soil microorganisms. The chemistry of arsenic in soil is to a certain extent similar to that of phosphorous; however, it is far more complex because changes in redox conditions and microbial reactions affect the oxidation state of the element (Sadiq, 1995). In natural conditions arsenic mostly occurs in association with sulphur and oxygen (Boyle and Jonasson, 1973). Its mobility, solubility and bioavailability depend on its oxidation and redox potential (Masscheleyn et al., 1991).

Many studies have been devoted to arsenic sorption on well-characterized minerals or soil clay particles, oxides of Al, Fe, Mn, calcium carbonates and/or organic matter (Sadiq, 1997; Smith et al., 1998).

REACTIONS THAT REMOVE ARSENIC FROM SOLUTION

The reactions resulting in removal of arsenic from the solution phase include adsorption into clays or coprecipitation into laterites or metal ion precipitation (McLaren et al., 1995; Carrillo and Drever, 1998; McGeehan et al., 1998; Bhattacharya, 2000; Davis, 2000). Arsenate, the stable form in aerobic water, may be removed by several mechanisms which showed that arsenate species coprecipitate with or adsorb onto hydrous iron oxides (Jana et al., 2000). Shnyukov (1963) observed that iron ores are always enriched with arsenic, owing to the high adsorptive capacity of the hydrous iron oxides, with arsenic nearly absent from manganese ores. The fact that iron oxide has a positive surface charge in most geologic environments and preferentially adsorbs anions while manganese oxide is negatively charged and adsorbs cations has been cited as an explanation of such distributions of arsenic. In addition, ferric arsenate is very insoluble while manganese arsenate and presumably manganous arsenate are much more soluble because arsenate species are adsorbed by aluminium hydroxide and by clays (Onishi and Sandell, 1955). Arsenate is chemically similar to phosphate and may isomorphously substitute and is enriched in phosphate minerals (Krauskopf, 1955). However, in floride phosphate pebbles, arsenic content is proportional to the iron content and inversely proportional to phosphate, indicating that the chemical affinity of arsenic for iron is predominant (Stow, 1969). Arsenite, +3 As, species may also be present in surface waters if the Eh is less than 0.1 V or if oxidation to arsenic +5 is incomplete. Arsenious acid species will absorb or coprecipitate with iron oxide (Gupta and Ghosh, 1953) in a manner similar to arsenic acid. As +3 has strong affinity for sulfur; it readily adsorbs or coprecipitates with metal sulfides. Report was found up to 3000 mg kg^{-1} arsenic in sedimentary pyrite, FeS_2 (Ferguson and Davis, 1972).

GROUNDWATER ARSENIC IN BANGLADESH AND WEST BENGAL, INDIA

The contamination of groundwater by arsenic in Bangladesh is the largest poisoning of a population in history, with millions of people exposed. Tube-wells were installed to provide "pure water" to prevent morbidity and mortality from gastrointestinal disease. The water from the millions of tube-wells that were installed was not tested for arsenic contamination. Studies in other countries where the population has had long-term exposure to arsenic in groundwater indicate that 1 in 10 people who drink water containing 500 ppb of arsenic per litre may ultimately die from cancers caused by arsenic, including lung, bladder and skin cancers. The rapid allocation of funding and prompt expansion of current interventions to address this contamination should be facilitated. The fundamental intervention is the identification and provision of arsenic-free drinking water. Arsenic is rapidly excreted in urine and for early or mild cases no specific treatment is required. Community education and participation are essential to ensure that interventions are successful; these should be coupled with follow-up monitoring to confirm that exposure has ended. Taken together with the discovery of arsenic in groundwater in other countries, the experience in Bangladesh shows that groundwater sources throughout the world that are used for drinking-water should be tested for arsenic (Smith et al., 2000b). Nine districts in West Bengal, India and 42 districts in Bangladesh have arsenic levels in groundwater above the World Health Organization maximum permissible limit of 50 ppb. The area and population of the 42 districts in Bangladesh and the nine districts in West Bengal are 92,106 km^2 and 79.9 million and 38,865 km^2 and 42.7 million, respectively. Analyzed water samples from 42 arsenic affected districts in Bangladesh and water samples from nine arsenic-affected districts in West Bengal shows 59 and 34%, respectively, contained arsenic levels above 50 ppb. Analyses of hair, nail, and urine samples from people living in arsenic-affected villages have shown arsenic above the normal level. Surveyed 27 of 42 districts in Bangladesh for arsenic patients shows patients with arsenical skin lesions in 25 districts. In West Bengal, patients with lesions in seven of nine districts were identified (Chowdhury et al., 2000).

The source of arsenic is geological. Analyzed water samples contain a mixture of arsenite and arsenate and in none of them methylarsonic or dimethylarsenic acid could be detected (Das et al., 1995). A regional groundwater quality survey from 20 tube wells in the Purbasthali area of the Burdwan district of West Bengal province (India) identified arsenic pollution in this area. Arsenic was detected in 19 cases at a concentration level 0.5 to 135.9 ppb. Speciation studies indicate that As(III) is present in only one sample and organo-arsenic compounds have not been detected. Iron, antimony and pH of such water samples were also studied to see if there is any correlation of the presence of arsenic and these parameters. A high concentration of iron (0.3 to 10.7 mg/L) has been detected. Antimony is present in all these water samples (0.03 to 0.9 ppb). The pH value of the groundwater in this area shows that it is more or less neutral (Nag et al., 1996). Safe water from a source having 2 ppb arsenic has been supplied for two years to five affected families comprising 17 members (eight of them with arsenical skin-lesions) of different age groups for impact assessment study in terms of loss of arsenic through urine, hair and nail. After minimizing the level of contamination, a noteworthy declining trend after eight months was observed in urine, hair and nails in all the cases, but not to that level observed in a normal population, due to prevailing elevated background level of arsenic in the area. The eight members, who had already developed skin lesions, are far from recovering completely, indicating a long-lasting damage (Mandal et al., 1998).

To determine the relationship of arsenic-associated skin lesions and degree of arsenic exposure, a cross-sectional study was conducted in Bangladesh, where a large part of the population is exposed through drinking water. A total of 430 subjects had skin lesions (keratosis, hyperpigmentation, or hypopigmentation). Individual exposure assessment could only be estimated by present levels and in terms of a dose index, i.e., arsenic levels divided by individual body weight. Arsenic water concentrations ranged from 10 to 2040 ppb, and the crude overall prevalence rate for skin lesions was 29/100. After age adjustment to the world population the prevalence rate was 30.1/100 and 26.5/100 for males and females, respectively. There was a significant trend for the prevalence rate both in relation to exposure levels and to dose index ($p < 0.05$), regardless of sex. This study shows a higher prevalence rate of arsenic skin lesions in males than females, with clear dose-response relationship.

The overall high prevalence rate in the studied villages is an alarming sign of arsenic exposure and requires an urgent remedy (Tondel et al., 1999). A cross-sectional survey was conducted between April 1995 and March 1996 to investigate arsenic-associated skin lesions of keratosis and hyperpigmentation in West Bengal, India, and to determine their relationship to arsenic water levels. Although water concentrations ranged up to 3400 ppb of arsenic, over 80% of participants were consuming water containing <500 ppb. The surprising finding of cases who had arsenic-associated skin lesions with apparently low exposure to arsenic in drinking water needs to be confirmed in studies with more detailed exposure assessment. Further research is also needed concerning susceptibility factors, which might be present in the exposed population (Guha Mazumder et al., 1998). A prevalence comparison of hypertension among subjects with and those without arsenic exposure through drinking water was conducted in Bangladesh to confirm or refute an earlier observation of a relation in this respect. These results suggest that arsenic exposure may induce hypertension in humans (Rahman et al., 1999). Animal studies have indicated that 2,3-dimercaptosuccinic acid can be used as an oral chelating agent. A prospective, double blind, randomized controlled trial was carried out to evaluate the efficacy and safety of 2,3-dimercaptosuccinic acid for chronic arsenicosis due to drinking arsenic-contaminated (> or = 50 ppb) subsoil water in West Bengal. Under the conditions of this study, 2,3-dimercaptosuccinic acid was not effective in producing any clinical or biochemical benefit or any histopathological improvement of skin lesions in patients with chronic arsenicosis (Guha Mazumder et al., 1998).

Since 1983 large number of people are being encountered with arsenic toxicity due to drinking of arsenic contaminated water (0.05-3.2 ppm) in six districts of West Bengal. There was no correlation between the quantity of arsenic taken through water and the level of arsenic in hair, nail, liver tissues and the degree of fibrosis (Mazumder et al., 1998). In order to assess possible risk to brain function by drinking such water, rats were given arsenic mixed (0.05, 0.1, 0.3 and 3.0 ppm) in drinking water at the above four concentrations for 40 days. There was increased lipid peroxidation at all doses of arsenic, including the 'permissible limit', decrease in glutathione level, superoxide dismutase and glutathione reductase activities, indicating the free-radical-mediated degeneration of brain (Chaudhuri et al., 1999). Other than West Bengal, information also emerged from Rajnandgaon district (Madhya Pradesh) that As contamination of groundwater was found (880 ppb) (Chakraborti et al., 1999). Recently, unpublished data also shows Brahmaputra and Barak river basin area around Assam has arsenic in groundwater (100 ppb).

NATURAL SOURCE OF ARSENIC IN MALDA DISTRICT, WEST BENGAL

An intensive field survey was carried out in one of the least affected districts—Malda—in Indian State of West Bengal and also other districts for Eh, pH and Arsenic measurements during January-February 2002. Malda district has about two million population with a reported (Das et al., 1996) arsenic concentration level, based on 222 samples of ground water ranging from 50 to 500 ppb. In this area, sub surface water is being extracted from wells that are drilled with simple indigenous technology by local community. Using this technique, it was possible to obtain subsurface aquifer cross sections, for eight locations over a 35 km stretch between two rivers—the Ganges and Mahananda that joins the Ganges farther south. Ground water was obtained from a minimum depth of about three metres near the Ganges River progressively increasing up to a maximum depth of about 100 metres near the Mahananda River.

Arsenic values range from a low of around 78 ppb to a high of around 473 ppb in the same direction while the Eh in the same locations vary from a high of +118 mV near the Ganges to a low of +40 mV towards the Mahananda. There are no spatial trends in Eh and arsenic levels in the entire Bengal Basin area as shown in Fig. 1. The lowest Eh value in West Bengal is around –30 mV whereas the values are reported to be as low as −400 mV in some regions of Bangladesh (Mukherjee and Bhattacharya, 2001) indicating that the redox processes responsible for arsenic contamination are more intensive in Bangladesh in comparison to West Bengal. In terms of increasing arsenic concentrations there is general north-south trend in West Bengal while it is West-East in Bangladesh; even some of the shallow wells in Bangladesh show strong reducing Eh (Mukherjee and Bhattacharya, 2001), indicating depth and not just Eh alone may be critical in the arsenic problem of this region. Incidentally, it was shown, based on more than 30,000 analyses (Welch et al., 2000) that even depth does not control arsenic levels in the ground waters of various physiographic and geological regions in the mainland part of USA. Many hypotheses on the source of arsenic have been postulated, both mineral and microbial. Extremely reducing conditions as observed for the district Meherpur (see Fig. 1) in Bangladesh does not contain very high arsenic whereas highest values of arsenic reported in the region actually show strongly oxidizing environment.

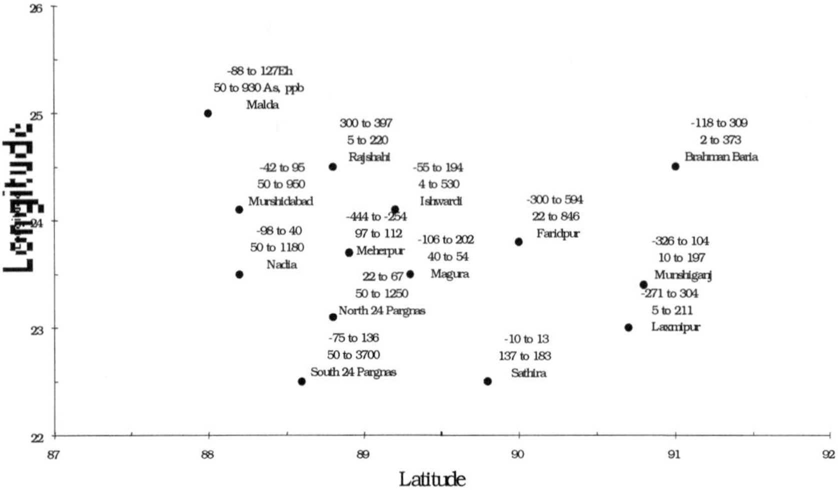

Figure 1. Range of Eh and arsenic concentration in the Bengal basin area in West Bengal (India) and Bangladesh.

The distribution of arsenic and also much talked about redox control do not show any well defined trend in this region or in many other regions of arsenic contamination spread throughout the world (Smedley and Kinniburgh, 2002). Thus, until the source of As is identified any simple chemical model is unlikely to help in the As problem.

SULPHIDE MINERAL MINING AS A SOURCE OF ARSENIC

Metal mines move, store, and dispose very large volumes of material. In order to get at an ore body, metal mining generates large quantities of waste rock, which typically contains heavy metals (copper, arsenic, lead, etc.). These previously buried materials are exposed to the elements upon excavation and become susceptible to leaching by rain and snow. Unless carefully controlled and monitored, the leaching process can lead to ground and surface water sources contaminated with heavy metals and other toxic chemical pollution that would not have occurred naturally.

Surface water samples were obtained from Zawar (Pb and Zn sulphides) and Khetri (Cu sulphide) mines and also groundwater from the surrounding residential areas were collected. It should be noted that some of the other mining areas have similar minerals and are therefore potential sites of arsenic contamination. Fresh grains of the sulphide ores, Pb, Zn and Cu concentrates and their tailings were also collected from the ore beneficiation plants at Zawar and Khetri.

Table 2. Total arsenic in groundwater (ppb), mine water (ppb) and in the sulphide ore, smelter concentrates and tailings (mg/kg) in the Khetri and Zawar region

Zawar	*Gw*	*MW^IN^*	*MW^OUT^*	*As in Pb & Zn ore*	*As in Zn Conc.*	*As in Pb Conc.*	*As in Pb & Zn tailings*
1	4.7	6.9	120	3262	6256	49	463
2	13.8	0.9	140	8410	15197	72	2575
3	1.4	0.9	–	5558	–	–	–
4	3.3	3.3	–	13501	–	–	–
Mean	5.8	3.0	130	7683	10727	61	1519
SD	5.5	2.8	14	4414	6322	16	1493

Khetri	*Gw*	*MW^IN^*	*MW^OUT^*	*As in Cu ore*	*As in Cu Conc.*	*As in Cu tailings*
1	tr	tr	3	70	7	202
2	tr	tr	2	202	16	2156
3	tr	tr	–	105	–	–
4	tr	tr	–	180	–	–
5	tr	tr	–	74	–	–
6	tr	tr	–	241	–	–
Mean	tr	tr	3	145	12	1179
SD	tr	tr	1	72	6	1382

Gw Groundwater
MW^{IN} Mine water inlet
MW^{OUT} Mine water outlet
tr, trace (< 1 ppb)

In groundwater (Zawar area) the total dissolved arsenic was between 1.4 and 13.8 ppb with an average of 5.8 ± 5.5 ppb. In the Khetri area the concentration was generally below <1 ppb (detection limit). The average total dissolved arsenic in Zawar and Khetri mine inlet water was 3.0 ± 2.8 ppb and <1 ppb, respectively while the mine outlet water had 130 ± 14 ppb and 3 ± 1 ppb arsenic, respectively. Oxidation of metallic sulphides under the action of inlet water produces sulphuric acid and the metals released are soluble in this acidic water thus releasing arsenic to the outlet water. Our analysis shows higher concentration of arsenic in Zawar and Khetri ore with an average of 7683 ± 4414 mg As/kg and 145 ± 72 mg As/kg of the sulphide ores, respectively. The average concentration of arsenic in Zawar zinc and Khetri copper tailings was 1519 ± 1493 mg As/kg and 1179 ± 1382 mg As/kg of tailings, respectively. In the metal sulphide smelter, in the zinc concentrate 10727 ± 6322 mg As/kg of the concentrate, in the Pb concentrate 61 ± 16 mg As/kg of the concentrate and in the copper concentrates 12 ± 6 mg As/kg of the concentrate were observed. Processing of Pb-Zn sulphide ores involves first the separation of Zn from the Pb concentrates through addition of chemicals such as potassium ethyl xanthate leading to enhancement of As in the Zn concentrates relative to the Pb concentrates. On the other hand, in the beneficiation of Cu ore, while the concentration of As in the Cu ore is less relative to that in Pb-Zn ores, even the small excess As is eliminated in liquid form in the tailings. Thus, the Cu beneficiation is more harmful to the environment due to direct mixing of As enriched tailing with the immediate aquatic systems compared to the Zn beneficiation where As is enriched in the Zn concentrates and hence is in solid form (Madhavan and Subramanian, 2000).

CONCLUSION

Thus our observation indicates that from the sulphide ores the mine water leaches out as under redox and acidic conditions and further as concentration takes place in the metal sulphide beneficiation activities. Thus no specific As-enriched mineral such as arsenopyrite is required in the neighbouring lithology but all sulphide mining and smelter sites are potential locations of future As contaminated areas and should not await an episodic event to unleash an environmental disaster. Arsenic studies should therefore precede mining/smelter operations as a precaution in all sulphide related mineral resources exploitation.

REFERENCES

Acharya, S.K., Lahiri, S., Raymahashay, B.C. and Bhaowmik, A. (2000). Arsenic toxicity of groundwater in parts of Bengal Basin in India and Bangladesh: The role of Quaternary Stratigraphy and Holocene sea-level fluctuation. *Environ. Geol.*, **39:** 1127–1137.

Acharyya, S.K., Chakraborty, P., Lahiri, S., Raymahashay, B.C., Guha, S. and Bhowmik, A. (1999). Arsenic poisoning in the Ganges delta. *Nature*, **401:** 545.

Adriano, D.C. (1986). *Trace Elements in the Terrestrial Environment.* Springer-Verlag, New York, 533 pp.

Adriano, D.C. and Doner, H.E. (1982). *In:* A.L. Page (ed.), *Methods of Soil Analysis Part 2*, Am. Soc. Agron. Inc, Madison, WI., 449-483.

Agrawal, V., Vaish, A.K. and Vaish, P. (1997). Groundwater quality: Focus on fluoride and fluorosis in Rajasthan. *Curr. Sci.*, **73**: 743-746.

Alabdula'aly, A.I. (1997). Fluoride content in drinking water supplies of Riyadh, Saudi Arabia. *Environ. Monit.Assess.*, **48**: 261-272.

Anawar, H.M., Akai, J., Mostofa, K.M.G., Safiullah, S. and Tareq, S.M. (2002). Arsenic poisoning in groundwater: Health risk and geochemical sources in Bangladesh. *Enviorn. Int.*, **27:** 597–604.

Anderson, M.A., Zelazny, L.W. and Bertsch, P.M. (1991). Fluoro-Aluminium complexes on model and soil exchangers. *Soil Sci. Soc. Am. J.*, **55**: 71- 75.

Arnesen, A.K.M. (1998). Effect of fluoride pollution on pH and solubility of Al, Fe, Ca, Mg, K and organic matter in soil from Ardal (Western Norway). *Water, Air and Soil Pollut.*, **103**: 375-388.

Arulanatham, A., Ramakrishna, T.V. and Balasubramanian, N. (1992). Studies on fluoride removal by coconut shell carbon. *Indian J. Environ. Protection*, **12**: 531-536.

Azcue, J.M. (1995). Environmental significance of elevated natural levels of arsenic. *Enviorn. Rev.*, **3:** 212-221.

Backer, D.E. and Chesnin, L. (1975). Chemical monitoring of soils for environment quality and animal and human health. *Adv. Agron.*, **27:** 305-374.

Barrow, N.J. and Shaw, T.C. (1977). The slow reactions between soil and anions: Effect of time and temperature of contact fluoride. *Soil Sci.*, **124**: 265-278.

Belzile, N. and Tessier, A. (1990). Interactions between arsenic and iron oxyhydroxides in lacustrine sediments. *Geochem, Cosmochim.Acta*, **54:** 103-109.

Bhargava, R.K., Saxena, S.C. and Thergaonkar, V.P. (1978). Ground water quality in Ajmer district. *Indian J. Environ. Hlth.*, **20**: 290-299.

Bhattacharya, P., Chatterjee, D. and Jacks, G. (1997). Occurrence of arsenic-contaminated groundwater in alluvial aquifers from the Delta Plain, Eastern India: Options for safe drinking water supply. *Water Res. Dev.*, **13:** 79-92.

Bhattacharya, P. (2000). Ridding West Bengal groundwater arsenic free. *Down to Earth*, **9:** 54-55.

Bond, W.J., Smith, C.J., Gibson, J.A.E. and Willett, I.R. (1995). The effect of sulfate and fluoride on the mobility of aluminium in soil. *Aust. J. Soil Res.*, **33**: 883-897.

Bowen, H.J.M. (1979). *Environmental chemistry of the elements*. Academic Press, New York. 333 pp.

Boyle, R.W. and Jonasson, J.R. (1973). The geochemistry of arsenic and its use as an indirect element in geochemical prospecting. *J Geoch. Expl.*, **2:** 251-256.

Brannon, J.M. and Patrick, W.H. (1987). Fixation, transformation, and mobilization of arsenic in sediments. *Environ. Sci. Technol.*, **21:** 450-459.

Brewer, R.F. (1965). *In:* H.D. Chapman (ed). *Diagnostic Criteria for Plants and Soils*, Quality Printing Co. Inc., Abilene, Tx., 180-196.

Carrillo, A. and Drever, J.I. (1998). Adsorption of arsenic by natural aquifer material in the San Antonio-El Triunfo mining area, Baja California, Mexico. *Envriron. Geol.*, **35:** 251-257.

Chakraborti, D., Biswas, B.K., Basu, G.K., Mandal, B.K., Chowdhury, U.K., Mukherjee, S.D., Gupta, J.P., Chowdhury, S.R. and Rathore, K.C. (1999). Arsenic groundwater contamination and sufferings of people in Rajnandgaon district, Madhya Pradesh, India. *Curr. Sci.*, **77:** 502-504.

Chaudhuri, A.N., Basu, S., Chattopadhyay, S. and Das Gupta, S. (1999). Effect of high arsenic content in drinking water on rat brain. *Indian J. Biochem. Biophys.*, **36:** 51-54.

Choubisa, S.L. (1998). Fluorosis in some tribal villages of Udaipur district (Rajasthan). *J. Environ. Bio.*, **19**: 341-352.

Chowdhury, U.K., Biswas, B.K., Chowdhury, T.R., Samanta, G., Mandal, B.K., Basu, G.C., Chanda, C.R., Lodh, D., Saha, K.C., Mukherjee, S.K., Roy, S., Kabir, S., Quamruzzaman, Q. and Chakraborti, D. (2000). Groundwater arsenic contamination in Bangladesh and West Bengal, India. *Environ Health Perspect.,* **108:** 393-397.

Craw, D., Falconer, D. and Youngson, J.H. (2003). Environmental arsenopyrite stability and dissolution: theory, experiment, and field observations. Chemical Geology, **199:** 71-82.

Das, D., Chatterjee, A., Mandal, B.K., Samanta, G., Chakraborti, D., Chanda, B. (1995). Arsenic in ground water in six districts of West Bengal, India: The biggest arsenic calamity in the world. Part 2. Arsenic concentration in drinking water, hair, nails, urine, skin-scale and liver tissue (biopsy) of the affected people. *Analyst*, **120:** 917-924.

Das, D., Samanta, G., Mandal, B.K., Roy Chowdhury, T., Chanda, C.R., Chowdhury, P.P., Bose, G.K. and Chakraborti, D. (1996). Arsenic in groundwater in six districts of West Bengal, India. *Environ. Geochem. Health*, **18:** 5-15.

Datta, P.S., Tyagi, S.K., Mookerjee, P., Bhattacharya, S.K., Gupta, N. and Bhatnagar, P.D. (1999). Groundwater NO_3 and F contamination processes in Puskar valley, Rajasthan as reflected from ^{18}O isotopic signature and ^{3}H recharge studies. *Environ. Monit. Asses.*, **56**: 209-219.

Davis, J. (2000). *Control of arsenic solubility by iron, barium and copper solids.* 31st International Geological Congress (Aug. 3-5, 2000), Rio de Janeiro, Brazil.

Deshmukh, A.N., Wadaskar, P.M. and Malpe, D.B. (1995). Fluorine in environment: A review. *Gondwana Geol. Mag.*, **9**: 1-20.

Dissanayake, C.B. (1991). The fluoride problem in the groundwater of Sri Lanka—Environmetnal management and health. *Int. J. Environ. Stud.*, **38**: 137-156.

Dzombak, D.A. and Morel, F.M.M. (1990). Surface Complexation Modelling—Hydrous Ferric Oxide. John Wiley, New York.

Ferguson, J.E. (1989). The heavy elements: Chemistry, environmental impacts and health effects, Oxford, Pergamon Press, pp. 614.

Ferguson, J.F. and Gavis, J. (1972). A review of the arsenic cycle in natural waters. *Wat. Res.*, **6:** 1259-1274.

Fleischer, M. and Robinson, W.O. (1963). Some problems of the geochemistry of fluorine. *Roy. Soc. Can. Spec. Publ.*, **6**: 58-75.

Flühler, H., Polomski, J. and Blaser, P. (1982). Retention and movement of fluoride in soils. *J. Environ. Qual.*, **11**: 461-468.

Fung, K.F., Zhang, Z.Q., Wong, J.W.C. and Wong, M.H. (1999). Fluoride contents in tea and soil from tea plantations and the release of fluoride into tea liquor during infusions. *Environ. Pollut.*, **104**: 197-205.

Gelogical Survey of India (GSI) (1963). *Annotated index of Indian mineral occurrences*. Part (II). Edited by Chatterjee, P.K., pp. 147-285.

Geological Survey of India (GSI) (1977). *Geology and mineral resources of the states of India*. Part XII—Rajasthan, Miscellaneous Publication No. 30, 75 pp.

Gilpin, L. and Johnson, A.H. (1980). Fluorine in agricultural soils of Southern Pennsylvania. *Soil Sci. Soc. Am. J.*, **44**: 255-258.

Guha Mazumder, D.N., Haque, R., Ghosh, N., De, B.K., Santra, A., Chakraborty, D. and Smith, A.H. (1998a). Arsenic levels in drinking water and the prevalence of skin lesions in West Bengal, India. *Int. J. Epidemiol.*, **27:** 871-877.

Gupta, R.K., Chhabra, R. and Abrol, I.P. (1982). Fluorine adsorption behaviour in alkali soils: Relative roles of pH and sodicity. *Soil Sci.*, **133**: 364-368.

Gupta, S.R. and Ghosh, S. (1953). Precipitation of brown and yellow hydrous iron oxide. III Adsorption of arsenious acids. *Kolloid-Z.*, **132:** 141-143.

Handa, B.K. (1975). Geochemistry and genesis of fluoride containing ground waters in India. *Groundwater*, **13:** 275-281.

Hasegawa, H. (1997). The behaviour of trivalent and pentavalent methylarsenicals in Lake Biwa. *Appl. Organomental. Chem.*, **11:** 305-311.

Hasegawa, H., Matsui, M., Okamura, S., Hojo, M., Iwasaki, N. and Sohrin, Y. (1999). Arsenic speciation including 'hidden' arsenic in natural waters. *Applied Organomental. Bhem.*, **13:** 113-119.

Hopkins, D.M. (1977). *J. Res USGS.*, **5:** 589-593.

Huang, P.M. and Jackson, M.L. (1965). Mechanism of reaction of neutral fluoride solution with layer silicates and oxides of soils. *Soil Sci. Soc. Am. Proc.*, **29**: 661-665.

Jana, J., Sahu, S.J., Roy, S., Bhattacharyya, R., Nath, B., De Dalal, S.S., Chatterjee, D., Bhattacharya, P. and Jacks, G. (2000). *Redox-induced arsenic mobilization in an anoxic groundwater environment—a field study in West Bengal, India.* 31st International Geological Congress (Aug. 3-5, 2000), Rio de Janeiro, Brazil.

Jinadasa, K.B.P.N., Dissanayake, C.B., Weerasooriya, S.V.R. and Senaratne, A. (1993). Adsorption of fluoride on goethite surfaces—Implications on dental epidemiology. *Environ. Geol.*, **21:** 251-255.

Jones, B.F., Eugster, H.P. and Rettig, S.L. (1977). Hydrogeochemistry of the lake Magadi basin, Kenya. *Geochim. Cosmochim. Acta.*, **44:** 53-72.

Jung, M.C., Thornton, I., Chon,Hyo-Taek (2002). Arsenic, Sb and Bi contamination of soils, plants, waters and sediments in the vicinity of the Dalsung Cu-W mine in Korea. *The Science of the Total Environment*, **295:** 81-89.

Kabata-Pedias (1989). *Trace Elements in Soils and Plants.* CRC Press, Inc. Boca Raton, Florida, 315 pp.

Karim, M.M. (2000). Arsenic in groundwater and health problems in Bangladesh, *Water Res.*, **14:** 1304-1310.

Kau, P.M.H., Smith, D.W. and Binning, P. (1998). Fluoride retention by kaolin clay. *J. Contaminant Hydro.*, **28:** 267-288.

Keerhisinghe, G., MeLaughlin, M.J. and Randall, P.J. (1991). Improved recovery of fluoride in plant material using a low temperature sealed chamber digestion technique in conjunction with a fluoride ion specific electrode. *Commun. Soil Sci. Plant Analy.*, **22:** 1831-1846.

Keller, E.A. (1979). *Environmental Geology*, Charles and Merril Publ. Co., Ohio, USA, 548 pp.

Killedar, D.J. and Bhargava, D.S. (1993). Effect of stirring rate and temeperature on fluoride removal by fishbone charcoal. *Indian J. Environ. Hlth.*, **35:** 81-87.

Krauskkopf, K.B. (1955). Sedimentary deposits of rare metals. *Econ. Geol.*, **50:** 411-463.

Krumgalz, B.S., Fainshtein, G., Gorfunkel, L. and Nathan, Y. (1990). Fluorite in recent sediments as a trap of trace metal contaminants in an estuarine environment. *Estuarine, Coastal and Shelf Sci.*, **30:** 1-15.

Kumpulainen, J. and Kovistoinen, P. (1977). *Residue. Rev.*, **68**: 37-57.

Larsen, S. and Widdowson, A.E. (1971). Soil fluorine. *J. Soil Sci.*, **22:** 210-221.

Latha, S.S., Ambika, S.R. and Prasad, S.J. (1999). Fluoride contamination status of groundwater in Karnataka. *Curr. Sci.*, **76:** 730-734.

Madhavan, N. and Subramanian, V. (1999). Uptake of fluoride by activated charcoal. *In:* Proceedings of the national seminar on fluoride contamination, fluorosis and defluoridation techniques. Gyani, K.C., Vaish, A.K. and Vaish, P. (eds.), SARITA, Udaipur, 146 pp.

Madhavan, N. and Subramanian, V. (2000). Sulphide mining as a source of arsenic in the environment. *Curr. Sci.*, **78:** 702-709.

Madhavan, N. and Subramanian, V. (2001). Fluoride concentration in river waters of south Asia. *Curr. Sci.*, **80:** 1312-1319.

Madhavan, N. and Subramanian, V. (2002). Fluoride in fractionated soil samples of Ajmer District, Rajasthan. *J. Environ. Moni.*, **4(6):** 82-822.

Maithani, P.B., Gurjar, R., Banerjee, R., Balaji, B.K., Ramachandran, S. and Singh, R. (1998). Anomalous fluoride in groundwater from westen part of Sirohi district, Rajasthan and its crippling effects on human health. *Curr. Sci.*, **74:** 773-777.

Mallick, S. and Rajagopal, N.R. (1996). Groundwater development in the arsenic affected alluvial belt of Went Bengal—Some questions. *Curr. Sci.*, **70:** 956-958.

Mandal, B.K., Chowdhury, T.R., Samanta, G., Mukherjee, D.P., Chanda, C.R., Saha, K.C. and Chakraborti, D. (1998). Impact of safe water for drinking and cooking on five arsenic-affected families for 2 years in West Bengal, India. *Sci. Total Environ.*, **218:** 185-201.

Masscheleyn, P.H., Delaune R.D. and Patrick W.H. (1991). Effect of redox potential and pH on arsenic specification and solubility in contaminated soil. *Env Sci Technol.*, **25:** 1414-1419.

Mazumder, D.N., Das Gupta, J., Santra, A., Pal, A., Ghose, A. and Sarkar, S. (1998). Chronic arsenic toxicity in West Bengal—the worst calamity in the world. *J. Indian Med. Assoc.*, **96:** 4-7.

McArthur, J.M., Ravenscroft, P., Safiullah, S. and Thirlwall, M.F. (2001). Arsenic in groundwater: testing pollution mechanisms for sedimentary aquifers in Bangladesh. *Water Res. Res.*, **37:** 109-117.

McGeehan, S.L., Fendorf, S.E. and Naylor, D.V. (1998). Alteration of arsenic sorption in flooded-dried soils. *Soil Sci. Soc. Am. J.*, **62:** 828-833.

McLaren, S.J. and Kim, N.D. (1995). Evidence for a seasonal fluctuation of arsenic in New Zealand's longest river and the effect of treatment on concentrations in drinking water. *Environ. Pollu.*, **90:** 67-73.

Mehrotra, R., Kapoor, B. and Narayan, B. (1999). Defluoridation of drinking water using low cost adsorbent. *Indian J. Environ. Hlth.*, **41:** 53-58.

Ming, L., Yi, S.R., Hua, Z.J., Lei, B.Y.W., Ping, L. and Fuwa, K.C. (1987). Elimination of excess fluoride in potable water by electrolysis using an aluminium anode. *Fluoride*, **20:** 54-63.

Mukherjee, A.B. and Bhattacharya, P. (2001). Arsenic in groundwater in the Bengal delta plain: slow poisoning in Bangladesh. *Environ. Rev.*, **9:** 189-220.

Nag, J.K., Balaram, V., Rubio, R., Alberti, J. and Das, A.K. (1996). Inorganic arsenic species in groundwater: A case study from Purbasthali (Burdwan), India. *J. Trace Elem. Med. Biol.*, **10:** 20-24.

Naidu, R., Summer, M.E. and Harter, R.D. (1998). Sorption of heavy metals in strongly weathered soils: An overview. *Environ. Geochem Health*, **20:** 5-9.

National Academy of Sciences (NAS), 1971. Fluorides. *In Comm. Biol. Effects of Air Pollut.*, NAS, Washington, DC., 295 pp.

National Research Council of Canada (NRCC) (1977). *In Environmental Fluoride*, NRCC No. 16081. Ottawa, Ontario, 151 pp.

Nickson, R.T., McArthur, J.M., Burgess, W.G., Ravenscroft, P., Ahmed, K.M. and Rehman, M. (1998). Arsenic poisoning of Bangladesh groundwater. *Nature*, **395:** 338.

Nickson, R.T., McArthur, J.M., Ravenscroft, P., Burgess, W.G. and Ahmed, K.M. (2000). Mechanism of arsenic poisoning of groundwater in Bangladesh and West Bengal. *Appl. Geochem.*, **15:** 403-413.

Nicolli, H.B., Suriano, J., Gomez Peral, M., Ferpozzi, L.H., Baleani, O. (1989). Groundwater contamination with arsenic and other trace elements in an area of the Pampa, Province of Corboda. *Environ. Geol. Water Sci.*, **14:** 3-16.

Omueti, J.A.I. and Jones, R.L. (1977). Fluoride adsorption by Illinois soils. *J. Soil Sci.*, **28:** 564-572.

Onishi, H. and Sandell, E.B. (1955). Geochemistry of arsenic. *Geochim. Cosmochim. Acta.*, **7:** 1-33.

Peterson, M.L. and Carpenter, R. (1986). Arsenic distribution in porewaters and sediments of Puget Sound, Lake Washington, the Washington coast and Saanich Inlet, B.C. *Geochim. Cosmochim. Acta*, **50:** 353-369.

Polomski, J., Flühler, H. and Blaser, P. (1982). Fluoride induced mobilisation and leaching of organic matter, iron and aluminium. *J. Environ. Qual.*, **11:** 452-456.

Rahman, M., Tondel, M., Ahmad, S.A., Chowdhury, I.A., Faruquee, M.H. and Axelson, O. (1999). Hypertension and arsenic exposure in Bangladesh. *Hypertension*, **33:** 74-78.

Rajwanshi, P., Singh, V., Gupta, M.K., Shrivastav, R., Subramanian, V., Prakash, S. and Dass, S. (1999). Aluminium leaching from surrogate aluminium food containers under different pH and fluoride concentration. *Bull. Environ. Contam. Toxicol.*, **63:** 271-276.

Robinson, W.O. and Edington, G. (1946). Fluorine in soils. *Soil Sci.*, **61:** 341-353.

Rose, A.W., Hawkes, H.E. and Webb, J.S. (1979). *Geochemistry in mineral exploration.* Academic Press, London. 658 pp.

Roychaudhury, T., Basu, G.K., Mandal, B.K., Biswas, B.K., Samanta Gautam, Chowdhury, U.K., Chanda, C.R., Lodh Dilip, Roy, S.L., Saha, K.C., Roy, Sibtosh, Kabir Saiful, Quamruzzaman Qazi, Chakraborti, Dipankar. (1999). Poisoning of Ganga Delta. *Nature*, **401:** 545-546.

Sadiq, M. (1995). Arsenic chemistry in soils: An overview of thermodynamic predictions and field observations. *Water Air Soil Pollut.*, **93:** 117-136.

Sahu, S.K., Pati, S.S. and Padapanda, R.K. (1998). Fluorine content in groundwater around an aluminium industry in Hirakund, Orissa, *Environ. Geol.*, **16:** 169-171.

Saifullah, S., Kabir, A., Tereq, S.M., Khan, M.M.K. and Alam, F.R. (1998). Removal of arsenic by composite porous materials based on Fe_2O_3-MnO_2-laterite soil. *J. Bangladesh Chem. Society.* **12(2):** 185-192.

Sangodoyin, A.Y. and Ogedengbe. K. (1991). Surface water quality and quantity from the standpoint of irrigation and livestock. *Intern. J. Environ. Studies*, **38:** 251-262.

Schneider, H.-J. and Möller, P. (1977). Fluorine contents in carbonate sequences and rare earths distribution in fluorites of Pb-Zn deposits in East Alpine Mid-Triassic. *Mineral Deposita (Berl.)*, **12:** 22-36.

Serrano, M.J.G., Sanz, L.F.A and Nordstrom, D.K. (2000). REE speciation in low-temperature acidic waters and the competitive effects of aluminum. *Chem. Geol.*, **165:** 167-180.

Shacklette, H.T. and Boerngen, J.G. (1984). *Elemental concentrations in soils and other surficial materials of the conterminous United States.* USGS Prof Paper 1270. US Govt. Printing Office, Washington, DC.

Shnyukov, E.F. (1963). Arsenic in the Cimmerian iron ores of the Azov-Black Sea region. *Geochemistry*, 7-93.

Singh, R.P., Singh, Y. and Swarrop, D. (2000). Defluoridation of groundwater in Agra city using low cost adsorbent. *Bull. Environ. Contam.Toxicol.*, **65:** 120-125.

Slavek, J., Farrah, H. and Pickering, W.F. (1984). Interaction of clays with dilute fluoride solutions. *Water, Air & Soil Pollut.*, **23:** 209-220.

Smedley, P.L., KinniBurgh, D.G. (2002). A review of the source, behaviour and distribution of arsenic in natural waters. *Applied Geochemistry*, **17:** 517-568.

Smith, E., Naidu, R. and Alston, A.M. (1988). Arsenic in the soil environment: A review. *Adv. Agron.*, **64:** 149-195.

Smith, A.H., Lingas, E.O. and Rahman, M. (2000b). Contamination of drinking-water by arsenic in Bangladesh: A public health emergency. *Bull. World Health Organ.*, **78:** 1093-1103.

Sparks, D.L. (1995). *Environmental Soil Chemistry.* Academic Press, San Diego. 267 pp.

Steinkoenig, L.A. (1919). *J. Indus Eng. Chem.*, **11:** 463-465.

Stevens, D.P., McLaughlin, M.J. and Alston, A.M. (1995). Limitations of acid digestion techniques for the determination of fluoride in plant material. *Commun. Soil Sci. Plant Analy.*, **26:** 1823-1842.

Stow, S.H. (1969). The occurrence of arsenic and the colour-causing components in Florida land-pebble phosphate rock. *Econ. Geol.*, **64:** 667-671.

Stummeyer, J., Marchig, V. and Knabe, W. (2002). The composition of suspended matter from Ganga-Brahmaputra sediment dispersal system during low sediment transport season. *Chem Goel.*, **185(1-2):** 125-147.

Subramanian, V. (1980). A geochemical model for phosphate mineralisation in marine environment. *Geological Survey of India. Misc. Publication* No. **44:** 308-313.

Susheela, A.K. and Kharb, P. (1990). Arotic calcification in chronic fluordie poisoning: Biochemical and electromicroscopic evidence. *Experimental and Molecular Pathology*, **53:** 72-80.

Susheela, A.K. and Kumar, A. (1991). A study of the effect of high concentrations of fluoride on the reproductive organs of male rabbits, using light and scanning electron microscopy. *J. Reprod. Fert.*, **92:** 353-360.

Susheela, A.K. (1999). Flurosis management programme in India. *Curr. Sci.*, **77:** 1250-1256.

Takamatsu, T., Kawashima, M. and Koyama, M. (1985). The role of Mn^{+2}-rich hydrous manganese oxide in the accumulation of arsenic in lake sediments. *Water Res.*, **19(8):** 1029-1032.

Tamaki, S., Frankenberger (1992). Environmental Biochemistry of arsenic. *Rev. Environ. Contam. Toxicol.*, **124:** 79-110.

Tareq, Shafi M., Safiullah, S., Anawar, H.M., Rahman, M. Majibur and Ishizuka, T. (2003). Arsenic pollution in groundwater: A self-organizing complex geochemical process in the deltaic sedimentary environment, Bangladesh. *The Science of Total Environment*, **313:** 213-226.

Tondel, M., Rahman, M., Magnuson, A., Chowdhury, I.A., Faruquee, M.H. and Ahmad, S.A. (1999). The relationship of arsenic levels in drinking water and the prevalence rate of skin lesions in Bangladesh. *Environ. Health Perspect.*, **107:** 727-729.

Underwood, E.J. (1977). *Trace elements in human and animal nutrition.* Academic Press, New York. 545 pp.

United States Environmental Protection Agency (US-EPA) (1999). *Current Drinking Water Standards.* OGWDW, 8 pp.

Vinogradov, A.P. (1959). The Geochemistry of Rare and Dispersed Chemical Elements in Soils, 2nd ed., New York, pp. 65-70.

Vora, J. and Joshi, J.D. (1998). Alum treatment process for fluoride reduction in potable water. *Curr. Sci.*, **75:** 338-339.

Wadia, D.N. (1994). *Geology of India*. Tata McGraw-Hill, New Delhi, pp. 508.

Wang, Y. and Reardon, E.J. (2001). Activation and regeneration of a soil sorbent for defluoridation of drinking water. *App. Geochemist.*, **16:** 531-539.

Wasay, S.A., Harson, Md. J. and Tokunaga, S. (1996). Adsorption of fluoride, phosphate and arsenate ions on lanthanum-impregnated silica gel. *Water Environ. Res.*, **68:** 295-300.

Wedepohl (1974). *Handbook of Geochemistry*. Springer-Verlage Berlin. Heidelberg. New York. **2(4):** 9K-1 pp.

Welch, A.H., Westjohn, D.B., Helsel, D.R. and Wanty, R.B. (2000). Arsenic in groundwater of the United States: Occurrence and geochemistry. *Groundwater*, **38:** 589-604.

WHO (1996). *Guidelines for drinking water quality. Recommendations*. World Health Organisation, Geneva, **1:** 188.

Yan, X.-P., Kerrich, R. and Hendry, M.J. (2000). Distribution of the arsenic (III), arsenic (V) and total inorganic arsenic in pore-waters from the thick till and clay-rich aquitard sequence, Saskatchewan, Canada. *Geochem. Cosmicha. Acta.*, **64:** 2637-2648.

Yan-Chu, Huang (1994). Arsenic Distribution in Soils in Arsenic in the Environment, Part-I: Cycling and Characterization. Jerome O. Nriagu (ed.), John Wiley & Sons. Inc.

Yang, M.M, Hashimoto, T., Hoshi, N. and Myoga, H. (1999). Fluoride removal in a fixed bed packed with granular calcite. *Wat. Res.*, **33:** 3395-3402.

Zobrist J., Dowdle, P.R., Davis, J.A. and Oremland, R.S. (2000). Mobilization of arsenite by dissimilatory reduction of adsorbed arsenate. *Environs. Sci. Technol.*, **34:** 4747-4753.

7

Characterization of Fracture Properties in Hard Rock Aquifer System

J.C. Maréchal, B. Dewandel and K. Subrahmanyam[1]

BRGM (Bureau de Recherches Géologiques et Minières), Water Department, 1039 rue de Pinville, 34 000 Montpellier, France
[1]National Geophysical Research Institute, Indo-French Centre for Groundwater Research, Uppal Road, Hyderabad, 500 007, India

PREAMBLE

The hydrodynamic properties of the weathered-fractured layer of a hard-rock pilot watershed in a granitic terrain are characterized using hydraulic tests at different scales. The interpretation of numerous slug tests leads to characterize the statistical distribution of local permeabilities in the wells. The application of flow meter profiles during injection tests determines the vertical distribution of conductive fracture zones and their permeabilities. It appears that the extension of the most conductive part of the weathered-fractured layer is limited down to 35 metres depth. The partition of drainage porosity between blocks (90%) and fractures (10%) is determined thanks to the interpretation of pumping tests using a double porosity model. The application of anisotropic and single fracture analytical solutions on pumping test data allows to determine, respectively, the degree of anisotropy of permeability ($K_r/K_z = 10$) and the radius (4 to 16 metres) of the horizontal conductive fractures crossed by the wells. Two different scales of fracture networks are identified: the primary fracture network (PFN), which affects the matrix on a decimeter scale by contributing to an increase in the permeability and storage capacity of the blocks, and the secondary fracture network (SFN), which affects the blocks at the borehole scale. SFN is composed of two sets of fractures. The main set of horizontal fractures is responsible for the sub-horizontal permeability of the weathered-fractured layer. A second set of less permeable sub-vertical fractures insures the connectivity of the aquifer at the borehole scale. The good connectivity of fracture networks is shown by fractional dimension flow solutions. The absence of scale effect in the study area suggests that the hydraulic conductivity at the borehole scale is laterally homogeneous. Finally, the analysis and synthesis of the hydrodynamic properties allow to propose a comprehensive hydrodynamic model of the fractured-weathered layer. Many geological and hydrogeological indicators suggest that a continuous and laterally homogeneous weathering process is responsible for the origin of the fractures and permeability encountered in the aquifer. These results confirm the major role played by weathering in the origin of fractures and on resulting hydrodynamic parameters in the shallow part of hard-rock aquifers.

INTRODUCTION

Hard rocks and their associated aquifers occur in many areas of the world. They are principally (but not always) exposed in large areas called shields, composed mainly of metamorphic and magmatic rocks of Precambrian or Archean ages, and form the most tectonically stable parts of the continents. The importance of hard-rock aquifers for hydrogeological and water management issues differs from place to place, depending on various factors, but mainly on the overall availability of water and the water demand. In arid and semi-arid regions, due to the lack of surface water resources, special attention has traditionally been paid to shallow groundwater. The existence of semi-arid or arid climates in hard-rock regions increases the interest of hydro-geologists for such aquifers namely in Africa, Brazil or India (Gustafson and Krasny, 1994). India, with hard rock constituting more than two-thirds of the total surface, is a great example of the nexus between water scarcity and the occurrence of hard-rock aquifers. The groundwater boom during the Green revolution of the seventies has led to a complete inversion of the irrigation scenario, with groundwater now sustaining almost 60 % of irrigated land (Roy and Shah, 2002). As recently pointed out by Bredehoeft (2002), sustainable groundwater development of an aquifer usually depends on the dynamic response of the system to development. The principal tool to evaluate the sustainability of management scenarios for hydro-systems is groundwater modeling. One of the major prerequisites for establishing the reliability of such models is an accurate knowledge of the geometry and the hydrodynamic properties of the aquifer.

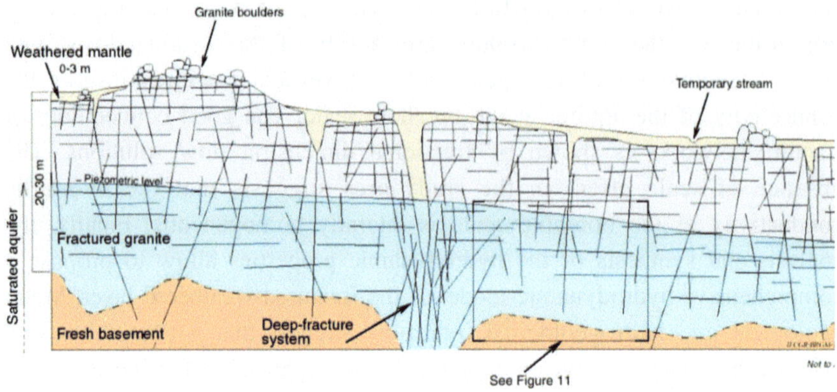

Figure 1. Simplified geological profile of a hard-rock aquifer, modified from Wyns et al. (2004), in accordance with the thickness and geometry of the layers corresponding to the case study.

Hard-rock aquifers generally occupy the upper tens of metres of the subsurface profile (Detay et al., 1989). The hydrogeological characteristics (e.g., hydraulic conductivity and storage) of the weathered mantle and underlying bedrock derive primarily from the geomorphologic processes of deep weathering and erosion (Taylor and Howard, 2000; Wyns et al., 2004). The classical weathering profile (see Fig. 1) is composed of the following layers (from top to bottom), which have specific hydrodynamic properties:

- Unconsolidated weathered mantle (saprolite or regolith), with a thickness from negligible to a few tens of metres, derived from prolonged in-situ decomposition of bedrock. Usually, this unconsolidated layer has a high porosity and a low permeability (Acworth, 1987). When saturated, this layer constitutes the reservoir of the aquifer.
- Fractured-weathered layer, generally characterized by a fracture density that decreases with depth (Houston and Lewis, 1988; Howard et al., 1992). These fractures could be caused by cooling stresses in the magma, subsequent tectonic activity (Houston and Lewis, 1988) or litho-static decompression processes (Davis and Turk, 1964; Acworth, 1987; Wright, 1992). Recent works (Lachassagne et al., 2001; Wyns et al., 2004) demonstrate that they also result from the weathering process itself. This layer mainly assumes the transmissive function in the aquifer and is pumped by most of the wells drilled in hard-rock areas.
- Fresh basement, which is permeable only locally where deep tectonic fractures are present. This deep part of the hard-rock aquifers has been investigated in numerous countries (Canada, England, Finland, Sweden,

Switzerland, United States) under various nuclear waste disposal programmes (Blomqvist, 1990; Walker et al., 2001; Pickens et al., 1987) and is not the subject of this paper, which focuses on the shallow aquifers of hard rock.

Given that the unconsolidated weathered mantle can be represented by a porous medium, its hydrodynamic properties can be easily measured in-situ and many data (see, for instance, Acworth, 1987) are available for modeling. On the other hand, due to the heterogeneity, discontinuity and anisotropy induced by the fracture networks, the hydrogeology of the fractured-weathered layer is more complex. The scientific community has long focused on the specific capacities (Summers, 1972) or yields of wells in crystalline rocks as a function of their depth (Davis and Turk, 1964), geology (Houston and Lewis, 1988) or weathering profile (Foster, 1984). Many authors have studied the evolution with depth of the transmissivity and the storage in the upper part of the weathered crystalline basement (Chilton and Foster, 1984). The optimization of the location of boreholes in tectonically fractured areas using geophysics (Barker et al., 1992; White et al., 1988), remote sensing (Sander, 1997; Mabee and Hardcastle, 1997) or geological data (Houston and Lewis, 1988) has been a major issue for rural water supplies, namely in Africa (Chilton and Smith-Carington, 1984). More recently, Taylor and Howard (2000) have proposed a tectono-geomorphic model of the hydrogeology of crystalline rock from Uganda, and have pointed out the shallow hydrogeology of such rocks in relation to their geological context. Most of these studies cite transmissivity estimates of the fractured zone of weathered crystalline rocks without considering the total hydrodynamic behaviour of the layer or the relationships between fracture networks and the rock matrix. Other related properties such as anisotropy, connectivity or distribution of storage between matrix and fractures are not explored. Therefore, this section presents the application of existing interpretation techniques specific to fractured media in order to propose a conceptual model of the geometry and physics of water flow in the weathered-fractured layer of hard-rock aquifers.

METHODOLOGY AND STUDY AREA

Extensive fieldwork has been carried out on a hard-rock aquifer in India (Maheshwaram watershed). Hydraulic tests at different investigation scales were used to determine the hydrodynamic properties of the weathered-fractured layer. Slug tests, injection tests, flow meter tests, and pumping tests are interpreted using specific techniques for fractured media. Hydraulic conductivity, storage coefficient, anisotropy, connectivity, fracture density, conductivity and radius of the fractures are investigated using various analytical solutions (Table 1).

Table 1. Characteristics of techniques used for the interpretation of hydraulic tests

Hydraulic test	*Interpretation method*	*u/c*[1]	*t/s*[2]	*Parameters obtained*	
Slug test	Solution for unconfined aquifers (Bouwer and Rice, 1976)	u	t	Local permeability	Scale effect
Injection test	Classical Dupuit (1848, 1863) solution	u	s	Permeability	
Flowmeter test	Classical Dupuit (1848, 1863) solution for confined aquifers corrected for unconfined aquifers	u	s	Permeability and density of CFZ[3]	
Pumping test	Double porosity (Warren and Root, 1963)	c	t	Bulk permeability and storage of blocks and fractures	
Pumping test	Anisotropy (Neuman, 1975)	u	t	Permeability and storage, degree of permeability anisotropy	
Pumping test	Single fracture (Gringarten, 1974)	c	t	Radius of fractures	
Pumping test	Fractional dimension flow (Barker, 1988)	c	t	Flow dimension, generalized transmissivity and storage	

[1]hypothesis on aquifer property in the analytical solution (u: unconfined, c: confined)
[2]analytical solution type (t: transient, s: steady-state)
[3]CFZ: conductive fracture zones

First, the interpretation of numerous slug tests in observation wells allows getting information on the statistical distribution of local permeabilities. The application of flow meter profiles during injection tests locates in depth zones of fractures and their permeabilities, including the vertical density of fractures. Pumping tests are then interpreted using a double porosity model in order to determine the respective contribution of blocks and fractures to storage coefficient and hydraulic conductivity. The application of anisotropic and single fracture analytical solutions allows determining, respectively, the degree of anisotropy of permeability and the radius of conductive horizontal fractures crossed by the wells. Then, connectivity of fractures networks is explored using fractional dimension flow solutions. Except flow meter profiles and injection tests, the results are interpreted using transient analytical solutions in order to improve the analysis precision. Both unconfined and confined solutions are used; this point will be justified later. The scale effect is characterized comparing previous results of various tests at different scales. Finally, the analysis and synthesis of the hydrodynamic properties introduce a hydrodynamic model of the fractured-weathered layer.

The Maheshwaram watershed is composed mainly of Archean granites. The surface area of the watershed, located around the village of Maheshwaram 30 kilometres from Hyderabad (Andhra Pradesh, India), is about 55 km^2 (Fig. 2a). The major part of the basin is constituted by biotite granite (centimetric grain-size) often including porphyritic K-feldspars. Leucogranites are observed in the southwestern part of the basin. The weathering profile is observable from numerous dug-wells previously used by the farmers for irrigation (Fig. 2b). The weathering profile is generally truncated by erosion; overlain by a few decimetres of red soil, the weathered mantle has quite a homogeneous thickness of less than three metres throughout the watershed. A high density of sub-horizontal fractures is observed in the underlying fractured-weathered layer. A few sub-vertical fractures are also present. This layer is estimated from bore wells observations (from cuttings) to be between 20 and 40 metres thick. Due to the over-exploitation of groundwater resources, water levels are far below ground level; thus, the weathered mantle is dry and only the deeper weathered-fractured layer is saturated. This feature was used to carry out hydraulic tests with the objective of characterizing the hydrodynamic properties of the weathered-fractured layer only.

A series of 34 observation wells were drilled in the watershed to depths ranging from 32 to 45 metres (Fig. 2a). The absence of a confining layer and the storage values obtained from hydraulic tests suggest that the aquifer is unconfined. The frequent observation during drilling that the final static water level lies above the level at which water is initially encountered is explained by the separation of the fracture systems into discrete flow horizons by low-permeability blocks.

LOCAL HYDRAULIC CONDUCTIVITY AND VERTICAL DISTRIBUTION OF FRACTURES

Thirty slug tests were achieved in the wells and interpreted using the Bouwer and Rice (1976) technique for unconfined aquifers with completely or partially penetrating wells. The distribution of hydraulic conductivity obtained (Table 2) constitutes a preliminary estimate of the hydraulic conductivity of the fractured-weathered layer in close proximity to the well. The values

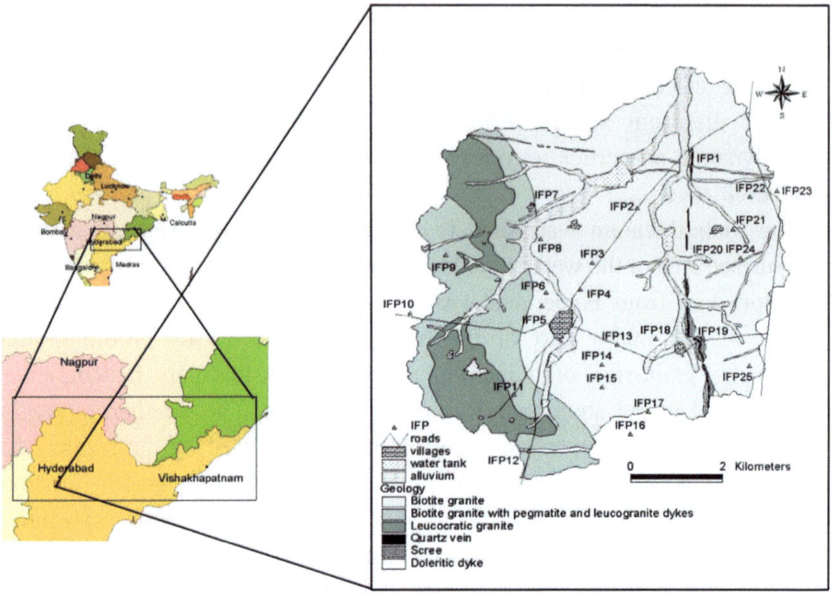

Figure 2(a). Simplified geology of the study area with locations of the observation wells (IFP-1...25) where the hydraulic tests were conducted (nine additional wells named IFP-1/1, IFP-1/2... are not located for map clarity reasons).

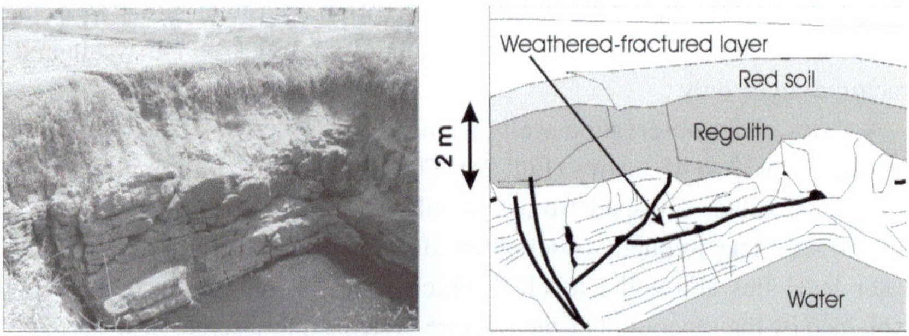

Figure 2(b). Photograph and geological interpretation of a dug well in the Maheshwaram area. The horizontal and sub vertical fracture networks are visible.

range from 2×10^{-8} m/s up to 5×10^{-4} m/s and exhibit a near-lognormal distribution with a geometric mean of 4.4×10^{-6} m/s (Fig. 3a). Within this distribution, a small population of three boreholes (IFP-5, IFP-19 and IFP-25) with permeabilities of less than 10^{-7} m/s corresponds to the less fractured granite.

Table 2. Characteristics of wells tested by slug and injection

Well	*Depth* m	*Casing depth* m	*Water-table depth* m	*Basement Depth*[3] m	*K, slug test* m/s	*K, injection* m/s	*Number of CFZ*[4]	*CFZ density*[5] m^{-1}	*Persistence*[6] m
IFP-1	42	8.8	16.7	37.5	7.4E-6	1.3E-5	4	0.14	7.08
IFP-1/1	45	5.75	16.7	35	2.1E-6	n/a[1]	n/a	n/a	n/a
IFP-1/2	35	11.95	18.1	30.2	2.7E-6	5.1E-7	2	0.08	12.29
IFP-1/3	35	5.65	18.5	29	4.4E-7	1.9E-7	u/k[2]	u/k	u/k
IFP-2	27.1	0.6	14.8	20	1.9E-7	1.1E-5	u/k	u/k	u/k
IFP-3	37	6.25	11.9	23.5	2.0E-6	2.7E-5	6	0.27	3.71
IFP-4	45	0.8	19	22	6.7E-7	6.0E-7	u/k	u/k	u/k
IFP-5	60	11.65	25.8	36	3.9E-8	1.7E-7	0	0.00	>25.5
IFP-6	42	21.3	21.9	32.2	3.4E-5	4.9E-5	6	0.44	2.25
IFP-7	38	14.8	20.5	18.5	5.1E-4	2.6E-4	3	0.80	1.25
IFP-8	38	12	20.4	22.86	2.6E-5	1.4E-4	5	0.43	2.30
IFP-9	42	12	28.4	33.53	3.8E-5	2.2E-5	4	0.18	5.47
IFP-9/2	42	11.95	28.1	n/a	9.1E-6	n/a	n/a	n/a	n/a
IFP-10	40	1.2	24.7	22	6.4E-6	9.7E-8	u/k	u/k	u/k
IFP-11	42	18.2	21.5	37.5	1.9E-5	6.4E-5	3	0.12	8.62
IFP-11/1	n/a	n/a	21.9	30	n/a	1.6E-5	u/k	u/k	u/k
IFP-11/3	42	19	21.9	35.2	5.0E-6	1.7E-6	1	0.04	23.12
IFP-11/4	39	<12	23.5	28.75	n/a	2.1E-5	5	0.47	2.14
IFP-12	42	21.1	21.0	37.5	3.6E-6	2.7E-6	4	0.14	7.13
IFP-13	50	15.1	23.4	45.73	2.0E-6	n/a	n/a	n/a	n/a
IFP-14	38	15	20.0	23.78	1.0E-5	2.1E-5	3	0.15	6.54
IFP-15	41	12	19.1	23.7	7.3E-5	1.8E-4	2	0.19	5.33
IFP-16	42	14.8	15.2	36.58	6.2E-5	2.0E-4	27	1.10	0.91
IFP-17	38	19.8	18.2	30.48	6.5E-6	2.1E-6	2	0.07	14.71
IFP-18	36.6	16	19.5	22.8	2.9E-5	n/a	n/a	n/a	n/a
IFP-19	41	5.8	13.2	41.14	5.9E-8	6.3E-7	0	0.00	>32.3
IFP-20	41	18.1	19.2	21.1	1.5E-5	8.2E-5	3	0.53	1.90
IFP-21	40	12	13.8	21.1	7.9E-5	4.6E-5	5	0.48	2.09
IFP-22	41	9.7	15.1	41	1.1E-6	9.2E-8	u/k	u/k	u/k
IFP-23	42	13.8	21.2	39	8.8E-7	7.6E-7	u/k	u/k	u/k
IFP-24	49	16.8	18.1	27.5	1.0E-5	1.3E-5	u/k	u/k	u/k
IFP-25	46	5.45	22.5	42	2.0E-8	2.4E-7	0	0	>23.4

[1]n/a = not available for practical reasons.
[2]u/k = unknown data in low-permeability wells.
[3]According to the geological log, also corresponds to the bottom of the weathered-fractured layer.
[4]0.5 metre-thick zones.
[5]CFZ density is the ratio between the number of conductive fracture zones and the thickness of the investigated part of weathered-fractured layer (equal to the difference between the final water level during the injection test and the bottom of the weathered-fractured layer).
[6]Persistence, equal to CFZ spacing, is the inverse of density.
K: hydraulic conductivity; CFZ: conductive fracture zones.

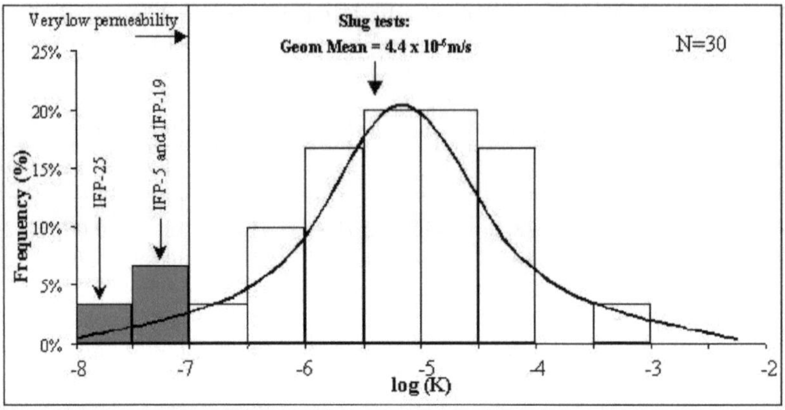

Figure 3a. Histograms on a logarithmic scale of the hydraulic conductivity obtained from the slug tests. Lognormal distribution of the permeability includes a population of very low-permeability wells (IFP-5, IFP-19 and IFP-25).

The location of hydraulically conductive fractures was obtained using vertical profiles of flow meter measurements. Given that pumping induces drawdown of water level in the well, it decreases the thickness of the zone where flows take place. In order to increase the vertical extension of the investigated zone where flow meter measurements can be done, the experiment was made during injection and not pumping. During injection at constant rate, after a pseudo-steady state had been reached, the flow meter, which measures the vertical flow within the screened portion of the well, was lowered almost to the bottom of the well and a measurement of the velocity was obtained. Velocity measurements were taken every 0.5 m. The final result is a series of data points giving the vertical discharge within the screen as a function of depth.

As is usually assumed (Molz et al., 1989), the aquifer is considered as a series of horizontal layers. The difference between two successive meter readings constitutes the net radial flow (q_i) entering each layer of the aquifer. Then, the Dupuit (1848, 1863) formula for horizontal flow to a well in a confined aquifer is applied to each layer, leading to an estimation of the layer hydraulic conductivity:

$$K_i = \frac{q_i}{2\pi s} \ln \frac{R_i}{r_w}$$

where K_i = the conductivity of the ith layer; s = the drawdown; R_i = the radius of influence; and r_w = the well radius.

A correction factor is applied to the obtained conductivities in order that the total conductivity of the well calculated from the sum of the layer conductivities (according to the principle of parallel layers: $K = \Sigma K_i$) is equal to the total conductivity obtained by applying the Dupuit (1848, 1863) formula for an unconfined aquifer to the injection test results:

$$K = \frac{Q}{\pi s(2s + b)} \ln \frac{R_i}{r_w}$$

where b = initial saturated thickness of aquifer; and Q = the total injection flow.

Seventeen tests were conducted in the most permeable observation wells according to the sensitivity of the flow meter, the total injection flow during the test being higher than 25 l/min. In the absence of core samples in such a fractured aquifer, it is not possible to determine if the identified conductive layers (which ones where the net radial flow is not nil) are constituted by a single fracture, several fractures or a fractured zone. Therefore, in order to avoid any confusion, the results are referred to "conductive fractures zones (CFZ)" without any hypothesis concerning their exact nature (single fracture, multi-fractures…). According to the geometry of the well and the observed drawdowns, the sensitivity of the flow meter limits the identification of fractures zones to those with a hydraulic conductivity higher than 1×10^{-5} m/s (hydraulic transmissivity $T > 5 \times 10^{-6}$ m²/s), corresponding to a net radial flow of about 5.5 l/m/min in a 0.5 metre-thick layer. This means that the technique gives information only on the most conductive fracture zones. This is illustrated in Fig. 3b where the distribution of permeability in the conductive fracture zones is assumed to be lognormal, even though information on the low-permeability conductive fracture zones is not available. The geometric mean of available data is $K_{CFZ} = 8.8 \times 10^{-5}$ m/s.

The vertical profile of fracture zones distribution (Fig. 4a) shows the occurrence of conductive fractures between 9 and 39.5 metres, with higher concentrations of fractures between 15 and 30 metres. However, observations are limited at shallow depths by the presence of unperforated casings in the boreholes, and by the bottom of the well at deeper depths. Thus, it is necessary in the interpretation to consider a statistical bias of the data introduced by unequal representation of the depth ranges for the investigated boreholes. The curve in Fig. 4b shows that observations between 15 and 42 metres can be considered as representative, the percentage of investigated wells being higher than 50% (for a given aquifer portion, ratio between the number of available observations and the total 17 possible measurements, Fig. 4b). The low number of observations at shallow depths (down to 15 metres) suggests that the apparent decrease in fractures density above 15 metres is an artifact due to a lack of observations (ratio decreasing much, Fig. 4b). Consequently, it can reasonably be assumed that the weathered-fractured layer extends above the upper limit of observations, which is consistent with field and well geological observations.

The bottom of the weathered-fractured layer is better constrained thanks to the quality of the available observations. The number of fractures starts to decrease between 30 and 35 metres (Fig. 4a) and is followed by an absence of fractures below 35 metres while the quality ratio of observations remains high down to 39 metres (Fig. 4b). This depth roughly corresponds to the top of the fresh basement as identified by geological observations during the drilling of the wells, and also to a sudden decrease in drilling rates at this depth. Figure 4c shows a profile of the arithmetic average of the hydraulic conductivities (including those less than 1×10^{-5} m/s, which are considered nil) obtained from flow metres for each 0.5 meter-thick layer in all tested wells. It demonstrates the effect of the number of fractures on hydraulic conductivity, with the more highly transmissive zone located mainly between 9 and 35 metres. These observations mean that in the watershed area the weathered-fractured layer has a limited vertical extent (e.g., from the bottom of the weathered mantle down to maximum 35 metres depth).

For each well, the vertical density of the fracture zones is the ratio of the number of conductive fracture zones to the thickness of the investigated weathered-fractured layer (equal to the difference between the final water level during the injection test and the bottom of the weathered-fractured layer as defined above). The density varies from 0.04 conductive fracture zones per metre in IFP-11/3, where only one conductive fracture zone was identified, up to 1.10 fractures per metre in IFP-16, where 27 conductive fracture zones were detected. While slug and injection tests investigate the full range of permeability, flow meter tests could be done only in wells with a total permeability in the range of 5.0×10^{-7} m/s up to 2.0×10^{-4} m/s. Thus, there is a lack of information for low-permeability wells with few fractures or with only low-permeability fractures. Consequently, the density of conductive fracture zones obtained from flow meter measurements is estimated from a biased data set and cannot be extrapolated to the entire data set. Therefore, we are able only to determine a range for the CFZ density according to the following hypothesis. As shown below, the three bore wells (IFP-5, IFP-19 and IFP-25 in Fig. 2a) with the lowest permeability were drilled in an un-fractured block of granite; it is assumed that their CFZ density is zero. This solves the problem for only the three wells concerned. The results of the other wells—where no flow meter test could be done because of the low total permeability, but where a single conductive fracture might be present—can be interpreted in two ways, introducing a range of values for the average density. (1) Neglecting these in the calculation of CFZ density will lead to its overestimation by up to 0.24 m^{-1}. (2) On the other hand, considering that they don't intersect any CFZ (e.g., assuming a nil density in these wells too impermeable for a flow meter measurement) will underestimate the density by as much as 0.15 m^{-1}. The true value is comprised between these two limits. This range of density of conductive fracture zones in such a weathered-fractured layer leads to a vertical persistence (e.g., the distance between two consecutive CFZ, equal to the inverse of the density) between 4.2 and 6.8 metres. Figure 5 shows a regular

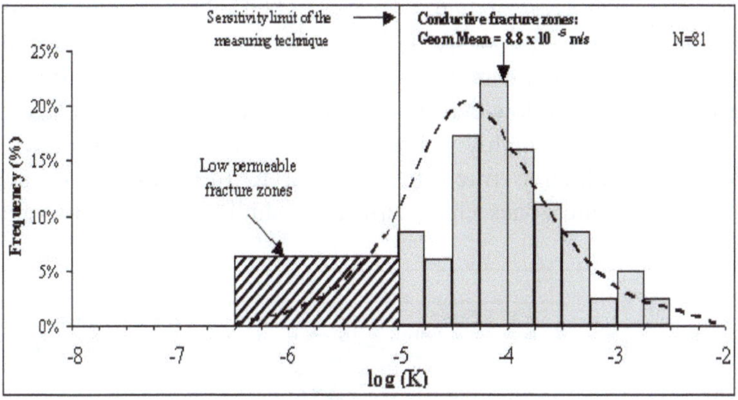

Figure 3b. Logarithmic histogram of the hydraulic conductivity in 0.5-metre-thick fracture zone obtained from flow meter injection tests.

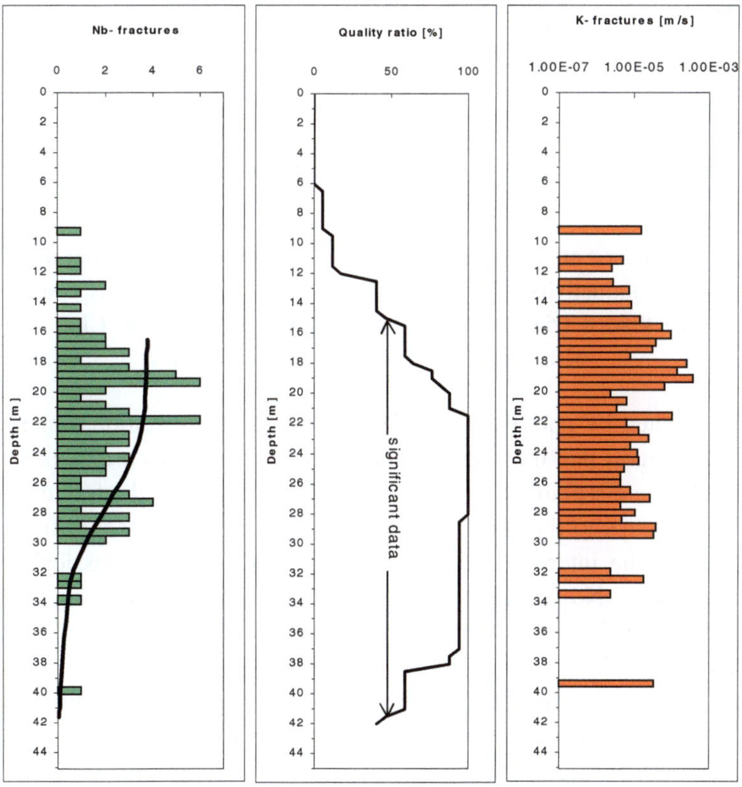

Figure 4. Interpretation of flow meter-measurement profiles in 0.5-metre-thick layers in the aquifer from 17 wells. (a) The number of hydraulically conductive fractures identified. (b) Quality of observation expressed as the ratio between the number of available observations for each aquifer portion and the total 17 measurements. (c) Averages of the fracture permeabilities.

increase of hydraulic conductivities for slug and injection tests with the density of CFZ intersected by the wells. This suggests that the total permeability of the aquifer is controlled mainly by the density of CFZ. As suggested by the absence of wells in the low-density of CFZ—high-conductivity zone (shaded triangle in Fig. 5), a highly permeable well does not correspond to one single very conductive fracture zone, but to the presence of several contributing fracture zones characterized by similar permeability values.

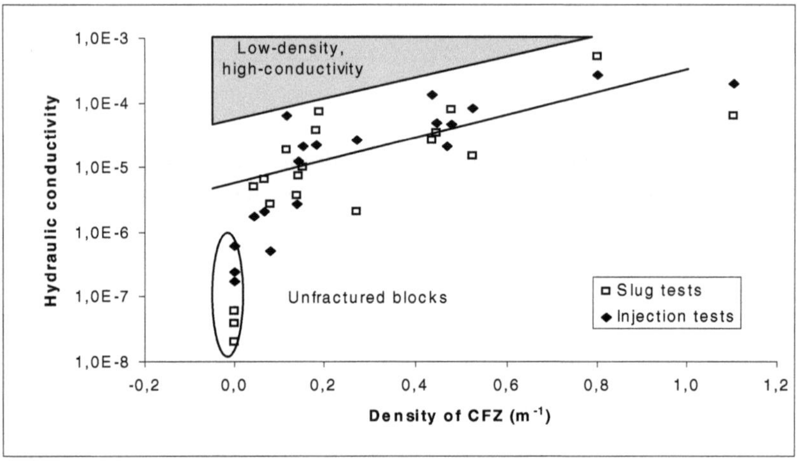

Figure 5. Influence of the density of CFZ (conductive fracture zones) on the hydraulic conductivity at a local scale (slug and injection tests results). The shaded triangle defines the zone of low-density of CFZ with high total hydraulic conductivity.

As illustrated by photograph in Fig. 2b and its interpretation, outcrop in dug wells suggest that the fractures intersected by the wells are mainly sub-horizontal. The distribution, anisotropy and connectivity of the fractures and their relationships to the blocks control the flow in the weathered-fractured layer making the interpretation of pumping tests using the classical Theis method difficult. No universal method exists to interpret pumping tests in fractured media, but the literature provides a package of specific methods that can reveal different aspects of the hydrodynamic properties of such aquifers (Maréchal et al., 2003a). These are applied below to available data using the information on the fracture distribution, providing a set of valuable information on the fractured layer.

DOUBLE POROSITY

Five long-duration pumping tests were conducted in five IFP wells located in the watershed (Fig. 2a). Another pumping test conducted by the Andhra

Pradesh Groundwater Department in the same weathered-fractured media (well BD9) a few kilometres away from the basin was also interpreted in order to increase the statistical significance of the results. Table 3 summarizes the characteristics of the tests.

Table 3. Characteristics of short- and long-duration pumping tests

Pumping well	*Radius r_w* m	*Observation wells*	*Pumping rate* m^3/s	*Duration minutes*
IFP-1	0.0825	IFP-1/1 (r = 28 m) IFP-1/2 (r = 27.5 m)	0.0013	1024
IFP-8	0.0825	n/o	0.0017	360
IFP-9	0.0825	IFP-9/1 (r = 30.7 m)	0.0007	720
IFP-16	0.0825	n/o	0.0058	1450
F-20	0.0825	IFP-20 (r = 14.6 m)	0.0009	370
BD-9	0.0825	BD-8 (r = 85 m) BD-10 (r = 50 m)	0.0053	369

n/o = no observation well

The vertical distribution of CFZ shows the existence between identified conductive fractures of blocks with a mean vertical thickness estimated between 4.2 and 6.8 metres. The double-porosity model (DP model), also known as the ‘overlapping continua’ or ‘double continuum’, was originally proposed by Barenblatt and Zheltov (1960) and Barenblatt et al. (1960). In this conceptual model, the fractured porous medium domain is represented by two distinct, but interacting, subsystems: one consisting of a network of fractures and the other, of the porous blocks. Each subsystem is conceptualized as a continuum occupying the entire investigated domain. Interaction phenomena between the two continua include the exchange of fluid between fractures and porous blocks, as described by Bear (1993) for a pumping test at a constant rate:

(i) Initially, fluid is removed primarily from the fractures, as a consequence of the much higher permeability of the fractures than of the porous blocks. During this phase, a straight line is obtained when the pressure decline in a well producing at a constant rate is plotted against the logarithm of time.

(ii) Gradually, flow at an increasing rate takes place from the porous blocks to the fractures. This appears as a curve of variable slope on the above plot and a U-shaped curve on the log-plot of the drawdown derivative (note that the “U” of derivatives can often be masked by well effects if they last for a long time).

(iii) Then equilibrium is reached between the two continua. This appears on the above plot as a straight line having the same slope as that describing the pressure decline during the initial period, but now the line is displaced parallel to the previous one. This indicates that the entire fractured porous rock behaves as an equivalent, single, homogeneous continuum. The equality of the slopes during the first and third stages indicates that the permeability in the latter is primarily that of the fractures.

In the Warren and Root (1963) double-porosity model, each system occupies the entire investigated domain and is characterized by its own hydrodynamic properties: K_f and S_f are the permeability and the storage coefficient of the fracture medium, respectively, and K_b and S_b are the permeability and the storage coefficient of the blocks, respectively. In a confined aquifer of thickness H, the flow radial to the pumping well is controlled only by the transmissivity of the fractures (flow from the blocks to the pumping well is nil, $K_f >> K_b$) and the fracture network drains the blocks in which the flow is stationary (spatial variation of the hydraulic head is neglected). Two main parameters characterize the geometry of fractures and blocks: N (entire and dimensionless) is the number of orthogonal fracture sets ($N = 1, 2$ or 3); l is the width of a block (or persistence).

This method is illustrated at pumping well IFP-16. The derivatives of drawdowns (derivatives with respect of logarithm of times, $ds/d(\ln t)$—in IFP-16 have a shape typical of a double-porosity aquifer (Figs. 6a and 6b):

(i) well effects and flow through fractures to pumping well;
(ii) transitional period: the "U" illustrates the contribution of the block flow through fractures to the pumping; and
(iii) flow in fractures and block.

The 'U' shape of the derivatives curve suggests that the application of the Warren and Root method is then justified for this data set (Fig. 6b). For the interpretation based on the flow meter measurements, the value l=0.9 metres was used for this well. The best matching of the observations was obtained with $N = 2$. The hydraulic conductivity of the fracture network appears to be $K_f = 5.9 \times 10^{-5}$ m/s and that of the block is $K_b = 2.6 \times 10^{-7}$ m/s. In Fig. 6a is also presented the interpretation using the Theis method, which cannot properly model the drawdown after 470 minutes of pumping.

The results of the application of this model to all the pumping tests are summarized in Table 4. The fracture network commonly used to match the observed data is a horizontal fracture set cross-cut by a vertical one ($N = 2$), suggesting the existence of a network complementary to the horizontal one as observed in the dug wells. The (geometric) mean of the hydraulic conductivity of the fracture network is 2.1×10^{-5} m/s, 400 times higher than that of the blocks (i.e. 5.1×10^{-8} m/s). This last value is similar to the lowest of the hydraulic conductivities estimated from slug-tests (Table 1). These low values were measured in three wells IFP-5, IFP-19 and IFP-25 in Fig. 3a,

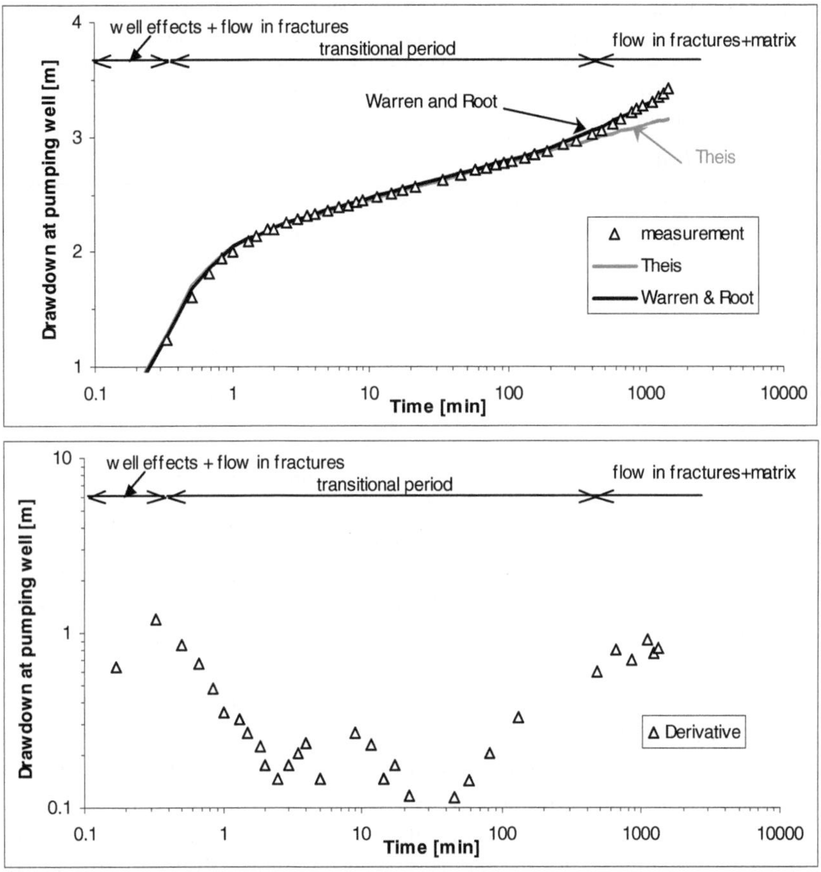

Figure 6. Interpretation at the pumping well IFP-16 using a double-porosity model. (a) Adjustment of drawdown using the Warren and Root double porosity and the Theis methods. (b) 'U'-shape of derivative.

suggesting that these wells were drilled in an unfractured area where only the hydraulic conductivity of the blocks contributes to the flow. However, it is quite probable that a primary network of fractures (PFN) also affects the matrix, accounting for the high value of the hydraulic conductivity (10^{-8} m/s) of the blocks compared to those measured on rock samples of the same lithology (matrix permeability K_m = 10^{-9} – 10^{-14} m/s, de Marsily, 1986). At this stage, the use of the term "primary" refers only to the small scale (block-level) of the fracture network and does not imply any consideration on the specific origin of the PFN. That point will be considered in the final discussion. Together, the horizontal and the vertical sets of fractures observed in dug wells and designated in the double-porosity model constitute the secondary fracture network (SFN) at the borehole scale. The total storage coefficient (specific yield in this unconfined aquifer, obtained

Table 4. Values of matrix and fracture media hydrodynamic properties calculated using the double-porosity (DP) model

Well	*Known parameters*		*Parameters calculated by adjustment on DP model*					
	l m	*N*	T_f m²/s	K_f m/s	S_f^2	K_b m/s	$S_b^{\ 2}$	$S_y^{\ 2}$
IFP-1	7.1	2	6.0E-5	2.8E-6	–	2.6E-8	–	–
IFP-1/1	7.1	1	2.3E-5	1.0E-6	9.0E-4	1.3E-7	3.4E-3	4.3E-3
IFP-1/2	7.1	2	2.6E-5	1.2E-6	1.0E-6	6.9E-10	2.6E-3	2.6E-3
IFP-8	2.3	2	1.2E-3	7.4E-5	–	1.7E-8	–	–
IFP-9	5.5	2	7.6E-4	1.0E-4	–	9.6E-9	–	–
IFP-9/1	5.5	2	7.3E-4	1.0E-4	3.0E-4	8.2E-7	5.0E-3	5.3E-3
IFP-16	0.9	2	1.4E-3	5.9E-5	–	2.6E-7	–	–
IFP-20	1.9	2	9.2E-4	5.1E-5	1.0E-4	3.9E-8	4.4E-3	4.5E-3
BD-8	2.4	2	1.8E-3	7.5E-5	2.0E-3	2.1E-7	1.8E-2	2.0E-2
BD-10	2.4	2	5.6E-4	2.3E-5	2.0E-4	1.7E-7	1.2E-3	1.4E-3
Mean[1]	3.7	–	3.5E-4	2.1E-5	5.8E-4	5.1E-8	5.7E-3	6.3E-3

[1]Arithmetic mean for storage and geometric mean for hydraulic conductivity
[2]S_b, S_f, and S_y: blocks, fractures and total storage coefficients ($S_y = S_b + S_f$), corresponding, in unconfined aquifer, to specific yields

from the sum of fracture and block storage coefficients) is equal to 6.3×10^{-3}, a value quite consistent with those evaluated from other interpretation methods: Neuman technique or the water-table fluctuation techniques (Maréchal et al., 2003b). Storage in the primary fractures network (PFN) affecting the matrix accounts for the bulk (91 %) of the total storage of the aquifer; the secondary fracture network (SFN) contributes the rest (9 %).

VERTICAL ANISOTROPY

In order to characterize the anisotropy induced by the horizontal set of the secondary fracture network (HSFN) observed in the dug-wells (Fig. 2b), drawdowns in observation wells were interpreted by the Neuman method (1975) while drawdown in the pumping wells were analyzed using the theory of the horizontal fracture developed by Gringarten and Witherspoon (1972).

Neuman Method for the Observation Wells

On a log-log plot (Fig. 7), the drawdown curves for the observation wells IFP-1/1 and IFP-1/2 during pumping tests at IFP-1 have a complex shape, which is difficult to interpret with classical methods, even considering an impermeable boundary. Drawdown curves are composed of three segments: an initial segment with a steep slope for a short time after the start, followed by an intermediate period during which the slope stabilizes, and a final segment the slope of which again steeping after a long time.

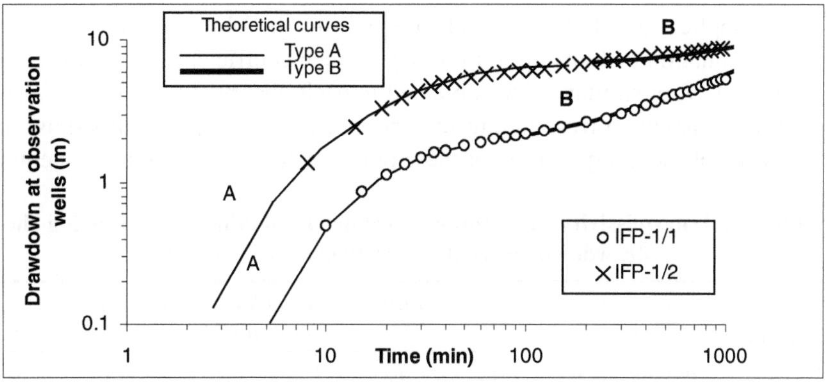

Figure 7. Adjustment of drawdown in observation wells IFP-1/1 and IFP-1/2 using the Neuman (1975) theoretical curves of types A and B.

This theory, initially developed by Boulton (1970), integrates the notion of "delayed yield from storage in unconfined aquifers" (Boulton and Pontin, 1971). It was improved by Neuman (1972, 1975) who developed an analytical solution adapted to anisotropic unconfined aquifers, where K_r is the radial horizontal permeability and K_z is the vertical permeability. The Neuman model (anisotropic (AN) model) considers an infinite unconfined homogeneous aquifer. When a completed well is pumped at a constant discharge rate, one part of the water comes from the storage in the aquifer and the other, from gravitational drainage at the free surface. The Neuman solution, plotted on type curves, provides reduced draw downs in an observation well located at a radial distance r from the pumping well, $S_{DN} = \frac{4\pi Ts}{Q}$ as a function of:

- Reduced time $t_s = \frac{Tt}{Sr^2}$ for a series of "type A" curves;
- Reduced time $t_y = \frac{Tt}{S_y r^2}$ for a series of "type B" curves;

where T = the transmissivity of the aquifer; S = the (elastic) storage coefficient; S_y = the specific yield; t = the time since the start of pumping; and s = the drawdown.

The application of this method consists in matching the observed drawdowns on the abacus for the two types of curves: a type-A curve for short times and a type-B curve for late times (Fig. 7). Both curves are characterized by the same parameter $\beta = \frac{r^2 K_D}{b^2}$, where the permeability anisotropy is $K_D = \frac{K_Z}{K_r}$ and b the thickness of the aquifer.

The application of this method (Table 5) to the observation wells leads to the evaluation of transmissivities (T), storage coefficients (S) and specific yields (S_y). Very similar values obtained in each observation well for T_A and T_B (transmissivity corresponding to curves A and B, respectively) show the coherence of the interpretation of these pumping tests using Neuman method.

Table 5. Transmissivity and storage parameters obtained by adjusting the drawdown using the anisotropic (AN) model

Observation well	*Known parameters*		*Parameters calculated by adjustment on AN model*					
	Pumping well	*r* m	T_A[1] m²/s	T_B[2] m²/s	T_A/T_B	T_{AB}[3] m²/s	*S*	*Sy*
IFP-1/1	IFP-1	28	1.8E-5	2.0E-5	0.90	1.9E-5	7.0E-5	1.6E-3
IFP-1/2	IFP-1	27.5	1.7E-5	1.8E-5	0.97	1.7E-5	3.7E-5	1.5E-3
IFP-9/1	IFP-9	30.7	5.5E-4	7.7E-4	0.73	6.5E-4	7.1E-4	3.4E-3
IFP-20	IFP-20F	14.6	6.5E-4	6.9E-4	0.90	6.7E-4	1.1E-3	7.8E-3
BD-8	BD9	85	1.6E-3	1.8E-3	0.90	1.7E-3	8.3E-4	9.7E-3
BD-10	BD9	50	3.2E-4	3.5E-4	0.90	3.4E-4	1.7E-4	1.9E-3
Mean[4]	–	–	1.9E-4	2.2E-4	0.88	2.1E-4	4.8E-4	4.3E-3

[1]T_A: transmissivity obtained by adjustment on type A curve.
[2]T_B: transmissivity obtained by adjustment on type B curve.
[3]T_{AB}: average of T_A and T_B.
[4]Arithmetic mean for storage and geometric mean for hydraulic conductivity.

The interpretation of the data from the observation wells (Table 6) gives rise to an anisotropy of the permeability tensor: the horizontal permeability is systematically higher (nine times on an average) than the vertical. This result is consistent with the observation of many horizontal fractures in dug-wells (Maréchal et al., 2003c). The average value of the vertical permeability ($K_z = 1.3 \times 10^{-6}$ m/s) is much higher than the permeability of the blocks obtained from the application of the double-porosity model. This confirms that the second set of sub-vertical fractures belonging to the secondary fracture network (VSFN) is less permeable than the horizontal and that it also affects the entire aquifer by ensuring its vertical permeability and drainage.

The advantage of the Neuman technique is the ability to determine both specific yield and elastic storage coefficient. The values obtained for the specific yield ($S_y = 5 \times 10^{-3}$ as an average) are consistent with an unconfined aquifer of low drainage porosity, as also suggested by water-table fluctuations analysis at the watershed scale (Maréchal et al., 2003b). The storage coefficient ($S = 5 \times 10^{-4}$) is one order of magnitude lower than the specific yield. Because the observed drawdowns are most of the times low (< 25%) compared to the thickness of the aquifer, the application of analytical solutions

Table 6. Permeability and degree of anisotropy determined at the observation wells using anisotropic (AN) model.

Observation well	*Known parameters*		*Parameters calculated by adjustment on SF model*				
	r m	*b* m	β	K_r m/s	K_z m/s	K_D	$1/K_D$
IFP-1/1	28	21.8	1.0	8.5E-7	5.2E-7	0.61	1.7
IFP-1/2	27.5	21.8	0.2	8.0E-7	1.0E-7	0.123	8.0
IFP-9/1	30.7	7.3	0.6	8.9E-5	3.0E-6	0.03	29.5
IFP-20	14.6	18	0.1	3.7E-5	5.6E-6	0.15	6.6
BD-8	85	23.9	0.8	7.0E-5	4.4E-6	0.06	15.8
BD-10	50	23.9	0.4	1.4E-5	1.3E-6	0.09	10.9
Mean[1]	–	–	–	1.1E-5	1.3E-6	0.1	8.7

[1]Geometric mean.

for confined aquifer to the unconfined aquifer of the study area is possible without introducing any inaccuracy. This is confirmed by the comparison between the values obtained for specific yields and hydraulic conductivity (Fig. 8) using double porosity model (confined aquifer solution) and anisotropic model (unconfined aquifer solution). The difference between the average values is less than 30% for both parameters, which is very low compared to uncertainties usually existing in the determination of storage and permeability. Consequently, although the aquifer is clearly unconfined, a few techniques for confined aquifers, are applied in this study.

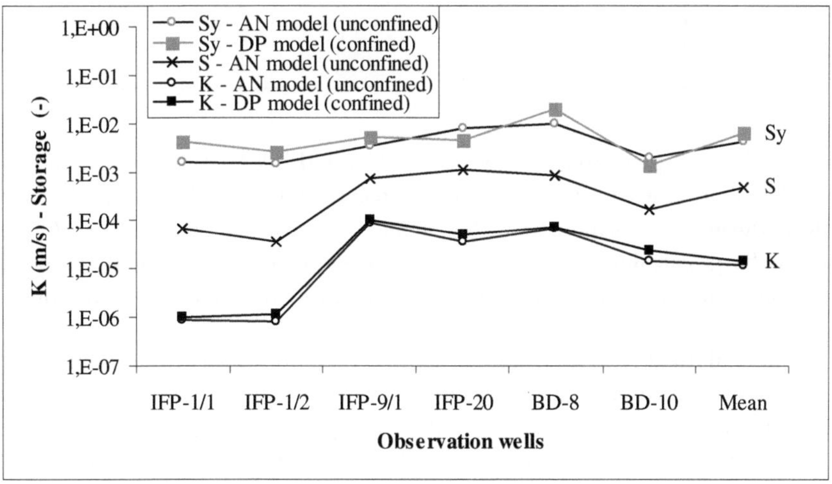

Figure 8. Comparison of storage coefficients and hydraulic conductivities calculated using Double Porosity (DP: confined aquifer solution) and Anisotropic (AN: unconfined aquifer solution) models.

Gringarten Method for Pumping Wells

The vertical flow meter profiles for wells IFP-1 and IFP-9 (Fig. 9) demonstrate that these wells intersect few conductive fracture zones: three CFZ (F1/1, F1/2 and F1/3) for the former and one CFZ (F9/1) for the latter. Actually, for IFP-1, only the deepest CFZ (F1/3 at 31.5 metres) was saturated during the whole pumping test. The low hydraulic-head losses estimated using the Jacob method (Jacob, 1945) for the step test suggest the desaturation of F1/1 and F1/2. In IFP-9, the only intersected CFZ (F9/1 at 29 metres) was also saturated during the whole pumping test. Moreover, analysis by the Neuman method documents the existence of hydraulic anisotropy due to the presence of horizontal fractures. Thus, the method developed by Gringarten and Ramey (1974) for a vertical well intersecting a horizontal fracture (single-fracture (SF) model) in an anisotropic aquifer, applicable to a pumping well, is well adapted to the hydrogeological context of IFP-1 and IFP-9 wells.

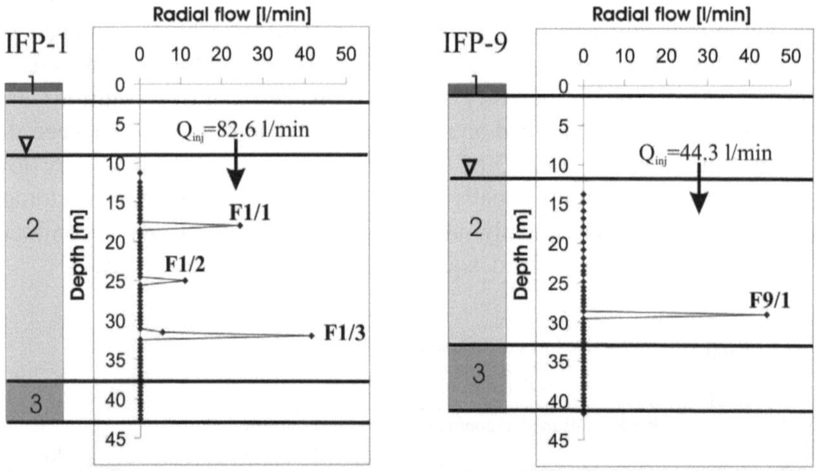

Figure 9. Geological profile of wells IFP-1 and IFP-9 (1, soil and regolith; 2, weathered- fractured layer; 3, fresh basement) and vertical profiles of radial fluxes during an injection test. The conductive fractures are identified. Q_{inj} is the discharge rate during injection and flow meter tests. The water table indicated in the figure corresponds to the level modified in the well by the injection.

The analytical solution requires an interpretation through the adjustment of observed drawdowns on theoretical type curves (Gringarten and Witherspoon, 1972) resulting in reduced draw-downs (s_{DG}) in a pumping well as a function of reduced time (t_{DG}) for various geometrical configurations represented by the parameter H_{DG}. When the fracture is located at the centre of the aquifer ($z_f/H = 0.5$),

$$t_{DG} = {K_r t}\Big/{S_s r_f^2}$$

$$s_{DG} = \frac{4\pi\sqrt{K_r K_z}\, r_f\, s}{Q}$$

and $$H_{DG} = \frac{H}{r_f}\sqrt{\frac{K_r}{K_z}}.$$

where z_f = the distance between the fracture and the bottom of the aquifer; H = the aquifer thickness; K_r= the permeability in the radial direction parallel to the fracture (which can be interpreted as the permeability increased by the existence of the horizontal fracture) K_z = the vertical permeability (which represents the block permeability); S_s = the specific storage coefficient; t = the time since the start of pumping; r_f = the radius of the horizontal fracture; s = the drawdown; and Q = the pumping rate.

Adjustments of observed drawdowns by the Gringarten theoretical curves lead to high values of H_{DG}, suggesting a high permeability anisotropy. The hydrodynamic properties have been evaluated only for the conductive fractures (Table 7), based on a knowledge of the geometry of the aquifer in each case (i.e. the thickness H of the aquifer), and the application of the anisotropy ratio value, K_r/K_z, determined previously by the Neuman method (average of 4.8 for IFP-1/1 and IFP-1/2 and 29.5 for IFP-9),with the assumption that the distance z_f between the bottom of the aquifer and the fracture is equal to 0.5*H. The radius of the intersected horizontal fractures is 16 and 4 metres at IFP-1 and IFP-9, respectively.

Table 7. Permeability, degree of anisotropy and radius of the horizontal fractures determined at pumping wells using the single-fracture (SF) model

Pumping Well	*Known parameters*				*Parameters calculated by adjustment on SF model*					
	H m	*Fracture*	z_f m	z_f/H	S_s m^{-1}	H_{DG}	r_f m	K_r m/s	K_z m/s	K_r/K_z
IFP-1	21.8	F1/3	6.0	0.28	–	3	15.9	3.4E-6	7.1E-7	4.8
IFP-9	7.3	F9/1	4.0	0.55	–	10	4.0	5.7E-5	1.9E-6	29.6

CONNECTIVITY OF FRACTURE NETWORK

Considering a typical hydraulic test in fractured rock, the problem arising when analyzing data is the geometry of the fractures system into which flow occurs. According to fracture density and its distribution, the dimension of flow will vary from 0 to 3. The GRF (Generalized Radial Flow) model generalizes the flow dimension to nonintegral values, while retaining the assumptions of radial flow and homogeneity (Barker, 1988).

The theory considers an n-dimensional radial flow into a homogeneous, confined and isotropic fractured medium, characterized by a hydraulic conductivity K_f and specific storage S_{sf}. This Generalized Flow Model introduces the concept of a fractional "dimension," n, for the flow, which characterizes the variation law of the flow section according to the distance from the pumping well. Values of n vary from 0 to 3: linear when n=1; cylindrical when n=2 (this corresponds to the Theis model); and spherical when n=3. Parameter n can take on any value, integer or not, indicating the geometrical complexity of the flow.

The results of the interpretation of long-duration pumping tests using this technique are summarized in Table 8.

Table 8. Hydrodynamic properties obtained from the interpretation of pumping tests using the Generalized Radial Flow (GRF) model.

	Known parameters			*Parameters calculated by adjustment on GRF model*				
	r m	b m	*Nr. of CFZ*[2]	n	K_f m/s	S_{sf} m^{-1}	$K_f b^{3-n}$ m^{4-n} s^{-1}	$S_{sf} b^{3-n}$ m^{2-n}
IFP-1	0.0825	21.8	4	2.2	5.0E-6	–	5.9E-5	–
IFP-1/1	28	21.8	–	1.2	9.5E-5	1.2E-5	2.5E-2	3.1E-3
IFP-1/2	30	21.8	2	2.5	9.5E-7	1.0E-6	4.4E-6	4.7E-6
IFP-8	0.0825	15.6	4	2.2	2.9E-5	–	2.6E-4	–
IFP-9	0.0825	7.3	4	2.5	6.2E-5	–	1.7E-4	–
IFP-9/1	30.65	7.3	–	2.0	1.4E-4	3.1E-4	1.0E-3	2.3E-3
IFP-16	0.0825	23.4	13	1.9	9.5E-5	–	3.2E-3	–
IFP-20	14.6	18.0	2	1.9	9.2E-5	7.1E-5	2.2E-3	1.7E-3
BD-8	85	23.9	–	2.0	2.6E-4	8.7E-5	6.3E-3	2.2E-3
BD-10	50	23.9	–	2.0	4.6E-5	1.4E-5	1.2E-3	3.7E-4
	Mean[1]		3.9	2.0	4.0E-5	8.3E-5	6.6E-4	1.5E-3

[1]Arithmetic mean for storage and geometric mean for hydraulic conductivity
[2]Inferior or equal to the number of CFZ in Table 2 because less fractures are investigated through pumping tests (water table drawdown desaturating CFZ) than through injection tests (water table increase saturating new CFZ)

The flow dimension, even though it is very difficult to interpret, is the parameter which provides information on the connectivity: a well-connected fracture network will have a higher value of flow dimension than a poorly connected fracture system (Black, 1994). The obtained flow dimensions are around 2, suggesting a cylindrical flow to a completed well in an isotropic system. This implies a quite good connectivity of the fracture network.

These values also suggest that in this anisotropic fractured rock, the horizontal fracture set (HSFN) is well connected, at the scale of pumping test, to a second sub-vertical fracture set (VSFN). Recent work (Kuusela-Lahtinen et al., 2003) on the evolution of the flow dimension with depth in wells in fractured media also suggest the predominance of dimension $n = 2$ at shallow depths. Nevertheless, according to Barker (1988), the GRF model does not permit the introduction of anisotropy in cases of a non-integer dimension; thus, the vertical anisotropy noted in the previous section cannot be confirmed or disproved by this model. Furthermore, there is no clear relationship between the flow dimension and the number of horizontal conductive fractures intersected by the well (Table 8), the connectivity being ensured by the vertical fractures set.

SCALE EFFECT

Various authors have observed that hydraulic conductivity is an apparent function of the scale of the measurement. The usual pattern is for hydraulic conductivity to increase with measurement scale up to the point defining the representative elementary volume, beyond which it is approximately constant over at least several orders of magnitude (Clauser, 1992; Rovey, 1998). Although a few authors (Butler and Healey, 1998) have considered it to be linked to artifacts and bias in small-scale field measurements, Clauser (1992), Neuman (1994) and Sanchez-Villa et al. (1996) link this scale dependence to heterogeneity.

From the data available on the weathered-fractured layer of Maheshwaram aquifer, five different scales of data sets are identified. At the smallest scale of investigation, the interpretation of pumping tests using the double-porosity model can describe the hydraulic conductivity of the blocks. Slug tests, injection and pumping tests characterize the hydraulic conductivity at the borehole level, while calibration of a numerical model with deterministic continuum conductivities produces an estimate of the same parameter at a higher scale (several kilometres). This classical finite-difference 2D-meshed model was calibrated on transient water table fluctuations observed during two years using the MARTHE code (Thiéry, 1993). MARTHE is a hydrodynamic modeling code, in transient regime with three-dimensional and/or multi-layer flow in porous media. The resolution method uses finite differences with a regular rectangular grid (in this case, a 53x53 grid of 200x200 metres square meshes). To quantify the dependency of hydraulic conductivity on scale, some length parameter must be associated with each test. In the numerical model, for calibration, the watershed was divided into several zones of equal permeability: the size of these zones is attributed as the length parameter for each calibrated permeability. For the pumping and injection tests, as suggested by Rovey and Cherkauer (1995), the radius of influence R_i can be estimated (accurately enough for the purpose) using a form of the Cooper-Jacob distance-drawdown equation:

$$R_i^2 = \frac{2.25\,Tt}{S}$$

For the slug tests, the radius of influence is given by Bouwer and Rice (1976). A scale-effect analysis for hydraulic conductivity of the blocks (K_b), obtained from the double-porosity model, was carried out by the introduction of an arbitrary length of 0.2 metre in order to avoid statistically the secondary fracture network.

The relationship between hydraulic conductivity and scale of measurement (R_i) is plotted on a log-log diagram (Fig. 10), along with an outline of the region corresponding to the data trend obtained by Clauser (1992) on a large data set of crystalline rocks. The data for Maheshwaram are located at the upper limit of that general trend (towards the highest permeabilities). This is most probably due to the near-surface origin of these data compared to the deeper origins of the crystalline rocks constituting the total database (data from tunnels, very deep wells, etc.). The impact of weathering-related fracturing in the basin increases the hydraulic conductivity by two orders of magnitude, as compared to the general data. Moreover, for the data from North America, the erosion by Quaternary glaciers of the superficial formations including the weathered-fractured layer contributes to a decrease in the general data trend of Clauser (1992).

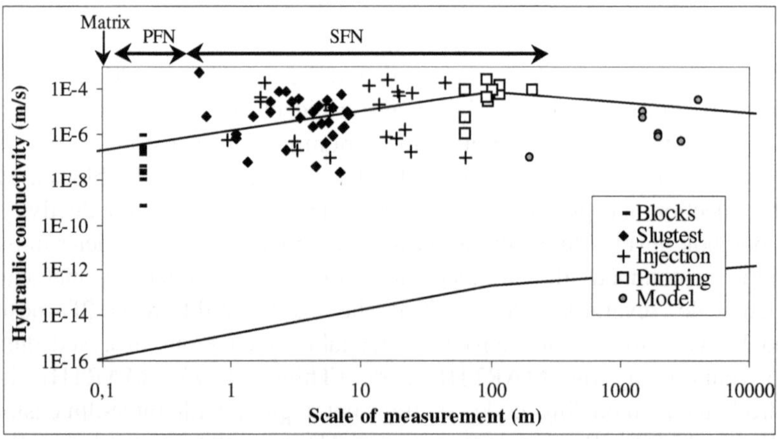

Figure 10. Permeabilities obtained in the study area as a function of the scale of the measurement. The limits of the general data from Clauser (1992) are shown. The permeability increases from the matrix-scale up to the block-scale due to the effects of the primary fracture network (PFN). The secondary fracture network (SFN) contributes to the increase of permeability from block-scale up to hydraulic tests scale.

As mentioned previously, the hydraulic conductivity increases from matrix (centimetre scale: K_m) to block (decimetre scale: K_b) size due to the contribution of the primary fracture network (PFN). It continues to increase from block scale up to borehole scale (metres to hundreds of metres), as observed in many cases, due to the effect of the horizontal (HSFN) and sub-vertical (VSFN) sets of secondary fracture network. At the borehole scale, the ratio between the average permeability from pumping tests and the average permeability from slug tests for wells tested by pumping tests is 3:1. This same ratio is observed between the average slug-test hydraulic conductivity (1.2×10^{-5} m/s) in wells investigated by long-duration pumping tests and the average slug-test hydraulic conductivity (4.4×10^{-6} m/s) for all wells. This means that the apparent permeability increase is linked to a bias in the pumping-test data, as has been previously noted for the flow metre data (Fig. 5). Thus, at the borehole scale, the hydraulic conductivity does not increase significantly.

Rovey and Cherkauer (1995) explained the increase in permeability from the pumping tests relative to the slug tests by proposing that at small scales the hydraulic conductivity and the groundwater flow (as measured by slug tests) would tend not to be influenced by the rare heterogeneities that increase the conductivity and flow rates over a regional scale (as measured by long duration pumping tests). Stated otherwise, the chances of a small-scale test encountering an extremely rare high-conductivity heterogeneity would be disproportionately small relative to the degree with which that heterogeneity increases regional hydraulic conductivity. This is not the case in our study area. Reversing Rovey's argument that the greater the heterogeneity the larger the scale effect (Rovey, 1998), the absence of a scale effect in the weathered-fractured layer of Maheshwaram suggests that the hydraulic conductivity at the borehole scale is laterally homogeneous. This is in agreement with the previous results showing the relative homogeneity of the fracture permeability, with the total permeability being controlled mainly by the frequency of medium-conductivity fractures. Finally, the large-scale data from the numerical model MARTHE exhibit a lower variability due to the longer-scale averaging of the parameters.

DISCUSSION

The information acquired in the study area introduces a conceptual hydrodynamic model for the weathered-fractured layer of the hard rock aquifers (Fig. 11).

Our data show that the weathered-fractured layer is mainly conductive from its upper surface to a depth of 35 metres; within which range conductive fracture zones with transmissivities greater than 5×10^{-6} m^2/s are observed. The lower limit corresponds to the top of the fresh basement, which contains few or no highly conductive fractures ($T > 5 \times 10^{-6}$ m^2/s) and where only local deep tectonic fractures are hypothesized to be significantly conductive at great depths as observed by studies in Sweden (Talbot and Sirat, 2001), Finland (Elo, 1992) or United States (Stuckless and Dudley, 2002).

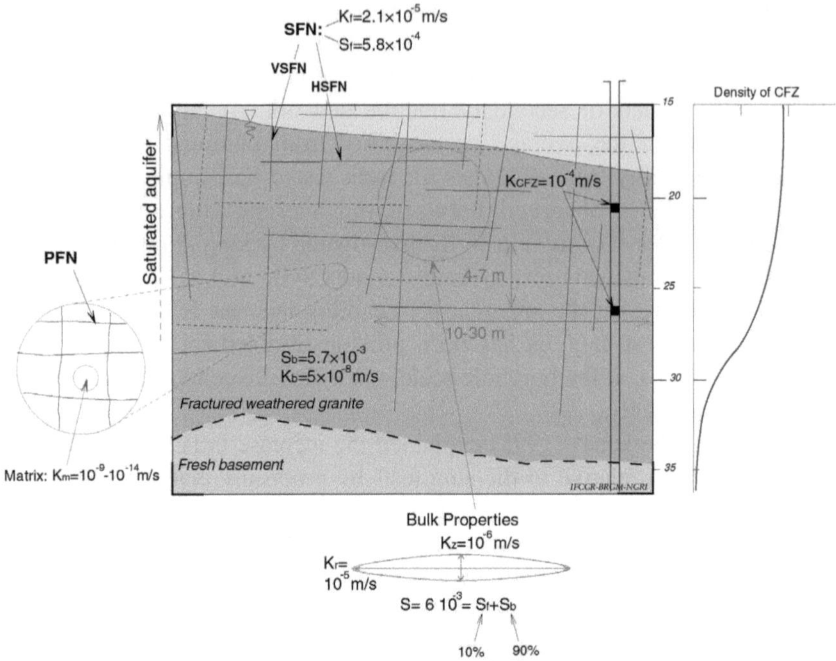

Figure 11. Hydrogeological model of the weathered-fractured layer of the hard-rock aquifer (This figure is a more detailed view of part of the complete weathering profile shown in Fig. 1). The dashed lines represent poorly conductive fractures unidentified by the flowmeter tests; solid lines, conductive fractures identified by the flow metre.

An unpublished study in the same area has statistically confirmed this result by the analysis of airlift flow rates in 288 boreholes, 10 to 90 metres deep. It is observed that the cumulative airlift flow rate ranges from 1.5 m^3/h to 45 m^3/h (Fig. 12), the mean value increasing drastically in the weathered-fractured layer between 20 and 30 metres deep. Below 30 metres, the flow rate is constant and does not increase with depth. In practice, drilling deeper than the bottom of the weathered-fractured layer (30-35 metres) does not increase the probability of improving the well discharge. The data confirm that the weathered-fractured layer is the most productive part of the hard-rock aquifer, as already shown elsewhere by other authors (Houston and Lewis, 1988; Taylor and Howard, 2000).

Two different scales of fracture networks are identified and characterized by the hydraulic tests: the primary fracture network (PFN) which affects the matrix at the decimetre scale, while secondary fracture network (SFN) affects the blocks at the borehole scale. The latter is described first below.

The secondary network is composed of two conductive fractures sets—a horizontal set (HSFN) and a sub vertical set (VSFN) as observed in outcrop. These are the main contributors to the permeability of the weathered-fractured layer. The average vertical density of the horizontal conductive set

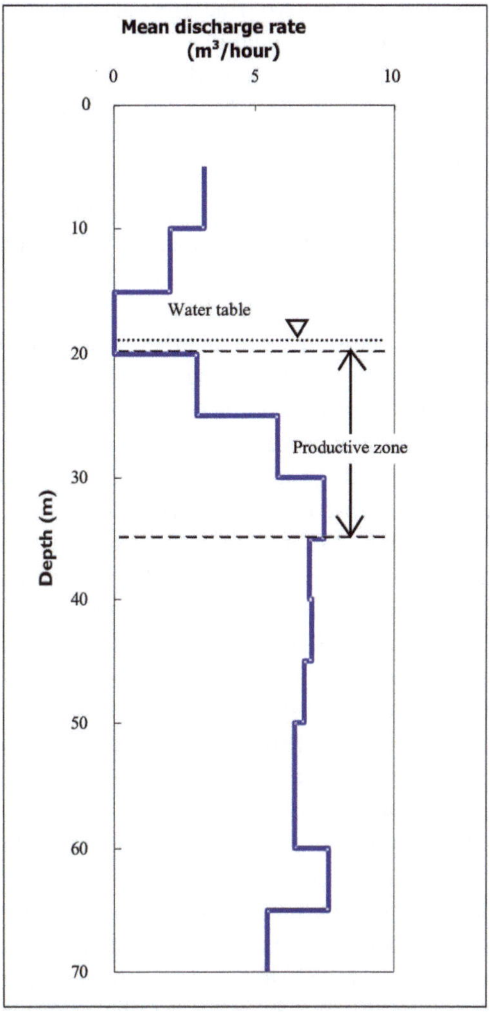

Figure 12: Mean cumulative airlift flow rate versus depth from the 288 wells in the study area. The major increase in flow is located between 20 and 30 metres.

ranges from 0.15 m^{-1} to 0.24 m^{-1}, with a fracture length of a few tens of metres (10 - 30 metres-diameter for the only available data). This corresponds to a mean vertical thickness of the blocks ranging from 4.17 (≈ 4) to 6.67 (≈ 7) metres.

The high dependence of permeability on the density of the conductive fractures indicates that individual fractures contribute more or less equally to the bulk horizontal conductivity ($K_r = 10^{-5}$ m/s) of the aquifer. No strong heterogeneity is detected in the distribution of the hydraulic conductivities of the fractures, and therefore no scale effect was induced at the borehole scale. The sub vertical conductive fracture set connects the horizontal network, insuring a vertical permeability (K_z = 10^{-6} m/s) and a good connectivity in the aquifer. Nevertheless, the sub vertical set of fractures is less permeable than the horizontal, introducing a horizontal-to-vertical anisotropy ratio for the permeability close to 10 due to the supremacy of horizontal fractures.

The existence of a secondary fractures network only in the upper part of the subsurface profile and the relative homogeneity of its characteristics are indicators that a superficial and laterally continuous process is responsible for the origin of the fractures. In the absence of glaciers and conditions favouring high erosion in the study area, another process than decompression must be invoked to explain the genesis of the primary and secondary fractures. Wyns et al. (2004) have shown that fracturing in hard-rock areas depends not only on the decompression processes, but also on the mineral weathering process. The weathering of the micaceous minerals, particularly biotites, to clays proceeds with an increase in volume, which induces cracks in the rock that will initiate fracturing. Where the rock texture is isotropic (e.g., in granite), the fractures are mostly sub-parallel to the contemporaneous weathering surface, as in the flat Maheshwaram watershed where they are horizontal (Maréchal et al., 2003c). In highly foliated rocks (metamorphic rocks), orientation of fractures can be also controlled by the rock structure (Pye, 1986). Resulting non-horizontal fractures are less affected by the closing effect of lithostatic constraint. This could explain the less rapid vertical decrease in permeability of these rocks compared with granitic rocks as reported by Havlík-Krásný (1998) and Krásný (1999). At a regional (and borehole) scale, the weathering process is much more homogeneous than tectonic phenomena (local deep-fault zones) and corresponds more closely to the observations in our study area. In the field, the bore-wells that were drilled with a quite homogeneous spacing throughout the watershed confirm this conclusion: of 288 wells, 257 (89%) were drilled deeper than 20 metres and 98% of these are productive. In any case, the probability of a vertical well crossing a horizontal fracture induced by such wide-scale weathering is very high.

The primary network of fractures (PFN), operating at the block scale, increases the original matrix permeability of $K_m = 10^{-14} – 10^{-9}$ m/s to K_b = 4×10^{-8} m/s. Regarding the matrix storage, it also should contribute to the storage coefficient in the blocks of $S_b = 5.7 \times 10^{-3}$. The storage in the blocks represents 91% of the total specific yield ($S_y = 6.3 \times 10^{-3}$) of the aquifer; storage in the secondary fracture network accounts for the rest. Such a high storage in the blocks (in both the matrix and the PFN) and the generation of the PFN would result from a first stage of the weathering process itself. The development of the SFN would be the second stage in the weathering process: that is why the words "primary" and "secondary" have been chosen to qualify the different levels of fracture networks. The obtained storage values are compatible with those of typical unconfined aquifers in the absence of an impermeable layer confining the aquifer.

CONCLUSION

The complementarity of various interpretation techniques specifically designed for fractured media is demonstrated in order to elaborate a comprehensive hydraulic model for example weathered-fractured zone of a hard rock aquifer. They constitute a useful toolbox for the interpretation of hydraulic tests in fractured hard rocks. A hydrodynamic model of the weathered-fractured layer of the aquifer is proposed based on the values of the properties obtained from hydraulic tests at different scales in the Maheshwaram watershed.

Many geological and hydrogeological indicators suggest that a continuous and laterally homogeneous weathering process is responsible for the origin of the fractures and permeability encountered in the aquifer. These results confirm the major role played by weathering in the origin of the fractures and on the resulting hydrodynamic parameters in shallow hard-rock aquifers.

The universal character of granite weathering and its worldwide distribution underlines the importance of understanding its impact on the hydrodynamic properties of the aquifers present in these environments. This generic model for permeability and storage in weathered granitic aquifers needs to be tested elsewhere.

Acknowledgements: This study has been carried out at the Indo-French Centre for Groundwater Research (BRGM-NGRI). The authors wish to thank the French Ministry of External Affairs and the Embassy in India Cooperation and Cultural Service for their support and the funding of missions of French scientists. The Indo-French Centre for Groundwater Research has also benefited from CNRS funding within the framework of the ACI Program "Water and Environment" and from funding from the IFCPAR (Indo-French Centre for the Promotion of Advanced Research). This paper has benefited immensely from the detailed comments and inputs provided by Patrick Lachassagne on various drafts and the research assistance provided by Jean-Baptiste Charlier, Riccardo Torri and Dewashish Kumar.

REFERENCES

Acworth, R.I. (1987). The development of crystalline basement aquifers in a tropical environment. *Q. J. Eng. Geol.,* **20:** 265-272.

Barenblatt, G.I. and Zheltov, I.P. (1960). Fundamental equations of filtration of homogeneous liquids in fissured rocks. *Soviet Dokl. Akad. Nauk.,* **13(2),** 545-548.

Barenblatt, G.I., Zheltov, I.P. and Kochina, I.N. (1960). Basic concepts in the theory of seepage of homogeneous liquids in fissured rocks. *Soviet Appl. Math. Mech.,* **24(5):** 852-864.

Barker, J.A. (1988). A generalized radial flow model for hydraulic tests in fractured rock. *Water Resources Research*, **24(10):** 1796-1804.

Barker, R.D., White, C.C. and Houston, J.F.T. (1992). Borehole siting in an African accelerated drought relief project. *In:* Hydrogeology of crystalline basement aquifers in Africa. E.P. Wright and Burgess, W.G. (eds.), 183-201, Geological Soc. Spec. Publ., **66**.

Bear, J.D. (1993). Modeling flow and contaminant transport in fractured rocks. *In: Flow and Contaminant Transport in Fractured Rock.* Bear, J., Chin-Fu Tsang and G. de Marsily, (eds.), 1-37, Academic Press, London.

Black, J.H. (1994). Hydrogeology of fractured rocks—A question of uncertainty about geometry. *Applied Hydrogeology,* **3:** 56-70.

Blomqvist, R.G. (1990). Deep groundwaters in the crystalline basement of Finland, with implications for nuclear waste disposal studies. *Geologiska Foereningen i Stockholm Foerhandlingar,* **112(4):** 369-374.

Boulton, N.S. (1970). Analysis of data from pumping tests in unconfined anisotropic aquifers. *J. Hydrol.*, **10:** 369-378.

Boulton, N.S. and Pontin, J.M.A. (1971). An extended theory of delayed yield from storage applied to pumping tests in unconfined anisotropic aquifers. *J. Hydrol.,* **14:** 53-65.

Bouwer, H. and Rice, R.C. (1976). A slug test for determining hydraulic conductivity of unconfined aquifers with completely or partially penetrating wells. *Water Resour. Res.,* **12(3):** 423-428.

Bredehoeft, J.D. (2002). The water budget myth revisited: Why hydrogeologists model. *Ground Water*, **40(4):** 340-345.

Butler, J.J. and Healey, J.M. (1998). Relationship between pumping-test and slug-test parameters: scale effect or artifact? *Ground Water*, **36(2):** 305-313.

Chilton, P.J. and Foster, S.S.D. (1984). Hydrogeological characterization and water-supply potential of basement aquifers in tropical Africa. *Hydrogeology J.*, **3(1):** 36-49.

Chilton, P.J. and Smith-Carington, A.K. (1984). Characteristics of the weathered basement aquifer in Malawi in relation to rural water supplies. *In:* Proc. Challenges in African hydrology and water resources Symp. IAH Series Publ., **144:** 15-23.

Clauser, C. (1992). Permeability of Crystalline Rocks. *Eos, Transactions, American Geophysical Union*, **73(21):** 233-240.

Davis, S.N. and Turk, L.J. (1964). Optimum depth of wells in crystalline rocks. *Ground Water*, **2(2):** 6-11.

Detay, M., Poyet, P., Emsellem, Y., Bernardi A. and Aubrac, G. (1989). Development of the saprolite reservoir and its state of saturation: Influence on the hydrodynamic characteristics of drillings in crystalline basement (in French). *C.R. Acad. Sci. Paris II,* **309:** 429-436.

Dupuit, J. (1848, 1863). Etudes théoriques et pratiques sur le mouvement des eaux dans les canaux découverts et à travers les terrains perméables, 1st and 2nd Ed. Dunod, Paris.

Elo, S. (1992). Geophysical indications of deep fractures in the Narankavaara-Syote and Kandalaksha-Puolanka zones. *Geological Survey of Finland*, **13:** 43-50.

Foster, S.S.D. (1984). African groundwater development—The challenges for hydrological science. IAHS Publication, **144:** 3-12.

Gringarten, A.C. and Witherspoon, P.A. (1972). A method of analysing pump test data from fractured aquifers. *In: Percolation through fractured rock.* I.A.o.R. Mechan (ed.), Deutsche Gesellschaft fur Red and Grundbau, Stuttgart, pp. T3B1-T3B8.

Gringarten, A.C. and Ramey, H.J. (1974). Unsteady-state pressure distribution created by a well with a single horizontal fracture, partially penetrating or restricted entry. *Trans. Am. Inst. Min. Eng.,* **257:** 413-426.

Gustafson, G. and Krásný, J. (1994). Crystalline rock aquifers: Their occurrence, use and importance. *Applied Hydrogeology,* **2:** 64-75.

Havlík, M. and Krásný, J. (1998). Transmissivity Distribution in Southern Part of the Bohemian Massif: Regional Trends and Local Anomalies. *In:* Hardrock Hydrogeology of the Bohemian Massif, Proc. 3rd Internat. Workshop 1998, Windischeschenbach. Münchner Geol. Hefte, **B8:** 11-18.

Houston, J.F.T. and Lewis, R.T. (1988). The Victoria Province drought relief project, II. Borehole yield relationships. *Groundwater,* **26(4):** 418-426.

Howard, K.W.F., Hughes, M., Charlesworth, D.L. and Ngobi, G. (1992). Hydrogeologic evaluation of fracture permeability in crystalline basement aquifers of Uganda. *Hydrogeology J.,* **1:** 55-65.

Jacob, C.E. (1947). Drawdown test to determine effective radius of an artesian well. *ASCE Trans.,* **112(232):** 1047-1064.

Krásný, J. (1999). Hard-rock hydrogeology in the Czech Republic. *Hydrogéologie,* **2:** 25-38.

Kuusela-Lahtinen, A., Niemi, A. and Luukkonen, A. (2003). Flow dimension as an indicator of hydraulic behaviour in site characterization of fractured rock. *Ground Water,* **41(3):** 333-341.

Lachassagne, P., Wyns, R., Bérard, P., Bruel, T., Chéry, L., Coutand, T., Desprats, J.F., Strat, P. Le (2001). Exploitation of high-yield in hard-rock aquifers: Downscaling methodology combining GIS and multicriteria analysis to delineate field prospecting zones. *Ground Water,* **39(4):** 568-581.

Mabee, S.B. and Hardcastle, K.C. (1997). Analysing outcrop-scale fracture features to supplement investigations of bedrock aquifers. *Hydrogeology J.,* **5(3):** 82-88.

Maréchal, J.C., Dewandel, B., Subrahmanyam, K. and Torri, R. (2003a). Review of Specific Methods for the Evaluation of Hydraulic Properties in Fractured Hard-rock Aquifers. *Current Science of India,* **85(4):** 511-516.

Maréchal, J.C., Galeazzi, L., Dewandel, B. and Ahmed, S. (2003b). Importance of irrigation return flow on the groundwater budget of a rural basin in India. *In:* Hydrology of the Mediterranean and Semiarid Regions, IAHS Publ, **278:** 62-67.

Maréchal, J.C., Wyns, R., Lachassagne, P., Subrahmanyam, K. and Touchard, F. (2003c). Anisotropie verticale de la perméabilité de l'horizon fissuré des aquifères de socle: concordance avec la structure géologique des profils d'altération, *C.R. Geoscience,* **335:** 451-460.

Marsily, G. de (1986). Quantitative Hydrogeology, San Diego, California. Academic Press.

Molz, F.J., Morin, R.H., Hess, A.E., Melville, J.G. and Guven, O. (1989). The impeller meter for measuring aquifer permeability variations: evaluation and comparison with other tests. *Water Resour. Res.,* **25(7):** 1677-1683.

Neuman, S.P. (1972). Theory of flow in unconfined aquifers considering delayed response of the water table. *Water Resour. Res.,* **8(4):** 1031-1045.

Neuman, S.P. (1975). Analysis of pumping test data from anisotropic unconfined aquifers considering delayed gravity response. *Water Resour. Res.*, **11(2):** 329-342.

Neuman, S.P. (1994). Generalized scaling of permeabilities: Validation and effect of support scale. *Geophysical Research Letters*, **21(5):** 349-352.

Pickens, J.F., Grisak, G.E., Avis, J.D., Belanger, D.W. and Thury, M. (1987). Analysis and interpretation of borehole hydraulic tests in deep boreholes: Principles, model development, and applications. *Water Resour. Res.*, **23(7):** 1341-1375.

Pye, K. (1986). Mineralogical and textural controls on the weathering of granitoids rocks. *Catena*, **13:** 47-57.

Rovey, C.W. (1998). Digital simulation of the scale effect in hydraulic conductivity, *Hydrogeology J.*, **6:** 216-225.

Rovey, C.W. and Cherkauer, D.S. (1995). Scale dependency of hydraulic conductivity measurements. *Ground Water*, **33(5):** 769-780.

Sanchez-Vila, X., Carrera, J. and Girardi, J.P. (1996). Scale effects in transmissivity. *J. Hydrol.*, **183:** 1-22.

Roy, A.D. and Shah, T. (2002). Socio-ecology of groundwater irrigation in India. *In*: Intensive Use of Groundwater: Challenges and Opportunities. Llamas, R. and Custodio, E. (Eds) A.A. Balkema.

Sander, P. (1997). Water-well siting in hard-rock areas: Identifying promising targets using a probabilistic approach. *Hydrogeology J.*, **5(3):** 32-43.

Stuckless, J.S. and Dudley, W.W. (2002). The geohydrologic setting of Yucca Mountain, Nevada. *Applied Geochemistry*, **17(6):** 659-682.

Summers, W.K. (1972). Specific capacities of wells in crystalline rocks. *Ground Water*, **10:** 37-47.

Talbot, C.J. and Sirat, M. (2001). Stress control of hydraulic conductivity in fracture-saturated Swedish bedrock. *Engineering Geology*, **61(2-3):** 145-153.

Taylor, R. and Howard, K. (2000). A tectono-geomorphic model of the hydrogeology of deeply weathered crystalline rock: Evidence from Uganda. *Hydrogeology J.*, **8(3):** 279-294.

Thiéry, D. (1993). Modélisation des aquifères complexes - Prise en compte de la zone non saturée et de la salinité. Calcul des intervalles de confiance. *Hydrogéologie*, **4:** 325-336.

Walker, D.D., Gylling, B., Strom, A. and Selroos, J.O. (2001). Hydrogeologic studies for nuclear-waste disposal in Sweden. *Hydrogeology Journal*, **9(5):** 419-431.

Warren, J.E. and Root, P.J. (1963). The behaviour of naturally fractured reservoirs. *Soc. Petroleum Eng. J.*, **3:** 245-255.

White, L.D., Houston, J.F.T. and Barker, R.D. (1988). The Victoria Province drought relief project, I. Geophysical siting of boreholes. *Groundwater*, **26(3):** 309-316.

Wright, E.P. (1992). The hydrogeology of crystalline basement aquifers in Africa. *In:* Hydrogeology of crystalline basement aquifers in Africa, Wright, E.P. and Burgess, W.G. (eds.), pp. 1-27, London Spec Publ, **66**.

Wyns, R., Baltassat, J.M., Lachassagne, P., Legchenko, A., Vairon, J. and Mathieu, F. (2004). Application of SNMR soundings for groundwater reserves mapping in weathered basement rocks (Brittany, France). *Bulletin de la Société Géologique de France*, **175(1):** 21-34.

8

Groundwater Models and Their Role in Assessment and Management of Groundwater Resources and Pollution

M. Thangarajan

Retired Scientist-G, National Geophysical Research Institute (NGRI)
Hyderabad 500 007, India

PREAMBLE

One can infer the past behaviour of an aquifer system from the available earlier field records, but there is no way of computing the future behaviour of the regional aquifer system unless it is actually subject to the stresses. A full-scale experiment would be prohibitively costly and time consuming. The only feasible recourse, therefore, is to appropriately simulate the aquifer and study its response to several input/output schemes and thereby evolve the optimal management schemes. A groundwater model is thus a simplified version of the real system that approximately simulates the input-output stresses and response relations of the system. One has to understand here that normally the real system is simplified to model the system as such there is no unique model for a given groundwater system. Normally, models are classified as predictive, interpretive and generic models. Predictive models are used to predict the future response of the aquifer, which needs a calibrated and validated model. Interpretive models are used for studying system dynamics and it is generally used for optimal data network design. Generic models are used to understand the flow dynamics in hypothetical situations.

H. Skibitzke and J. Karplus during sixties introduced the analogy between groundwater flow in porous medium and flow of electric current in resistance-capacitance (R-C) network and simulated the groundwater flow system.

Since then there were a lot of advancement in the modeling technology as well as understanding the flow dynamics in complex aquifer systems. Early eighties saw the era of numerical simulation technique in replacing the time consuming electric analog modeling techniques. The time old Hele-Shaw physical model to simulate density dependant flow was replaced by a powerful numerical model. Numerous numerical methods (Finite Difference, Finite Element and Integrate-finite Difference, etc.) were developed in eighties and nineties to simulate the realistic condition of groundwater flow system in porous medium. Though, a good perfection could be achieved in simulating continuous porous medium, we have no answer for non-continuum medium such as weathered fractured zone. Future attention has to be focused on the development of easily workable non-continuum models and apply them to assess the regional groundwater potential and contaminant migration in fractured medium. Another area of modeling technology to be perfected is the simulation of soil moisture and contaminant migration in the vadose zone and the simulation of interactions between surface water and groundwater.

INTRODUCTION

In many developing countries, groundwater plays a major life support to mankind, as it is the major source to support domestic needs and irrigation purposes. Groundwater occurs in a wide range of rock types and usually requires little or no treatment; therefore it is often the cheapest and simplest water supply option. However, the rising demand for water worldwide, mostly for irrigation, can lead to problems of over exploitation of these resources and conflicts with competing demands, especially with community potable supplies. In early days, abstraction from the shallow aquifer has been limited, mainly because water-lifting devices were animal-powered. However, since the 1950s groundwater abstraction in India and elsewhere has increased substantially, both as a consequence of the increase in the number of wells, and use of energized pumps capable of lifting water from deeper aquifer up to a depth of 200 m and more with much higher yields. This rapid development of groundwater resources had resulted in the declining of water table/levels rapidly in many parts of the Afro-Asian countries causing shallow wells to dry up with a particular impact on those rural poor farmers unable to deepen their wells to chase the declining water levels. In coastal areas, declining water levels are also associated with the ingress of saline water, leading to reduced crop yields, loss of drinking water supplies and ultimately loss of both fertile land and water supply wells. These problems are very acute in those areas underlain by hard rocks, since the hard rock aquifers has limited storage capacity and stores only limited quantity. Further, in the semi-arid regions, the climatic conditions of low and variable rainfall limit recharge and make these aquifers susceptible to drought.

The increased groundwater abstraction in many parts of India and elsewhere for the last many decades have resulted in a long-term continuous decline of water level. This has resulted in the deterioration of water quality and the widespread drying-up of wells in monsoon climatic regions due to 'failure' of the monsoon. Deepening of wells does not appear to be a viable option as most wells already fully penetrate the shallow weathered aquifer. *This has resulted only in debt trap of rural farming community. This needs effective groundwater management schemes.*

Effective groundwater management requires, firstly, a good understanding of the aquifer system, secondly, that practical measures to control abstraction can be identified and, thirdly augment groundwater resource through artificial recharge. It is therefore essential to quantify the response of the aquifer under study to different input output stresses. *Groundwater modeling study has proved to be a potential tool to study the aquifer response and thereby to evolve appropriate management schemes.* There are a quite number of modeling studies reported for porous medium in developed and developing countries but practically there is only one case study reported using fractured flow modeling technique.

Groundwater pollution is another dimensional hazardous movement. The increasing demand for water to meet the needs of domestic, industrial and agriculture is placing greater emphasis on the development of groundwater resources. The over-exploitation of groundwater resources at many developing countries induces degradation of groundwater quality and the discharge of untreated industrial effluents are adding contaminants to groundwater system. Globally, pollutions are identified as microbiological pollution, organic pollution, Stalinization, and total suspended solids. These are related to human (anthropogenic) activities such as agriculture, urbanization, industrialization, mining, land use change and climate change etc. The pollution of groundwater regime is not only due to sub-surface waste disposal, but is also attributable to the seepage of contaminants from the rivers and lakes, impoundment of toxic waste on unlined surfaces, indiscriminate spraying of insecticides, pesticides and excessive use of chemical fertilizers etc. The pollutant mass migrates with the groundwater flow and manifests itself even at places where one may least expect any contamination.

Some of the less hazardous contaminants include nitrates, sulphates, chlorides, suspended solids, mineral oil etc. However, if their quantities increase much beyond permissible limits, groundwater is rendered unusable. Still more toxic pollutants which are much more hazardous and have long term effects, are the toxic chemicals like arsenic, barium, boron, cadmium, chromium, cyanide and lead. There are several industries in the country, whose wastes containing such toxic chemicals have been causing serious groundwater pollution. Some of the biological pollutants include microorganisms, which cause diseases like infective hepatitis, gastro-enteritis, typhoid, dysentery etc. Radioactive nuclides such as ^{3}H, ^{90}Sr, ^{86}Rb, ^{133}Cs,

^{226}Ra etc., is yet another class of contaminants which enter the groundwater system from the nuclear waste disposal.

The contaminants gravitate and disperse in the unsaturated soil zone. Subsequently, they spread in the shallow saturated zone through advective (convective) migration and hydrodynamic dispersion. This process depends both on the groundwater flow velocities and the gradient of contaminant concentration. In deeper aquifers, where the groundwater velocities are relatively small, the rate of migration is slower and is governed primarily by the concentration gradient. This situation is prevalent in softer geological formations such as alluvial aquifers. In hard rock formations, the presence of fractures, fissures, joints etc., add another dimension to the transport of the pollutant mass. These discontinuities may provide easy conduits for fast migration of pollutants. Under these circumstances the groundwater becomes vulnerable to contaminants. The term "vulnerability of groundwater to contamination" was introduced by Margat in the year 1968 to describe how aquifer becomes very sensitive to contaminants.

With the increasing use of groundwater for drinking purposes, it is imperative to study the movement of contaminants in an aquifer so as to predict their migration and work out suitable remedial measures. Based on a well-planned network for collection of geo-hydrological and hydro chemical data, a deterministic numerical model of mass transport in groundwater regime may be prepared which can help assess the rate and extent of pollutant migration. It can also provide guidelines for future planning of disposal operations and for controlling the existing contaminant plumes.

TYPES OF MODELS

A model should be constructed as to reasonably portray the behaviour of the full-scale system (aquifer) and simulate all the relevant physical parameters, which describe the significant characteristics of the latter. There are enormous amount of models which have come either to predict the future behaviour or study the system dynamics or generic types. Since, there are various types of models reported in the past, it is desirable to classify them according to the techniques used, type of aquifer simulated and the dimension of the problem. The following classification have been followed in this book:

Aquifer Type

- Un-confined aquifer model
- Semi-confined aquifer (leaky) model
- Confined aquifer model

Dimensional Type

- 1-D (one dimensional model)
- 2-D (two-dimensional model)

- 3-D (three-dimensional model)
- Multi-layer (leaky) model

Techniques Type

- Physical Models
- Analog Models
- Mathematical Models

Problem (flow-conditions) Type

- Saturated groundwater flow models
- Un-saturated groundwater flow models
- Solute transport models

Models of Yester-Years and To-day

We will describe here the various modeling technology used in the past and now to solve groundwater flow and mass transport equation.

Physical Models

Forchheimer (1886) from Austria was the first to demonstrate the groundwater flow phenomenon by making use of physical models (sand tank model). The momentum of applying physical models to study the groundwater problem had picked up during 1930-1950, till the electric analog models came in force. In sand tank model, the actual field dimension was scaled down (three-dimensional) to the laboratory scale and the appropriate aquifer materials were introduced in the box and the model was simulated by incorporating appropriate pumping of water from the model and injection of water in to the model and with appropriate boundary conditions. Sand tank model was the well known among the Civil Engineers to study the well response and regional flow problem. The modeling technique actually flourished during thirties and forties in the hands of civil engineers. As mentioned above, a sand tank model is a scale down model of the actual system. Models of various shapes (rectangular, cube etc.) and various dimensions have been constructed to simulate porous flow condition. The author had seen a sand tank model of as big as 20 m × 10 m × 5 m at Radiometric Study and Hydraulic Laboratory, Munich, Germany during his visit to that laboratory in 1974. A channel flow model to represent the radial fracture flow to the well system was attempted by (Late) Er. P. Kumarasamy at the Institute of Hydraulics, Poondi near Chennai, India during seventies. Normally, the water levels are measured using manometers and the flow pattern is viewed through dye streams. The permeability pattern could be achieved by selecting various sizes of coarse sand. Laying various grain sizes of sands can create anisotropic permeability. Even during seventies many researchers had used the sand tank models to study the dispersion, artificial recharge and seawater intrusion etc. and one can refer the following for more details (Bruch, 1970;

Cahill, 1973; Kraijenhoff, 1962; Kimbler, 1970 and many others). Regional groundwater flow simulation by making use of sand tank model would be prohibitively costly and time consuming and as such the analog models took over the importance of physical model during sixties.

Analog Models

An analog model utilizes the similarity of the two physical systems and the one, which is easier to handle, is used as a model of the other. Various analog-modeling techniques used in the past to simulate groundwater flow had been reviewed by Prickett (1975). The prominent analog modeling techniques in vogue for regional groundwater flow studies have been (i) electric analog models and (ii) viscous fluid models. A brief description of viscous fluid model and a detailed discussion of Resistance-Capacitance (R-C) modeling approach will be focussed in this section.

Viscous Fluid Models

A two-dimensional groundwater flow can be analogous to the movement of a viscous liquid flow between two parallel plates. It is Mr. Hele-Shaw (1897), who had demonstrated the above principle known as Hele-Shaw model and widely used to study the density dependent flow problem (sea-water intrusion). By assuming laminar flow between two parallel plates, it can be shown that the flow lines form a two dimensional potential field. By making use of Navier-Stokes equation of fluid motion and Dupuit assumption for vertical flow condition, one can derive the equation of mean velocity as:

$$v_m = -\frac{b^2 \rho_m g}{12\mu_m}\frac{dh}{dx} \tag{1}$$

where μ_m = fluid viscosity (model); ρ_m = fluid density (model); b = inter-space between plates; g = acceleration due to gravity; and dh/dx = hydraulic gradient.

If we compare equation (1) with the formula for Darcian velocity or specific discharge, then one can arrive at the hydraulic conductivity of the model K_m as:

$$K_m = \frac{b^2 \rho_m g}{12\mu_m} \tag{2}$$

It is clearly evident from equation (2) that one can obtain the desired hydraulic conductivity values by varying spacing between plates, changing the fluid of different density and viscosity. According to Todd (1980), models of this type are constructed from two sheets of glass or plastic spaced a fixed distance apart. Reservoirs to control fluid flow between the plates are attached at the sides or ends of the model. Oil or glycerin is generally used to represent the fluid. Dye added to the fluid defines the free surface for unconfined condition. In the past, extensive study had been carried out by a number of researchers to study the bank storage near stream, seepage and drainage and seawater intrusion by making use of vertical viscous fluid model. The following references can be pursued for more details on viscous fluid model construction and application (Columbus, 1966; Collins et al., 1972; Marino, 1967; Todd, 1954; Yen and Hasie, 1972).

Electric Analog Models

Karplus (1958) had laid foundation for electric analog simulation to study the physical behaviour of the system, which obey the second order partial differential equation. Here, the flow of electric current in a solid conductor (Ohm's law) is analogous to the flow of water in a porous medium (Darcy's law). This can be expressed mathematically as:

$$I = -\sigma\, dV/dx \text{ (Ohm's law)} \tag{3}$$

where, I is the electric current per unit area, σ is the specific electrical conductivity of the material and dV/dx is the voltage (potential) gradient.

$$q = -Kdh/dx \text{ (Darcy's law)} \tag{4}$$

where, q is the Darcy velocity otherwise called specific discharge and is analogous to the electric current described in equation (3). Electrical specific conductivity is analogous to the hydraulic conductivity and electric potential V corresponds to hydraulic head h. These analogous quantities form the basis for the analog models.

There are two types of analog models viz. simulation of continuous medium through continuous electric conductor and the other one where the aquifer is discretised in space and simulated through resistance-capacitance network. In this type of model, time is continuous one whereas space is discretised as per the availability of field parameters.

Resistance-Capacitance Analog Modeling

The Resistance-Capacitance (R-C) analog modeling which had been widely used in sixties and seventies for aquifer simulation, is based on the analogy between the flow of water in a porous medium and the flow of electric current in a R-C network. Skibitzke (1960) propounded the idea of using R-C analog modeling technique to study the groundwater flow problems. Subsequently, Bermes (1960), Brown (1962), Stallman (1963), Walton and Prickett (1963), Patten (1965), Cole and Blair (1967), Thangarajan (1975, 1983) etc. applied this technique and demonstrated that R-C analog modeling is a potential tool for studies in groundwater management. Emery (1966) predicted the stream depletion due to excess pumping in an aquifer by making use of this technique. Scientists from USGS (US) and other parts of the world have carried out several investigations on complex groundwater basins. To mention a few, Bear and Schwarz (1966), Moore and Wood (1967), Anderson (1968, 1972), Reed (1972), Jorgenson and Donald (1975),

Gupta and others (1979, 1980) and Thangarajan (1983) etc. have made significant contributions in this particular field. During eighties, the work carried out by Rushton and Bannister (1970) Rushton and Wederburn (1973), Rushton and Ash (1975) and Batterman and Boochs (1976) in refinement of modeling technique to simulate complex boundary conditions such as aquifer changing between confined to unconfined and vice-versa, starting conditions of the model simulation and solving non-linear problem using computer controlled R-C network, are worth mentioning.

Basic Principles

The application of R-C analog modeling technique in simulating the groundwater flow regime in more realistic manner greatly depends on the sophisticated electronic instrumentation system used for the study.

Finite Difference Approximation of Groundwater Flow Equation

Groundwater models are the simplified representations of the subsurface aquifer systems. The calibrated and validated models may be used to predict aquifer response to pumping stresses. Groundwater flow in three dimensions in a porous media of constant density can be expressed by the following partial differential equation (Rushton and Redshaw, 1979):

$$\frac{\delta}{\delta x}\left(K_{xx}\frac{\delta h}{\delta x}\right)+\frac{\delta}{\delta y}\left(K_{yy}\frac{\delta h}{\delta y}\right)+\frac{\delta}{\delta z}\left(K_{zz}\frac{\delta h}{\delta z}\right)=S_s\frac{\delta h}{\delta t}\pm W \tag{5}$$

where K_{xx}, K_{yy} and K_{zz} = the hydraulic conductivity along x, y and z co-ordinates which are assumed to be parallel to the major axes of hydraulic conductivity (LT^{-1}); h = potentiometric head (L); W = volumetric flux per unit volume and represents sources and/or sinks of water (T^{-1}); S_s = the specific storage of the porous material (L^{-1}); and t = time (T).

Equation (5) describes groundwater flow under non-equilibrium conditions in a heterogeneous and anisotropic medium, provided the principal axes of hydraulic conductivity are aligned with the x-y Cartesian co-ordinate axes. The groundwater flow equations together with specification of flow and/or initial head conditions at the boundaries constitute a mathematical representation of the aquifer system. Numerical methods or electric analog are used in general to solve the groundwater flow equation. Analog methods were in vogue in sixties but mathematical methods had overtaken it during seventies. Advanced numerical methods are in use in solving the partial differential groundwater flow and mass transport equation. Partial derivatives are replaced by the finite difference approximations. The fundamental definition of derivatives and the Taylor series expansion (Rushton and Redshaw, 1979) of a function about a point together form the finite differential approximation (FDM). If $f(\chi)$ represents a function, then the Taylor series expansion of the function at point x_0 can be written as:

$$f(\chi) = f(x_0) + (\chi - \chi_0)f'(\chi) + \frac{(\chi - \chi_0)^2}{2I} f''(x^*) \tag{6}$$

where $x_0 < x^* < x$. By making use of this, one can approximate the first derivative by

$$f'(\chi) = \frac{f(\chi) - f(\chi_0)}{\chi - \chi_0} \tag{7}$$

The first derivative provides an error due to truncation of second derivative as given by

$$E_T = -\frac{(x - x_0)^2}{2I} f''(\chi^*) \tag{8}$$

where E_T is a truncation error.

If we assume that $x = x_0 + h$, then equation (7) becomes

$$f'(\chi) = \frac{f(\chi_0 + h) - f(\chi_0)}{h} \tag{9}$$

The above is called the forward difference approximation of the derivative of function $f(x)$. Similarly, by assuming $x = x_0 - h$ in equation (7), we can get backward difference approximation of the first derivative as:

$$f'(\chi) = \frac{f(\chi_0) - f(\chi_0 - h)}{h} \tag{10}$$

In the same way, one can arrive at the expression for the second derivative of $f(x)$ as:

$$f''(\chi) = \frac{f'(\chi_0 + h) - f'(\chi_0)}{h} \tag{11}$$

As reported by Bobba and Singh (1995), in order to avoid a biased condition, the forward difference scheme is used for $f'(\chi_0 + h)$ and backward difference scheme is used for $f'(\chi_0)$ and substituting these two in equation (9), one gets:

$$f''(\chi) = \frac{f(\chi_0 + h) - 2f(\chi_0) + f(\chi_0 - h)}{h^2} \tag{12}$$

In finite difference approximation, a continuous medium is effectively replaced by a discrete set of nodes at an interval of Δx in the x-direction and Δy in the y-direction. The finite difference of the medium is shown in Fig. 1a and corresponding R-C network in Fig. 1b.

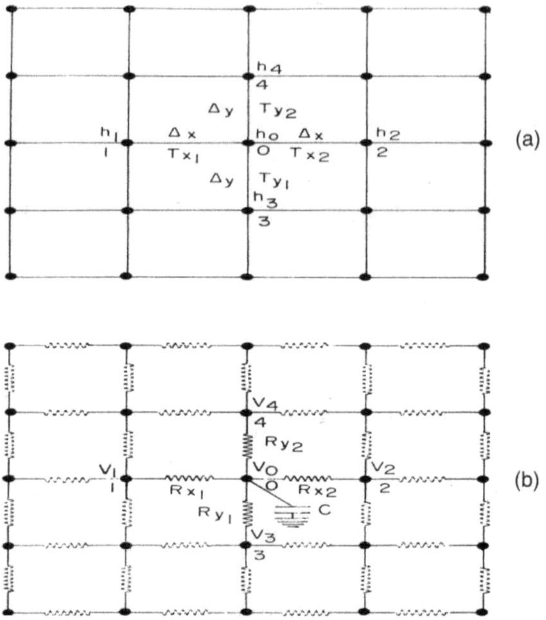

Figure 1. Discretization of a continuous medium in to (a) discrete set of nodes and (b) the corresponding R-C network

Here each node represents an element of aquifer of uniform aquifer parameters. The hydraulic head in the adjoining nodes are represented by h_0, h_1, h_2, h_3 and h_4 respectively.

If the groundwater flow system happens to be two-dimensional one, wherein the vertical component of flow is either very small or negligible and saturated thickness of aquifer does not vary with time in a given case, then the relevant partial differential equation (Pinder and Bredehoeft, 1968) describing groundwater flow may be written as:

$$\frac{\partial}{\partial x}\left(T_x \frac{\partial h}{\partial x}\right) + \frac{\partial}{\partial y}\left(T_y \frac{\partial h}{\partial y}\right) = S\frac{\partial h}{\partial t} \pm W \tag{13}$$

where T_x, T_y = the transmissivity values along x and y co-ordinates that are assumed to be parallel to the major axes of the transmissivity tensor ($L^2\ T^{-1}$); h = hydraulic head (L); S = storage coefficient of the porous material; t = time (T); $\pm W$ = volumetric flux per unit area, +ve for out flow and –ve for inflow; and x, y = the Cartesian Co-ordinate axes.

Equation (13) describes groundwater flow under non-equilibrium condition in a homogenous medium provided the principal axes of transmissivity are aligned with x-y Cartesian co-ordinate axes. Usually, it is difficult to find an exact solution of equation (13) and one has to resort to numerical technique for obtaining their approximate solutions.

A finite difference approximation of equation (13) can be written based on equation (12) as (truncated Taylor series expansion):

$$T_x \frac{h_1 + h_2 - 2h_0}{\Delta x^2} + T_y \frac{h_3 + h_4 - 2h_0}{\Delta y^2} = S \frac{\partial h}{\partial t} \pm W \tag{14}$$

Similarly, the equation for flow of electric current in a resistance-capacitance network (Fig. 1b) can be expressed as:

$$\frac{V_1 - V_0}{R_{X1}} + \frac{V_2 - V_0}{R_{X2}} + \frac{V_3 - V_0}{R_{Y1}} + \frac{V_4 - V_0}{R_{Y2}} = C \frac{\partial V_0}{\partial t} \pm J \tag{15}$$

where R_x, R_y are the resistance values in ohm along x and y direction respectively, C is the capacitance in farad, V's are the electric potentials in volts and J is the current in ampere withdrawn or injected.

If $R_{X1} = R_{X2}$, and $R_{Y1} = R_{Y2}$ for isotropic condition, we get

$$\frac{V_1 + V_2 - 2V_0}{R_X} + \frac{V_3 + V_4 - 2V_0}{R_Y} = C \frac{\partial V_0}{\partial t} \pm J \tag{16}$$

One can notice that equations (14) and (16) are analogous. Transmissivity is reciprocal of resistivity otherwise it corresponds to conductance, the storativity corresponds to capacitance, the hydraulic head corresponds to electric potential, and the groundwater flow corresponds to electric current. The instrumentation system used to simulate groundwater flow system is shown in Fig. 2.

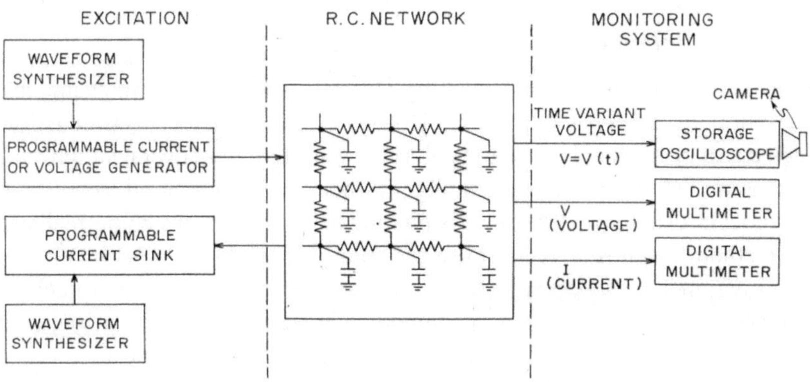

Figure 2. R-C analog model system

The measurement accuracy of this instrumentation system mainly depends on the components tolerance and the accuracy of measuring equipment viz. storage oscilloscope, and digital multi-meter. The maximum accuracy that can be obtained by using the oscilloscope could be 5%. This can be improved by using digital computer. It is possible to improve the accuracy of 0.01% by incorporating data logging system by using CMOS multiplexes as scanners.

R-C analog modeling technique has the following advantages over the other:

- Time is not discretised and thus associated errors are obviated, and
- There is an easy comprehension of the involved physical processes through visual display.

Though R-C analog modeling system was applied by many researchers during sixties and seventies to solve the complex groundwater flow problems, it could not withstand the advent of the advanced mathematical modeling, which could solve the much more complex problems with ease and fast.

Mathematical Modeling

Mathematical models may be deterministic, stochastic (statistical) or a combination of both. Stochastic models provide a range of solutions based on probabilities of occurrence. Stochastic models have found application in rainfall and flood forecasting. Deterministic models are based on cause and effect relationship of known systems and processes. Deterministic models are widely used for solving regional groundwater problems. Deterministic models can be further classified as (i) analytical and (ii) numerical. Numerical models can be further classified as Finite Difference Approximation Method (FDM) and Finite Element Method (FEM). Analytical element method model is another class of model in solving groundwater flow problem.

Analytical Modeling

An analytical model aims at obtaining an exact solution of a mathematical description of a physical process. However, groundwater flow equation, which could be amenable to analytical techniques, requires several simplifying assumptions of the system including the boundary and initial conditions. This process usually renders the system under study far from being realistic. This method has mostly been in use for pump test analysis for estimation of aquifer parameters, scooping calculations, inverse solutions for interpretation of flow tests and verification of numerical models. The earliest work on analytical solutions for groundwater flow can be found in Carslaw and Jaeger (1959) and the latest one on the above is from Strack (1989). Toth (1962, 1963) also had given two analytical solutions for the boundary value problem representing steady state flow in a vertical, two-dimensional (x-z), saturated, homogeneous, isotropic flow field bounded on top by a water table and on the other three sides by impermeable boundaries. Quickflow 2D analytical model developed by Rumbaugh (1991) is based on the Strack's work. Willis and Yeh (1987) have summarized the most common analytical response functions.

Numerical Modeling

Numerical modeling techniques have gained momentum during late seventies and eighties. Basically, numerical modeling employs approximate methods to solve the partial differential equation (PDE), which describes the flow in porous medium. The emphasis here is not on obtaining an exact solution but on obtaining reasonably approximate solution. Finite differences, finite elements, integrated finite differences, the boundary integral equation method and analytic element methods are the important numerical methods, which are presently in vogue.

Finite Difference

In the finite difference method (FDM), a continuous medium is replaced by a discrete set of points called nodes and various hydro-geological parameters are assigned to each of these nodes. Accordingly, difference operators defining the spatial-temporal relationships between various parameters replace the partial derivatives. A set of finite difference equations, one for each node is, thus, obtained. In order to solve a finite difference equation, one has to start with the initial distribution of heads and compute heads at later time instants. This is an iterative process and fast converging iterative algorithms have been developed to solve the set of algebraic equations obtained through discretization of equation (5). Some of the prevalent iterative numerical methods are (i) Successive Over-Relaxation (SOR) method (Watts, 1971, 1973; Aziz and Settari, 1972); (ii) Alternating Direction Implicit (ADI) Procedure (Peaceman and Rachford, 1955; Bredehoeft and Pinder, 1970; Remson et al., 1971; Rushton, 1974); (iii) Modified Iterative Alternating Direction Implicit Method (Prickett and Lonnquist, 1971); (iv) Strongly Implicit procedure (Stone, 1968; Weinstein et al., 1970; Trescott et al., 1976; etc.). Trescott and Larson (1977) had given comparative statement on various iterative methods used to solve two-dimensional groundwater flow problems. Shamir and Dagan (1971) developed a set of simultaneous partial differential equations to define motion of saltwater body in the coastal aquifer. They have introduced an implicit scheme to solve the algebraic equation simultaneously. The solutions resulted in an easily solvable seven-diagonal matrix. Bobba and Singh (1995) had brought out the important FDM models developed by various groups to solve different types of groundwater problems including seawater intrusion problem.

We will be concentrating here on "A modular three-dimensional finite difference groundwater flow model" developed by McDonald and Harbaugh (1988). The above model is also called as MODFLOW and presently many research scientists all over the Globe are using it. The latest version is called MODFLOW-2000.

As discussed in 'Finite Difference', the continuous medium described in equation (5) is replaced by a finite set of discrete points in space and also time is discretised. The partial derivatives are replaced by terms calculated from the differences in head values at these points. The process leads to systems of simultaneous linear algebraic difference equations. One can obtain head values at specific points and time by solving the above linear algebraic equations. The solutions obtained by this method are only an approximate one and not an exact one, as one would have got by analytical methods. An illustrative discretized aquifer of three dimensions is given in Fig. 3. Here, the aquifer system is discretized into mesh of blocks or cells, the location of which is described in terms of (i,j,k). 'I' represent rows, 'J' represents columns and 'K' represents layers. As per convention, the width of the cell in the row direction at a given column 'J' is designated as Δr_j, the width of the cell in the column direction, at a given row 'I' is designated as Δc_i, and the thickness of cells in a given layer 'K' is designated Δv_k. It is inferred from the schematic diagram that the cell with indices (i, j, k) has a volume of $\Delta r_j \times \Delta c_i \times \Delta v_k$.

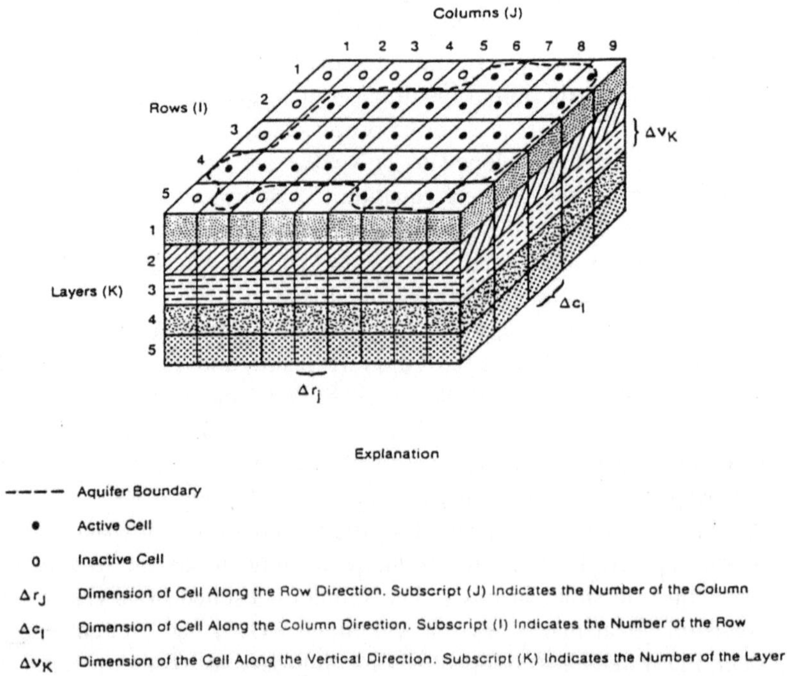

Figure 3. Illustration of discretization of continuous medium into finite difference cells (from McDonald and Harbaugh, 1988)

The continuous medium is discretized either as a mesh centered node (point centered) or block centered node as shown in Fig. 4 to solve groundwater flow equation (5). The head 'h' is a function of both space and time and it is required to discretise the continuous medium in to discrete nodes and time. Time is discretised in to stress periods and time steps and is shown in Fig. 5. One can refer McDonald and Harbaugh (1988) for the derivation of finite difference approximations and various iteration techniques to solve the matrix equation evolved during finite difference approximation.

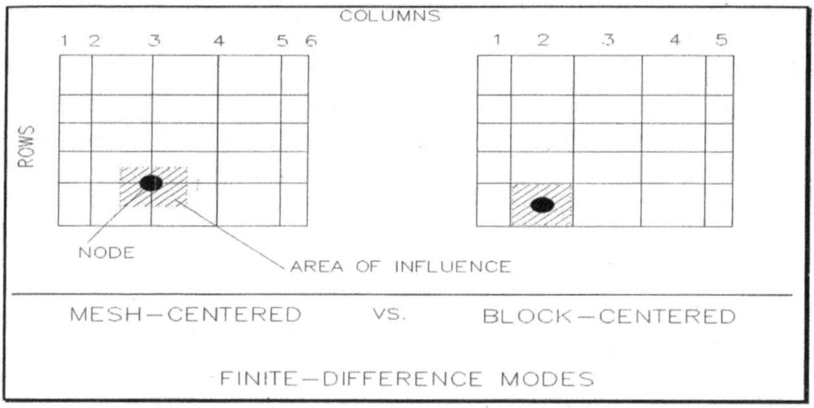

Figure 4. Illustration of two-dimensional space discretization (a) Mesh centered grid and (b) block centered grid (from Groundwater Vistas, 1998)

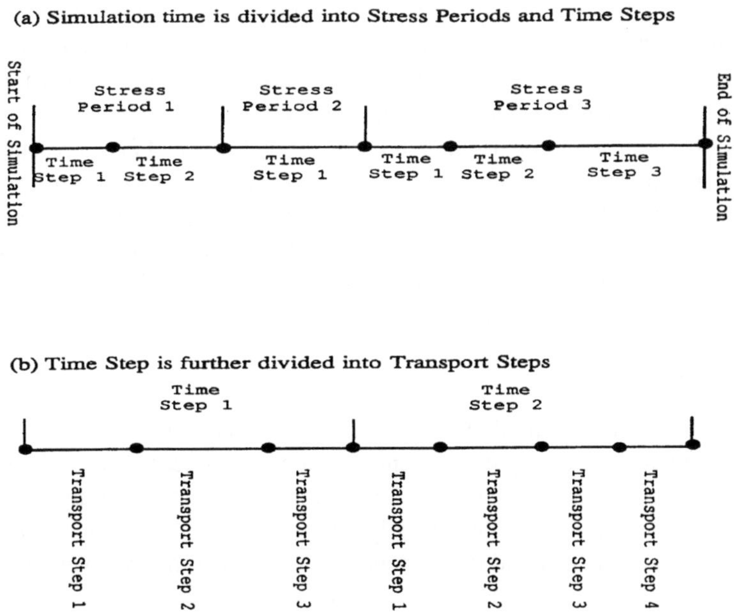

Figure 5. Illustration of time discretization in to various stress periods and time steps (from McDonald and Harbaugh, 1988)

The strengths of the finite difference method are that it is fairly simple to understand and to programme, has well known numerical properties, is economical and is quite easy to get accustomed to for those who have used difference methods for solving groundwater flow problems. One disadvantage is that the exact irregular boundary cannot be approximated with this method. This method does not work for advection-dominated solute transport problem due to numerical dispersion. Of course FDM is used to solve the hydrodynamic dispersion part of the mass transport equation 17 as given below.

The partial differential equation describing mass transport in flowing groundwater was derived by Reddell and Sunada (1970), Bear (1972), Bredehefot and Pinder (1973), Konikow and Grove (1977) and Javandal et al., 1984). Mass transport equation in a three-dimensional field can be expressed as:

$$\frac{\partial C}{\partial t}=\frac{\partial}{\partial x_{\mathrm{i}}}\left(D_{\mathrm{IJ}}\frac{\partial C}{\partial x_{\mathrm{j}}}\right)-\frac{\partial}{\partial x_{\mathrm{i}}}(v_{\mathrm{i}}C)+\frac{q_{\mathrm{s}}}{\theta}C_{\mathrm{s}}+\sum_{k=1}^{N}R_{\mathrm{k}} \tag{17}$$

where i, j = 1, 2, 3; C = the concentration of contaminants dissolved in groundwater, ML^{-3}; t = time, T; X_{i} = is the distance along the respective Cartesian coordinate axis, L; D_{ij} = the hydrodynamic dispersion coefficient along i and j direction, L^2T^{-1}; v_{j} = the seepage or linear pore water velocity, LT^{-1}; q_{s} = the volumetric flux of water per unit volume of aquifer representing sources (positive) and sinks (negative), T^{-1}; C_{s} = the concentration of the sources or sinks, ML^{-3}; θ = the porosity of porous medium dimensionless; and R_{k} = the chemical reaction term, $ML^{-3}T^{-1}$.

The first term in the right hand side of equation (17) represents the change in concentration due to hydrodynamic dispersion, while the second term gives the effect of advective transport. The third term represents source/sink while the last one represents the chemical reaction. The advection phenomenon is due to groundwater flow velocity, which in turn can be computed by solving groundwater flow equation (5).

Finite Element

The finite element method (FEM) is another powerful numerical technique used for modelling an aquifer system. The aim of numerical modelling is to evaluate the flow rate in a small-volume element by integration in space and time. This method first mathematically defines the potential surface over any small region of interest and then evaluates the gradient in any given direction by evaluating directives in that direction. In this method, the continuous flow field is discretised into a number of elements, which are used for interpolation of the field parameters such as the piezometric head and hydraulic conductivity. The basic idea is to transform the governing equations into integral form and to carry out piecewise integration over the elements. An example of finite element representation of aquifer system is given in Fig. 6.

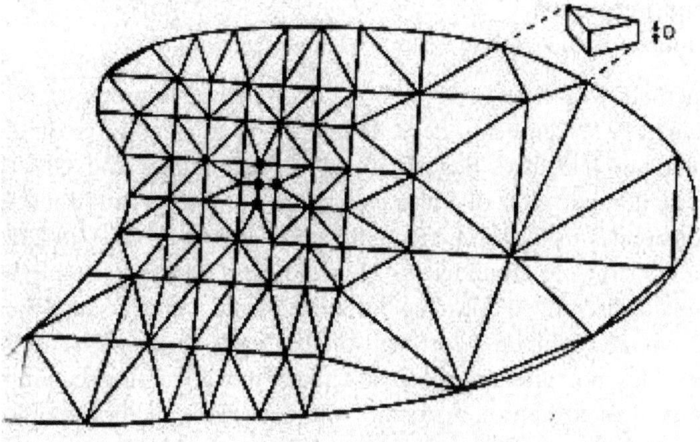

Figure 6. Illustration of aquifer discretization into finite elements (from Bobba and Singh, 1995)

The elements may have both different spatial dimensions and shapes. The order of the underlying interpolation scheme may typically be linear, quadratic or cubic. Continuity may be prescribed not only for the variables themselves but also for their derivatives. The procedures to be followed as given by UNESCO (1999) are given below:

(i) Discretization of the flow domain into a set of elements, where each element is defined by a number of nodes, for instance 3- or 6-node triangles, 4-, 8- or 9-node quadrilaterals in 2D or 4-node hexahedrons in 3D etc.

(ii) Expression of field parameters such as piezometric head, hydraulic conductivity etc., in the following form:

$$h\ (x,\ y,\ z) = \sum_{J=1}^{N} h_{\mathrm{j}}\Psi_{\mathrm{j}} \tag{18}$$

where h is piezometric head, (x, y, z) is Cartesian Coordinates, N is the number of nodal points in the discretized element grid and Ψ is the interpolation function or otherwise called shape functions.

(iii) Formulation of the groundwater flow equation (Partial Differential Equation or PDE) in integral form.

(iv) Element-wise integration of the integral form of the groundwater equation.

(v) Assembly of the algebraic matrix equations that result from the integration step into global system of linear equation of the form

$$[M]\left\{\frac{dh}{dt}\right\} + \{K(h)\}[h] = [f] \tag{19}$$

where $[M]$ denotes a matrix, and $\{\}$ denotes column vector.

(vi) Time integration
(vii) Solution of the global system of linear equations.

This method was first used in civil engineering (structural) problems (Zeinkiewicz, 1971) and has been found to be a potential tool to solve groundwater problem also. Segerlind (1976), and Pinder and Gray (1977) have presented application of finite element method to groundwater hydrology. Various codes using finite elements are available (Voss, 1984; Yeh and Huff, 1983) to simulate groundwater flow and mass transport. In this method, the domain is discretized into regular or irregular elements and heads or concentrations are computed for each of the nodes in the flow domain. In FEM, one does not attempt to approximate the governing equation, but rather the goal is to approximate the solutions. This method employs the mathematical shape or basis functions. Linear combination of the basic functions is used to represent the head or concentration distribution at each time step. The most common method of FEM for groundwater problem is the Galerkin method of weighted residuals. The residual is nothing but the difference between the true solution and that given by discrete approximation of the solution. The Galerkin method minimizes the weighted sum of the residuals for each of the finite elements and by doing so it attains the best approximate solution for the finite element mesh.

Boundary conditions can be tackled effectively by this method due to flexibility in choosing the shape of the elements. Another advantage of this method is that in its standard formulation it has the immediate property that each element can have its own value for the physical parameters such as hydraulic conductivity and storativity. One disadvantage by using finite element method could be that it is computationally difficult and computer memory requirements are larger than other methods.

Analytical Element Method

Strack (1989) had developed a new method called analytical element method. The basic idea of this method is that solutions of the basic equations can be obtained by superimposition of standard solutions of various problems, which may include singular solutions for homogeneities in the permeability, the infiltration rate etc. Since the solution is, in principle, an analytical one, it has all the advantage related to the analytical character. The solution is exact and it can be used to study the minute details of flow problem in a small region carved out of a big domain. This method was applied for multi-layer and heterogeneous system but results are not as on the expected line and, therefore, this method is ideally suited for only homogeneous and steady state flow problem.

The process of aquifer modeling, in general, consists of the following activities:

- Identification of parameters characterizing the physical framework of the aquifer and stress acting on it,
- Field estimation of the relevant hydrogeological parameters at as many control points as possible, particularly those at boundaries,
- Interpolation/extrapolation of these parameters to characterize the entire area under study,
- Integration of the entire hydrological data to conceptualize and resurrect the full-scale natural system,
- An appropriate mathematical description of the conceptual model giving spatial-temporal relationship between those parameters constrained by the fundamental laws of hydraulics,
- A solution of mathematical equations describing the groundwater regime in terms of observable field parameters such as groundwater levels or concentrations of pollutants etc.,
- Calibration of the model for steady and transient condition,
- A sensitivity analysis of the model to identify those parameters, which need to be estimated more accurately, and also to decipher the error bounds,
- The refinement of the model to progressively bring in plausibility and compatibility between field estimates of the various geohydrological parameters through the process of model calibration and validation, and
- The prognosis of the aquifer to evolve efficient management options for optimal utilization of the groundwater resources.

COMPUTER CODES FOR GROUNDWATER FLOW AND MASS TRANSPORT IN POROUS MEDIUM

A large number of computer codes had been developed to simulate groundwater flow and mass transport in late seventies and eighties either by using finite difference or finite element techniques. Among finite difference method, modular three-dimensional finite difference groundwater flow model is referred as MODFLOW and it is a well-documented one. United States Geological Survey (USGS, McDonald and Harbaugh, 1988) has developed this code and it has been received very well among the model users. Apart from MODFLOW, some other important codes viz. MT3D, MOC and MODPATH are briefed here for the benefit of the readers.

Modflow (Modular Three-dimensional Flow Model)

As mentioned above, MODFLOW is a versatile code to simulate groundwater flow in multi-layer porous aquifer. The model simulates flow in three dimensions using a block-centered finite difference approach. The groundwater flow in the aquifer may be simulated as confined, unconfined or the combination of both. MODFLOW consists of a main programme and a number of sub-routines called modules. These modules are grouped into various packages viz. basic, river, recharge, block centered flow, evapotranspiration, well, general head boundaries, drain, strongly implicit procedure (SIP), successive over relaxation (SOR), slice successive over relaxation (SSOR) and preconditioned conjugate gradients (PCG) etc. USGS is continuing its activities to expand the capabilities of MODFLOW and one can expect a lot more changes from the present Modflow-2000 version in coming years.

MT3D (Modular Three-dimensional Transport Model)

The MT3D transport model was developed by Zheng (1995) for use with any block-centered finite difference flow model such as MODFLOW and is based on the assumption that changes in concentration field will not affect the flow field significantly. This model can be used to simulate changes in concentration of single species miscible contaminants in groundwater considering advection, dispersion and chemical reactions with various types of boundary conditions as mentioned above and external sources or sinks. The chemical reactions included in the model are equilibrium-controlled linear or non-linear sorption and first order irreversible decay or biodegradation. MT3D makes use of the following spatial discretization capabilities and transport boundary conditions: (i) confined, unconfined or variably confined/unconfined layers; (ii) inclined model layers and variable cell thickness within the layer; (iii) specified concentration or mass flux boundaries and (iv) the solute transport effect of external sources and sinks such as wells, drains, rivers, areal recharge and evapotranspiration.

The model uses a mixed Eulerian-Lagrangian approach to the solution of advective-dispersive-reactive equation, based on combination of the method of characteristics and the modified method of characteristics. One can refer Thangarajan (2004) for more information regarding the solution techniques used to solve the mass transport equation (17).

MOC (Method of Characteristics)

Konikow and Bredehoeft (1978) developed a technique to solve the advective part of mass transport equation based on method of characteristics (MOC) to simulate changes in concentration of single species miscible contaminants in groundwater. This method consists in introducing a set of moving points that can be traced with reference to the stationary coordinates of finite difference grid. Each particle has a concentration associated with it and is moved in the flow field in proportion to the fluid velocity. The moving particle represents the advective transport. The change in concentration due to dispersion is computed by solving an explicit finite difference equation. This model was very much in use in USGS and other parts of the world, till MT3D was released during 1990. Presently MT3D makes use of both MOC (forward particle tracking) and modified version of MOC called

MMOC (backward particle tracking) for solving advective part of mass transport equation.

Modpath

It is a three-dimensional particle-tracking model, which works with MODFLOW and it is also developed by USGS. The initial version supported only steady state model but the later version accommodates transient condition also. This software helps one to view the groundwater flow direction and also computes flow velocity, which in turn will be used for capture zone analysis.

Sutra

The computer software SUTRA was developed by Voss (1984) and it simulates the groundwater flow and solute transport in the saturated aquifer and in the unsaturated vadose zone. This model employs a two-dimensional hybrid finite element and integrated finite difference method to approximate the governing equation of fluid density dependent saturated groundwater flow and solute transport in it. Though it can be used to simulate in two-dimensional horizontal aquifer systems, it is mainly used as x-z profile model. Also it can be used to simulate density dependent flow to study seawater intrusion.

OTHER CODES

There are a number of commercial codes, which use the MODFLOW, MT3D and MODPATH as the core software and supplement it with pre- and post-processors for easy inputting and analysis of the simulation results. Some of the user-friendly computer softwares are described here for the benefit of the readers.

Visual Modflow

Visual Modflow is the most common software used by the field hydrologists and researchers to simulate three-dimensional groundwater flow and contaminant transport simulation. This software supports MODFLOW, MT3D and MODPATH and interface the graphical software. The visual MODFLOW interface has been specially designed to increase modeling productivity and decrease the complexities in building the three-dimensional groundwater flow model. The interface is divided into three separate modules, the Input Module, the Run Module and the Output Module. It is this facility, which makes the user an easy going.

Groundwater Vistas

Groundwater Vistas (GV) is a sophisticated Windows graphical user interface for 3D groundwater flow and mass transport modeling, calibration and optimization. This software supports MODFLOW 2000, MT3D'99, MODPATH, RT3D, MT3DMS etc. This software is developed and marketed by Environmental Simulations International (ESI) Ltd., UK.

Special Features

- Automatic Calibration procedure and supports PEST ASP and UCODE
- Integration with ArcView and Equis
- Detailed parameter sensitivity analysis
- Importing and exporting various files
- Displays both plan and cross-sectional views
- Use of grid independent boundary conditions
- High tech and low price (1200 US $)

Among the abovesaid two softwares, the first one is very user friendly software with slightly higher cost (around 1600 US $), whereas the second one has more capability particularly automatic calibration and sensitivity analysis worth mentioning here and it is not that user friendly as the first one but it has less price than the first one. The modeler can choose any one of them depending upon his skill and need.

Nammu

Nammu is groundwater-modeling software from AEA Technology, UK. It is a finite-element software package for modeling groundwater flow and transport in porous media. It has the following capabilities:

- Groundwater flow in saturated and unsaturated conditions
- Coupled groundwater flow and solute transport with density dependent on concentration
- Coupled groundwater flow and heat transport
- Contaminant transport, including the effects of advection, dispersion and sorption
- Radioactive decay chains
- Steady state and transient simulations

It was reported by AEA that the above software is developed under a quality system that conforms ISO 9001 and Tick IT. One can get licences for NAMMU from AEA Technology Environment, UK.

PEST

PEST is used for non-linear parameter estimation through inverse modeling. Dr. John Doherty of Watermark Numerical Computing Company has developed this code. This code has been interfaced with most commercial codes viz. Groundwater Vistas (GV) and Visual MODFLOW.

WinFlow

WinFlow is an analytical groundwater flow model for Windows. It is a powerful yet easy-to-use groundwater flow model. It is an interactive, analytical model that simulates two-dimensional steady state and transient groundwater flow (both confined and unconfined aquifers) with wells, uniform recharge, elliptical recharge/discharge areas and the line source or sinks. The model depicts the flow field using streamlines, particle traces and water level contours. The above software can be obtained from ESI, UK

Application

- Assessing groundwater abstraction and aquifer drawdown
- Designing and interpreting pumping tests
- Predicting simple dewatering schemes and impacts
- Evaluating groundwater remedial options
- Delineating well capture zones
- Determining groundwater flow velocities and travel times

Flowpath (II)

Flowpath was widely used by hydrologists to determine groundwater flow direction, flow velocity and capture zones for wellhead protection. Waterloo Hydrogeologic Inc. has now extended its capabilities and termed as Flowpath II. This software has been released very recently and has the following capabilities:

- Solute transport simulation using Random walk solution process
- Animation of contaminant plumes and particle tracking
- Cell-by-Cell anisotropy for hydraulic conductivity properties
- Assignment of constant heads, constant flux, recharge, rivers, lakes, ditches and leakage
- Plot concentration vs. distance at each output time step

COMPUTER CODES FOR GROUNDWATER FLOW AND MASS TRANSPORT IN FRACTURED MEDIUM

There may be a number of commercially available codes to simulate the fracture flow system but the author is aware of only three codes and their salient aspects are presented here.

FracMan Software

The FracMan software, developed and marketed by Golder Associates Inc., Redmond, USA (Dershowitz et al., 1991), provides an integrated set of tools for simulation of discrete fracture network (DFN) analysis i.e., fractured and non-fractured heterogeneous rock masses. FracMan includes tools for discrete fracture data analysis, geologic modeling, spatial analysis, visualization, flow and transport, and geo-mechanics. FEM approach is used to simulate groundwater flow and particle tracking method for solute transport.

The following modules are incorporated in this software:

(i) FracSys (Data Analysis),
(ii) FracWorks (3D DFN Structural Modeling and Visualisation),
(iii) Mafic (Finite Element and Transport Modeling),
(iv) PA Works (Pathways and Analysis),
(v) FracCluster (Cluster Analysis), and
(vi) FracView (Visualisation and Software Linkage).

As per FracMan operating manuals the following specifications are provided:

- FracSys provides a unique set of tools to transform geological and well testing data into quantitative parameters necessary for discrete feature network modelling.
- FracWorks generates 3-dimensional realisations of discrete feature geology using the parameters derived with FracSys. Fifteen different spatial models are supported. FracWorks also includes the ability to validate models through simulated exploration.
- Mafic (Matrix and Fracture Interaction Code) uses the finite element method to solve the flow and transport through FracWorks geological models. Mafic idealises fractures using triangular finite elements and provides a dual porosity interaction using quadrahedral finite elements approximation. Mafic uses a pre-conditioning conjugate gradient solver, and has been applied for network of up to 100,000 fractures.
- PA Works and FracCluster are powerful tools to support analysis of the flow behaviour of discrete feature network, without requiring the solution of flow equations. PA Works determines flow pathways within a FracWorks model using graph theory methods, to identify most important pathways and their properties. FracClusters defines the geometry and properties of fracture network cluster and the block defined by fracture and fracture network.

Frac3DVS Software

Frac3DVS software is a 3D finite element model for steady/transient, variably saturated flow and advective-dispersive solute transport in porous and/or discretely-fractured porous media. Hydrogeologic Inc., Canada, had developed this software to simulate fracture flow system and its applicability and capability are given below:

- Frac3DVS is a 3-dimensional finite element model for simulating steady state or transient, variably-saturated groundwater flow and advective-dispersive solute transport in porous or discretely fractured porous media.
- Frac3DVS can use either an 8-node block or 6-node prism finite element or 7-point finite difference discretization of the porous medium and rectangular or triangular elements to represent fractures if present. Block elements can be sub-divided into tetrahedra so that grid composed of non-orthogonal block elements can be accommodated. Separate pre- and post-processors are included for easy input and output analysis.
- Frac3DVS uses numerical approach, which can be a control-volume finite element, Galerkin finite element or finite different method. The flow solute has the following features: An exact treatment of the saturation term by using a mixed, mass-conservation formation, full Newton-Raphson interaction for robustness in handling non-linearities in Richard's equation describing variably saturated flow.

NAPSAC

NAPSAC is a software package to model flow and transport through fractured network. This code was developed for more than 20 years jointly by scientists of UK, Germany, Sweden, and Canada for identifying suitable sites to dispose nuclear wastes. Presently, M/s AEA Technology Environment, UK, markets this code. Due to the extreme difficulty of quantifying actual fracture networks with a given rock mass, a stochastic approach was taken with NAPSAC. Rather than the user explicitly modeling actual fractures, statistical data on the dips, dip azimuths, and fracture lengths is collected from the field. Using this data set, NAPSAC will then generate a network of rectangular planar fractures with properties specified by the field statistical data. Each generated fracture centre will then have a rectangular fracture placed around it. Once all the fractures have been generated, those outside test volume are discarded. For those that remain, the intersections between fractures are calculated and the finite element nodes are assigned across the network. Flow can then be simulated at steady state, transient condition and particles tracked through the network. A 'random number seed', a value that is specified in the input data set, will control the random network that is generated.

NAPSAC can simulate flow either in parallel plate flow mechanism (each fracture is characterized by a single 'aperture' value) or specified distribution of local apertures (aperture variation over the entire plane). Though the second one is more realistic, it is very difficult to measure local aperture, hence one can opt for the first one even though it is unrealistic.

The models are based on a direct representation of the discrete fractures making up the flow-conducting network. NAPSAC uses a stochastic approach to generate networks of planes that have the same statistical properties as those that are measured in the field. The software is based on a very efficient finite element method.

Some of the capabilities of NAPSAC (Hartley, 1998) are as follows:

- Simulation of steady state or transient fluid flow in a fracture network. Steady state calculations use an efficient FEM scheme to enable calculations on very large networks. Transient calculations use domain decomposition to give good accuracy where changes take place rapidly without introducing high run times for the rest of the network.
- Prediction of transient pressures and drawdowns at well bores for various types of pump tests.
- Simulation of tracer transport through network using a stochastic particle tracking method.
- Simulation of mass transport for a variable density fluid. This also can be used to model both groundwater flow and salt transport.
- Generation of stochastic fractures from a wide variety of probability distribution functions.
- 'D' visualization of results using AVIZIER software.

MODELING AS A TOOL TO EVOLVE PRE-DEVELOPMENT MANAGEMENT SCHEMES: A CASE STUDY IN KUNYERE RIVER VALLEY IN BOTSWANA

Development of groundwater resources to meet the increased demand for drinking, industries, irrigation and other purposes is of prime importance in developed and developing countries. The over development of groundwater resources leads to the decline of water level causing socio-economic and environmental degradation. It is, thus, imperative to manage the groundwater resources in an optimal manner. Management schemes can be evolved, only if the groundwater potential is assessed in more realistic manner. Mathematical modeling in conjunction with detailed field investigations have been proved to be a potential tool for this purpose. Evolving pre-development management schemes still works out to be better choice. One such study was carried out in Kunyere River valley, Okavango Delta, Botswana.

A model with six layers flow regime was conceptualized by making use of available data. Fourth layer of the model is the main fresh water bearing aquifer and the bottom layer is brackish one. Long duration pumping test carried out in this area indicated the leaky nature of the aquifer system. Mathematical model of the basin was constructed and calibrated for steady state condition by using Visual Modflow computer software. Two prognostic runs were made and an optimal one was identified which will ensure minimum upward leakage from the bottom saline unit to the pumping aquifer. This simulation study indicates that substantial development of groundwater potential is possible in this area.

The Kunyere River valley is located in Okavango Delta (Fig. 7), Botswana (Southern Africa). The Okavango River, which originates in Angolan highlands, reaches in northwestern Botswana and terminates as a huge inland delta, namely the Okavango Delta. The Kunyere River valley runs along the Kunyere Fault and drains the basin to Lake Ngami (Fig. 8). Maun, a small town located on the bank of river Thamalakane, is emerging as a famous wild life tourist centre in Botswana and the demand for more water is ever increasing. The demand during the year 1997 was about 3540 m^3/day and it is likely to increase to about 5000 m^3/day during the year 2006. The

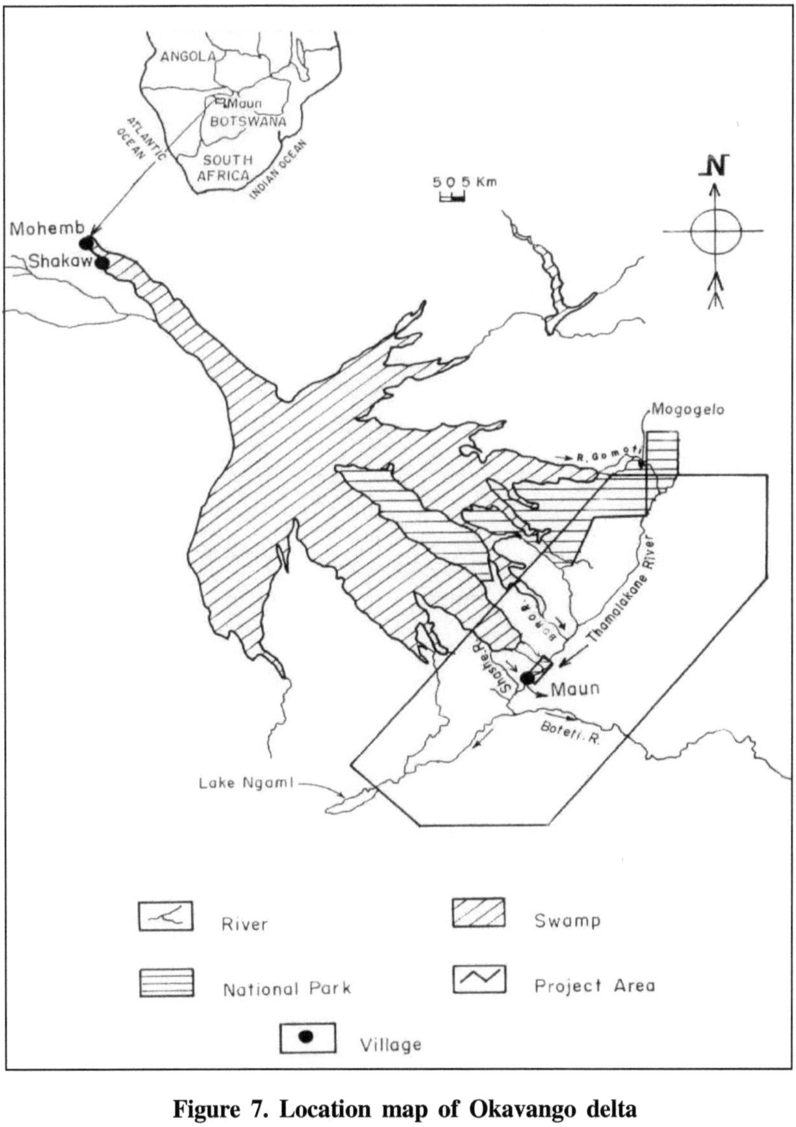

Figure 7. Location map of Okavango delta

present demand is met from the well fields located in the valleys of Shashe and Thamalakane and the future demand has to be met from new resources. Therefore, Government of Botswana formulated a project in the year 1996 to explore and assess the new groundwater structures through integrated geophysical and geohydrological studies in the neighborhood of Maun. The Kunyere River valley (Fig. 8) covering an area of about 120 km^2 is one such system wherein geohydrological, geophysical and chemical quality studies were combined to find suitable structures for exploiting groundwater. The

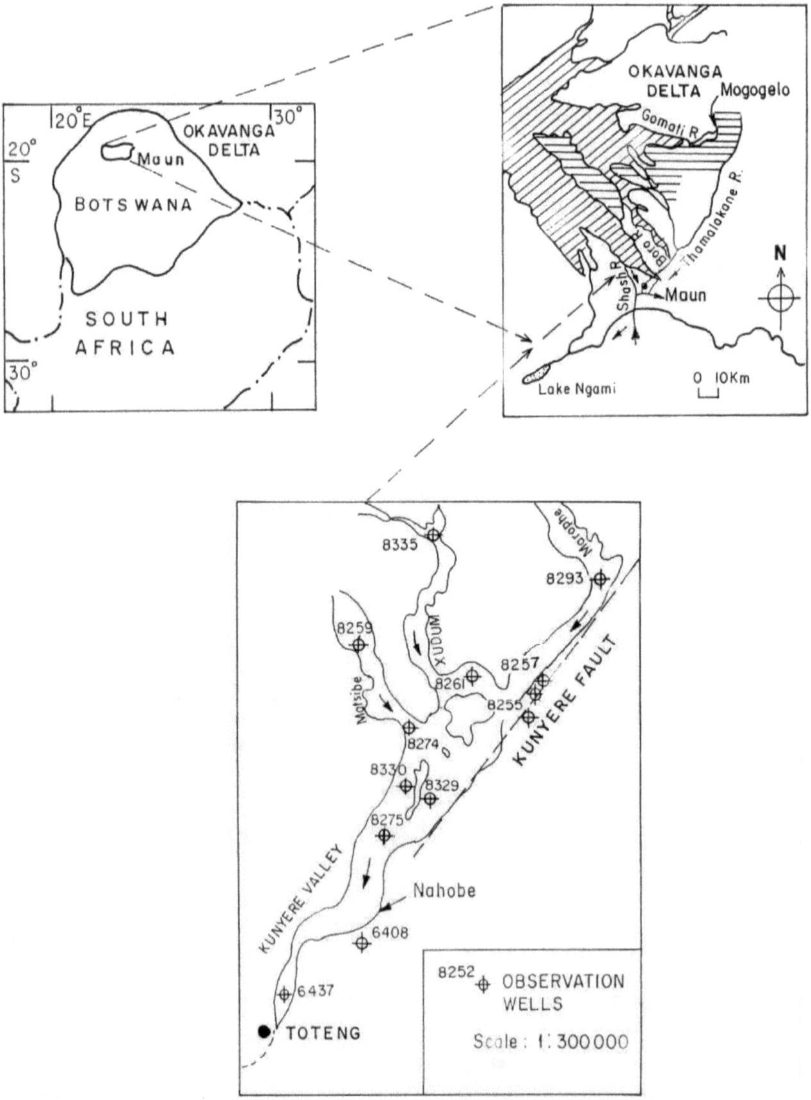

Figure 8. Location of Kunyere river valley

data collected under this project was used to conceptualize the groundwater flow regime and a preliminary mathematical model was developed to study the aquifer response and thereby to evolve pre-developmental management schemes. The model was constructed and calibrated for steady state condition. The model calibration has clearly indicated that upper and middle aquifers are semi-confined in nature. The calibrated model was then used to study the aquifer response under two possible scenarios for evolving optimal well field locations in the upper semi-confined aquifer.

Hydrogeological Setting

The aquifer system belongs to Kalahari beds. Three exploration boreholes and one water-level observation borehole have been drilled within the Kunyere valley. The vertical subsurface geological section along the valley is shown in Fig. 9. The lithology encountered during the drilling of these boreholes indicates a thin surficial cover of silts ranging in thickness from less than one metre in the south west (BH8275) to about seven metres in the northeast at the location of BH8255/BH8257. Clayey fine sands and silty fine sands underlie these silts with a thickness ranging from 15 to 20 m. The unit is in turn underlain by medium sand and fine to medium sand with a thickness between 11 and 19 metres. The medium sand and fine to medium sand is underlain by sandy to silty clay and silty fine sand with a thickness varying between 8 and 15 m. This unit is underlain by fine to medium sand inter-

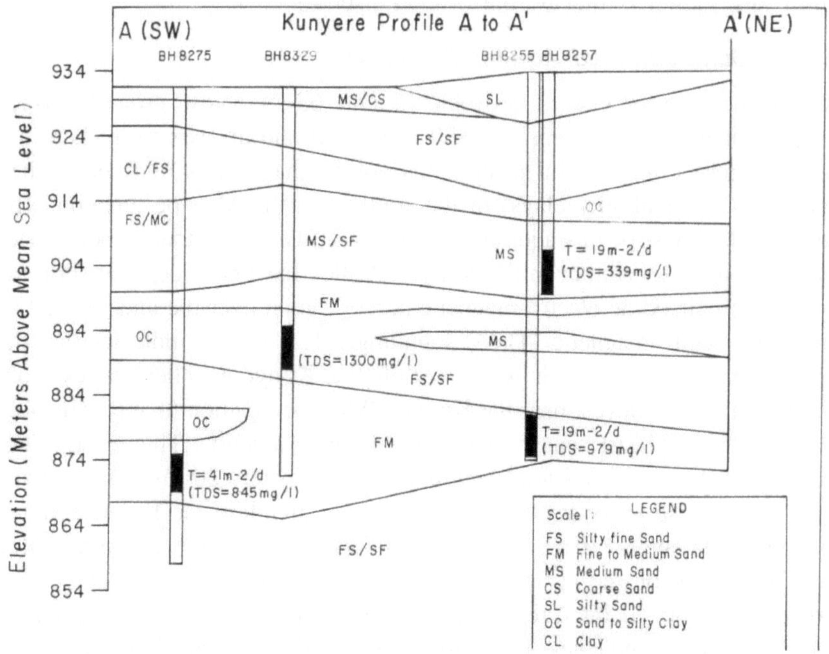

Figure 9. Subsurface Geological section

bedded with clay lenses (BH8275), which has a thickness varying from 7 to 21 metres. The fine to medium sand increases in thickness from the northeast to southwest down the Kunyere Valley and is underlain by fine silty sand. There are two main aquifer systems in the river valley (Fig. 9). The top zone (surface soil) is the low permeable zone with average thickness of 10 m. The top two aquifers are fresh water bearing units and the bottom one is saline. An outline of the hydrogeological conditions in each of these units is given below.

Upper Semi-confined Aquifer (Layer 2)

The upper semi-confined aquifer consists of fine to medium grained sand. The upper aquifer has a more or less uniform thickness throughout the reach of the valley between BH8255 and BH8275. This upper aquifer unit occurs at approximately 20 m below ground surface (approximately 910 m (amsl)) and the bottom occurs at around 36 m below ground surface (approximately 895 m (amsl)). The thickness of this aquifer is about 16 m. Recharge to the semi-confined aquifer has been predominantly from river water infiltration as downward leakage. Major rainfall events also recharge the top aquifer to a limited extent and it is derived from model calibration as 1.8 mm per annum. This is well within the range of value recommended by the recharge committee of Botswana. The mean annual rainfall for 1986-1996 from the Maun airport weather station is about 380 mm. Presently the main discharge from this aquifer is that by the evapotranspiration.

Lower Semi-confined Aquifer (Layer 4)

The lower aquifer occurs between 45 and 68 m below ground surface (approximately 890 m (amsl) to 880 m (amsl)) and appears to thicken in a southwestern direction. In general this aquifer has lenses of finer grained material (silts and silty clays) in the south near the Matsibe junction. The average thickness of this aquifer is about 24 m. Lateral extent or boundaries of the fresh water aquifer in the Kunyere Valley have been delineated on the basis of an air-borne electro-magnetic survey flown over the area as well as transient electro-magnetic (TEM) soundings. This survey indicates that the lateral extent of the fresh water within the Kunyere valley varies from 5.5 km in the southwest near the Kunyere/Matsibe junction to approximately 3 km in the northeast near the Kunyere/Marophe junction. This reach of the Kunyere valley covers a total length of approximately 30 km. The average area of the fresh water aquifer covered by this stretch of the Kunyere valley is approximately 127 km^2. The depth of fresh water in the Kunyere valley was delineated on the basis of TEM soundings conducted in the valley. A contoured depth to interpreted fresh water/saline water interface based on TEM soundings indicates that fresh water occurs to a maximum depth of 70 m and minimum depth of approximately 50 m.

The Transmissivity (T) values were estimated through single well pumping tests in the upper and lower semi-confined aquifers. T value of 19 m^2/day was estimated in layer 2 and 41 m^2/day in layer 4. The storability values were inferred and not estimated.

Bottom Aquifer (Brackish, Layer 6)

The aquifer, which occurs approximately between 70 and 75 m below the ground level, comprised fine to medium sands and contains brackish to saline water. There is not much information available as regards hydraulic characteristics of this unit. The hydraulic conductivity value is assumed of 7 m/d and its storativity value as 2.9×10^{-4}. The bottom aquifer is hydraulically connected with the overlying middle semi-confined aquifer.

Model Design

The computer software programme Modflow developed by the United States Geological Survey (USGS: McDonald and Harbaugh, 1988) was used for the present study. In this programme, using the Block Centered Finite Difference Approach solves groundwater flow equation. A pre- and post-model processor viz. Visual-Modflow developed by Nilson Guigner and Thomas Franz of Waterloo Hydrologic Software Inc., Canada (1996) was used for graphical data input, and for analysis and presentation of the output data.

Conceptual Model

The Kunyere River valley aquifer system was conceptualized as a six-layer system with three aquifers separated by two confining layers (confining/semi-confining) as follows:

Topsoil with low permeable zone
Upper semi-confined aquifer
Upper confining unit
Lower semi-confined aquifer
Lower semi-confining unit,
Bottom brackish/saline aquifer

The confluence points of three tributaries viz. Matsibe, Xudum and Morphe were taken as the inflow boundaries. The lateral boundary represents the fresh water zone in the upper semi-confined and lower semi-confined aquifers. This delineation is an approximate one and based on available data. The lateral boundaries were delineated based on the lithology of recently drilled lateral boundary definition boreholes, the airborne EM survey and TEM sounding data. The area considered for this model study is 130 km^2. The study area was divided into square grids and the map is shown in Fig. 10.

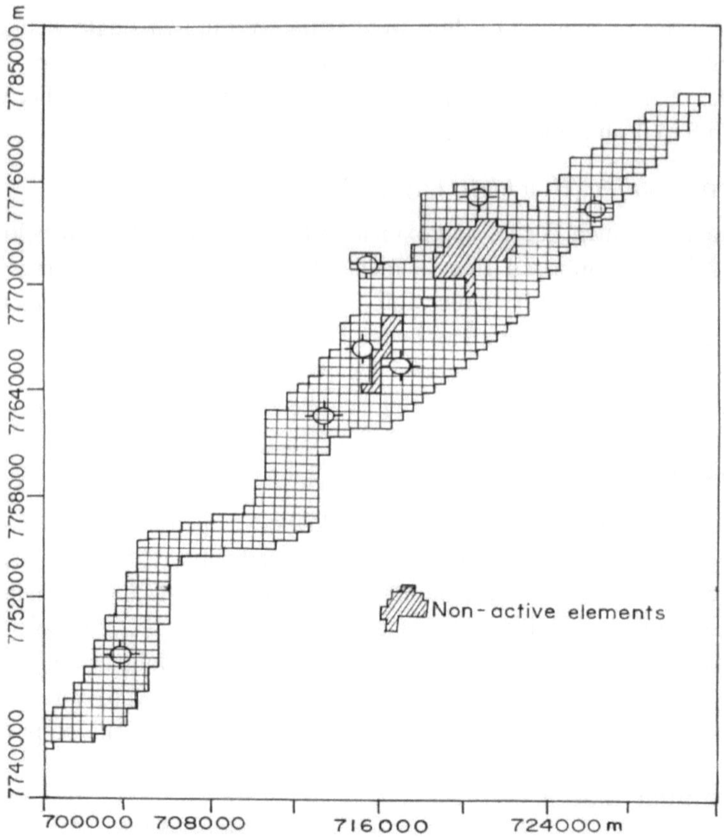

Figure 10. Grid map of the study area

Initial Condition and Boundary Conditions of the Model

The confluence points of three tributaries were taken as the inflow boundaries in the upper and lower semi-confined aquifers. The quantum of inflow flux was calculated by using Transmissivity values and the hydraulic gradient. It was estimated that about 1100 m^3/d is received in the upper aquifer and lower aquifer. Layers 3 and 5 (silty sands and clays) were taken as aquitards. Since bottom aquifer (saline unit) is laterally extended, the northeast and southwest boundaries were assumed as inflow and outflow boundaries respectively. The eastern and western lateral boundaries were treated as no flow boundaries as the flow is predominantly from north-east to south-west. The subsurface outflow towards the southwestern direction was simulated as fixed heads near Toteng. The water levels monitored in the presently drilled bore wells are very useful in fixing the boundary heads in all the three layers.

The aquifer parameters hydraulic conductivity (K, m/day), Specific yield (Sy) and Specific storage (Ss, L^{-1}) were assigned zonewise for each layer. Storativity value of 0.00029 was uniformly assumed for the upper and lower semi-confined aquifers. The vertical permeability for each layer is assigned as one tenth of horizontal conductivity. The upper semi-confining unit and lower confining unit were assumed to have storativity value of 0.001. The hydraulic conductivity value of 0.052 m/day was set for both the upper confining and lower confining units.

Model Calibration

Steady State Calibration

The aquifer condition of March 1997 was assumed to be the initial condition for the steady state model calibration. The model could not be initiated at an early date due to the non-availability of water level data before January 1997. Minimizing the difference between the computed and the field water level for each observation point started the steady state model calibration. Number of trial runs were made by varying the input/output stresses and the hydraulic conductivity values of the top and middle aquifers in order to keep the root mean square (RMS) error below 0.2 m and mean error below 0.1 m. The computed versus observed head for selected observation points are shown in Fig. 11. This figure indicates that there is a fairly good agreement between the calculated and observed water levels. The computed water level contours in layer 4 is shown in Fig. 12. The calibrated zonal hydraulic conductivity (K) values for the upper, lower and bottom aquifers and upper aquitard are shown in Table 1. Due to non-availability of historical water level data, transient state calibration could not be initiated in the present study. The calibrated steady state model was then used to prognoses the aquifer response.

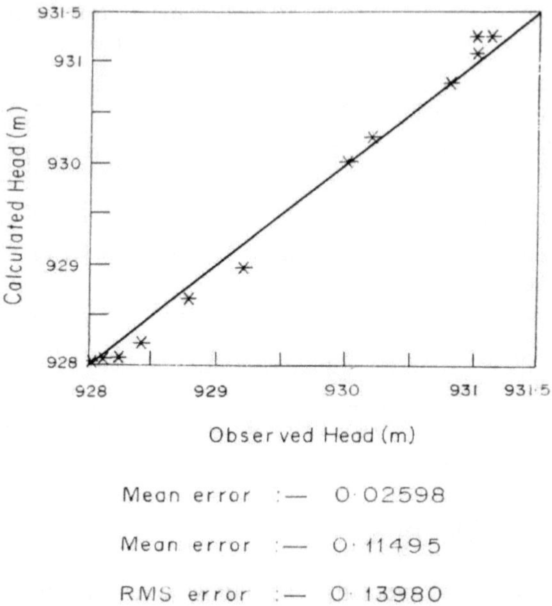

Figure 11. Scatter plot of computed vs. observed head for steady state model.

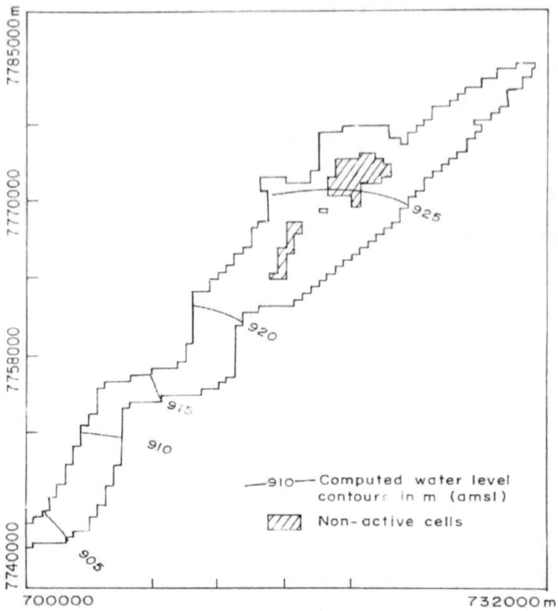

Figure 12. Computed water level contours for steady state model in layer 4.

Table 1: Model Parameters-Steady State Condition

Layer	*Unit Description*	*Average*	*Field Derived Hydraulic Conductivity (K in* m/d)	*Model Calibrated Horizontal Hydraulic Conductivity* (K_x *in* m/d)	*Storativity*
1	Semi-Confining Unit	20	NA	0.052	0.001
2	Semi-Confined Fresh Water Aquifer	15	K_x = 3.2	2.1 To 3.2	0.00029
3	Semi-Confining Unit	7	NA	0.052	0.001
4	Semi-Confined Fresh Water Aquifer	24	K_x = 3.2 To 7	3.5 To 9.5	0.00029
5	Semi-Confining Unit	2	K_x =0.0055	0.052	0.001
6	Semi-Confined Brackish Aquifer	50	NA	7	0.00029

K_x = Horizontal Hydraulic Conductivity. K_x = Vertical Hydraulic Conductivity, Assumed to be 0.1 of K_x.

Prediction Scenario Runs

Two preliminary model prediction scenarios were run utilizing simulated production boreholes completed in the lower semi-confined aquifer.

Prediction Scenario 1

In Scenario 1, 10 simulated production boreholes (Fig. 13) were placed over an area of 93 km^2 (3 km spacing) in the lower semi-confined aquifer, each pumping at a rate of 4800 m^3/d over a 10-year period. Table 2 summarizes the computed inflows and outflows to the lower semi-confined aquifer.

The bulk of inflow to the lower semi-confined aquifer is through downward vertical leakage from the upper semi-confining unit and the upper semi-confined aquifer (Column 6 of Table 2). Contribution to pumping from aquifer storage in the lower semi-confined aquifer is insignificant (Column 8). There is a net vertical downward leakage to the lower brackish/saline aquifer from the lower fresh water semi-confined aquifer up until Year 8 in the simulation (Column 9). After Year 8, there is an 8 to 9 percent contribution to pumping in the lower semi-confined aquifer due to upconing of water from the brackish/saline aquifer (Column 12). Calculated water quality impact due to this upconing in Year 10 of the simulation is an increase in TDS from a baseline of 800 mg/l to 1147 mg/l.

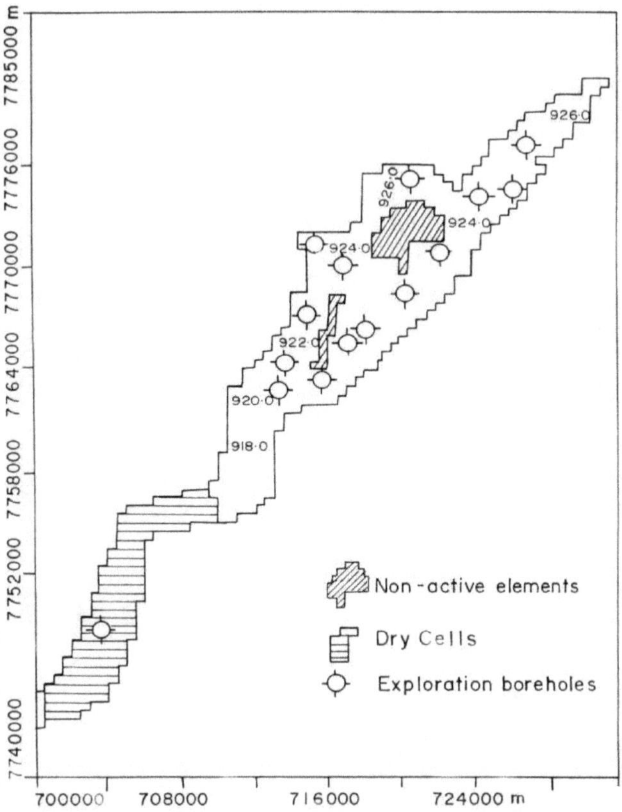

Figure 13. Predicted water level contours in m for January 1998 in layer 1

Table 2. Model Prediction Scenario 1: Pumpage of 4800 m^3/d from Lower Semi-Confined Aquifer with no River Recharge

Year	*Pumping Rate from Lower Semi-Confined Aquifer* (m^3/d)	*Horizontal Outflow* (m^3/d)	*Downward Leakage between lower semi-confined aquifer and brackish/ saline aquifer* (m^3/d)	*Total Outflow from the lower Semi-Confined aquifer* (m^3/d)	*Inflow from Downward Leakage from Upper Semi-Confining Layers and Upper Semi-Confined Aquifer* (m^3/d)	*Horizontal inflow* (m^3/d)	*Contribution from Aquifer Storage in the Semi-Confined Aquifer* (m^3/d)	*Upward leakage between Lower Semi-Confined Aquifer and Brackish/ Saline Aquifer* (m^3/d)	*Total inflow to the Low Semi-Confined Aquifer* (m^3/d)	*Percent Contribution to Pumping from Lower Semi-Confined Aquifer Storage*	*Percent Contribution to Pumping from Upconing*
1	4800	2288	3397	10,485	6179	1147	95	0	10,485	2	0
2	4800	2754	3578	11,132	6944	1310	108	0	11,132	2	0
4	4800	3169	1602	9571	6744	1535	94	0	9571	2	0
8	4800	4465	0	9265	6138	1852	68	394	9265	1	8
10	4800	4394	0	9194	5528	1974	50	431	9194	1	9

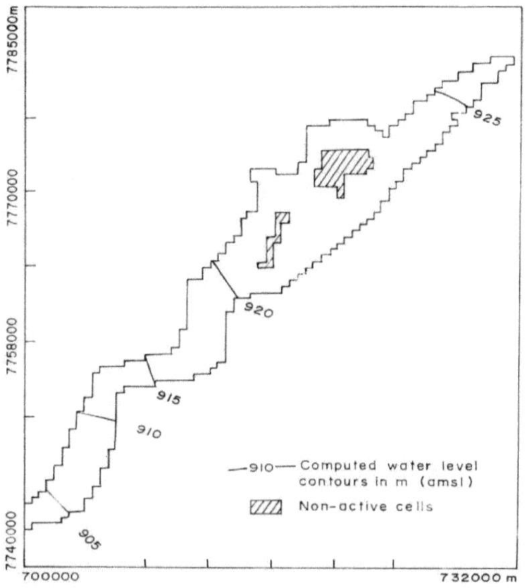

Figure 14. Predicted water level contours in m for January 1998 in layer 4

Figures 13 and 14 show predicted water-level contours in the top surface layer (Layer 1) and lower semi-confined aquifer (Layer 4) for January 1998 (First year); Figs 15 and 16 show the same for January 2007 (Tenth year). These figures indicate that the dewatering of the top surface layer progresses northeastward over the simulation period. The water-level contours indicate an approximate 5 m decline over the 10-year simulation.

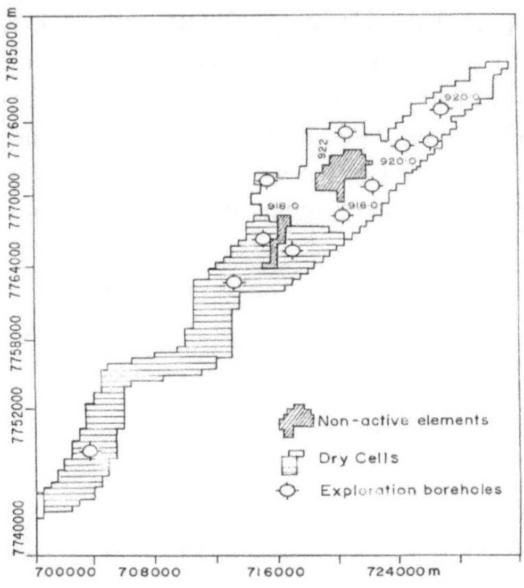

Figure 15. Predicted water level contours in m for January 2007 in layer 1

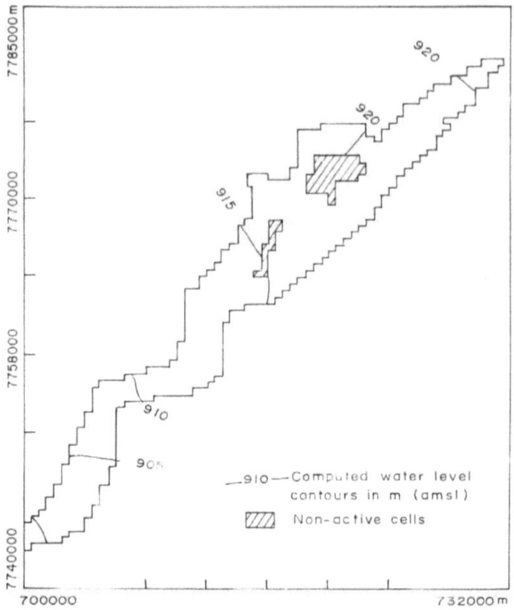

Figure 16. Predicted water level contours in m for January 2007 in layer 4

Figures 17 and 18 are computed water-level hydrographs for boreholes 8255, 8257 and 8274, 8275 respectively. Except one borehole 8255, all other boreholes are completed in the lower semi-confined aquifer (Layer 4) and the first one in the upper semi-confined aquifer (Layer 2). Water levels in the lower semi-confined aquifer appear to drawdown by 7 m at BH 8255 (northeastern portion of the valley) and at BH 8275 (southwestern portion of valley) after 10 years in this abstraction scenario. Also, there appear to be 7 m of drawdown within the upper semi-confined aquifer (Layer 2) at BH 8257. The top layer (Layer 1) becomes dewatered from the Nhabe/Kunyere confluence to the lower end of the wellfield. The correlation of drawdown in the upper and lower aquifer systems implies significant leakage between these two units.

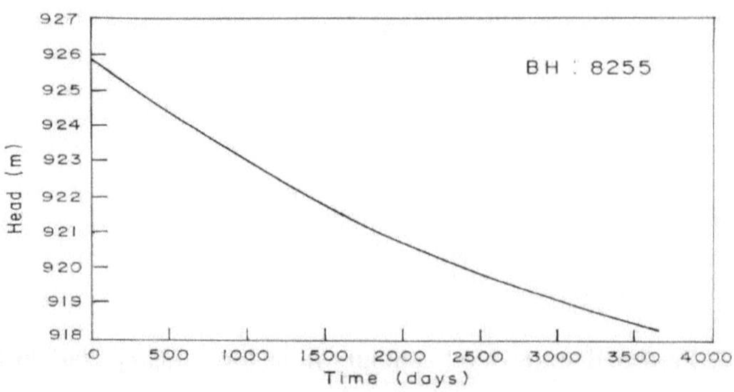

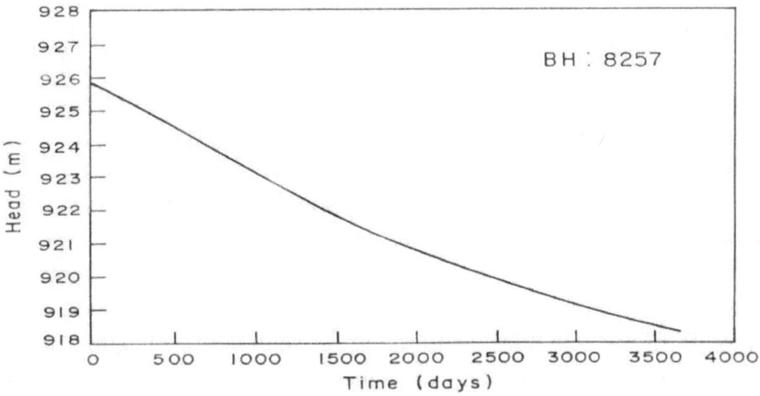

Figure 17. Predicted well hydrograph for boreholes (8255 and 8257)

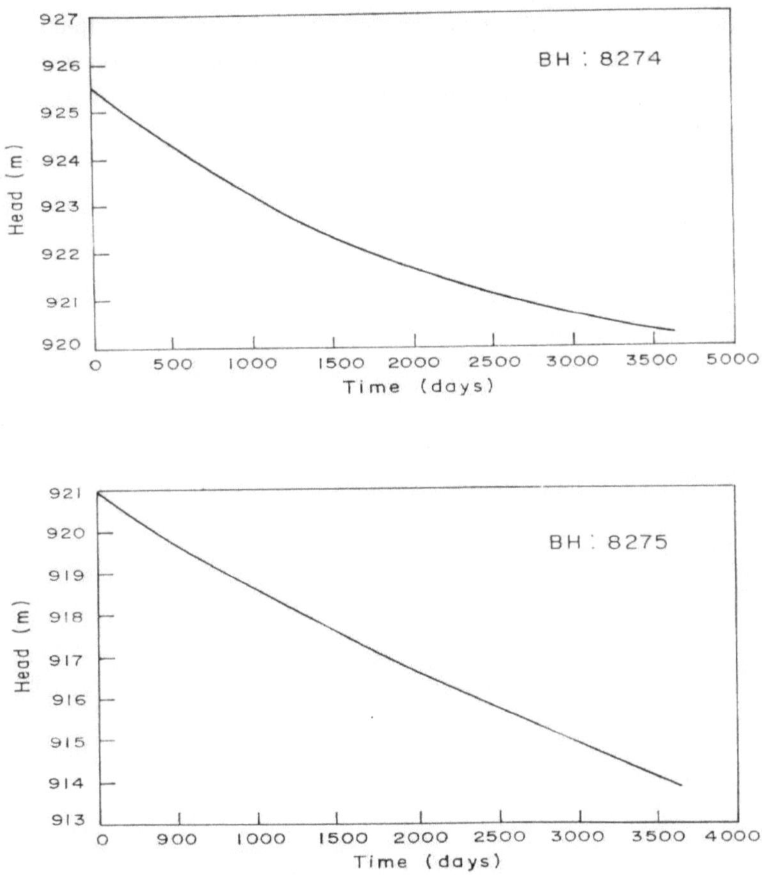

Figure 18. Predicted well hydrograph for boreholes (8274 and 8275)

Prediction Scenario 2

In Prediction Scenario 2, the number of production boreholes placed in the lower semi-confined aquifer within the 93 km^2 area was doubled to 20 boreholes (1.5 km spacing), each pumping at 480 m^3/d over the 10 year simulated period (Fig. 19). The inflows and outflows from the semi-confined aquifer for this scenario are summarized in Table 3.

As in Scenario 1, the bulk of the water supplied to pumping is via vertical downward leakage from the upper semi-confining layer and upper semi-confined aquifer (Column 6 in Table 3). The contribution from aquifer storage in the semi-confined aquifer is still insignificant (Column 8). Contribution from horizontal inflow to the lower semi-confined aquifer increases in this scenario due to lowering the hydraulic head as a result of the higher rates of pumping (Column 7). There is a net downward leakage from the lower semi-confined aquifer to the brackish/saline aquifer in Year 8 of the simulation period (Column 9). After Year 8, there is a 5 to 6 percent contribution to pumping in the lower semi-confined aquifer due to up coning of water from the lower brackish/saline aquifer (Column 12). The calculated water quality impact due to this upconing in Year 10 of the simulation is similar to that in Scenario 1.

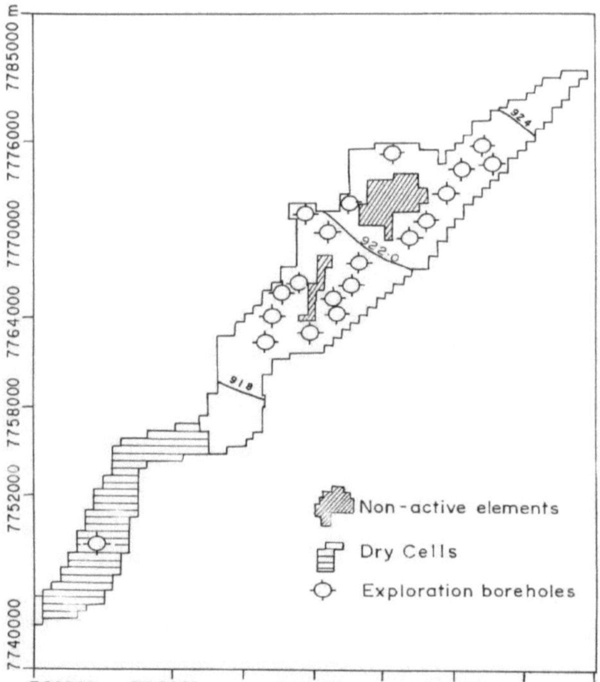

Figure 19. Predicted water level contours in m for January 1998 in layer 1

Table 3. Model prediction scenario 2: Pumpage of 9600 m^3/d from Lower Semi-Confined Aquifer with no river recharge

Year	*Pumping Rate from Lower Semi-Confined Aquifer* (m^3/d)	*Horizontal Outflow* (m^3/d)	*Downward Leakage between lower semi-confined aquifer and brackish/ saline aquifer* (m^3/d)	*Total Outflow from the lower Semi-Confined aquifer* (m^3/d)	*Inflow from Downward Leakage from Upper Semi-Confining Layers and Upper Semi-Confined Aquifer* (m^3/d)	*Horizontal inflow* (m^3/d)	*Contribution from Aquifer Storage in the Semi-Confined Aquifer* (m^3/d)	*Upward leakage between Lower Semi-Confined Aquifer and Brackish/ Saline Aquifer* (m^3/d)	*Total inflow to the Low Semi-Confined Aquifer* (m^3/d)	*Percent Contribution to Pumping from Lower Semi-Confined Aquifer Storage*	*Percent Contribution to Pumping from Upconing*
1	9600	2286	33125	15,011	9877	4948	186	0	15,011	1	0
2	9600	2735	3339	15,674	10,194	5295	185	0	15,674	1	0
4	9600	3153	1420	14,173	9342	4674	157	0	14,173	1	0
8	9600	4417	0	14,017	7925	5454	105	533	14,017	1	5.5
10	9600	4326	0	13,926	7060	6214	79	573	13,926	1	6

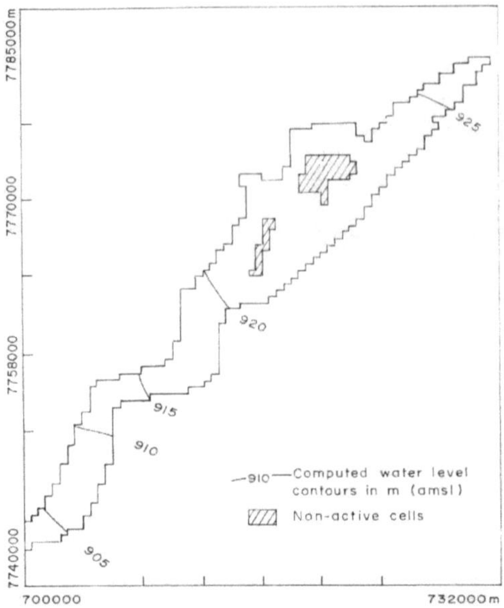

Fig. 20. Predicted water level contours in m for January 1998 in layer 4

Figures 19 and 20 show predicted water-level contours in the top low permeable aquifer (Layer 1) and lower semi-confined aquifer (Layer 4) for January 1998 (Year 1). Figures 21 and 22 show the same for January 2007 (Year 10). The entire length of top layer was de-watered after 10 years of pumping. Water-level declines in the lower semi-confined aquifer appear to be of the order of 10 to 15 m.

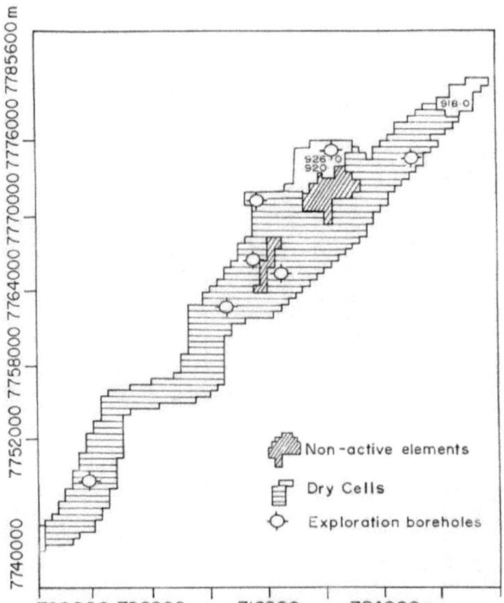

Figure 21. Predicted water level contours in m for January 2007 in layer 1

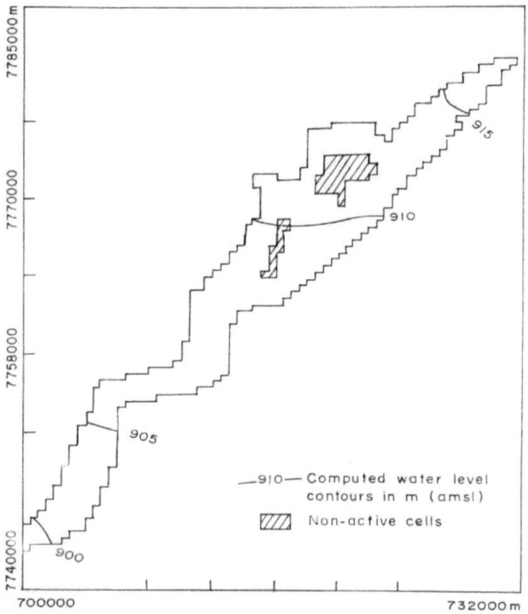

Figure 22. Predicted water level contours in m for January 2007 in layer 4

Figures 23 and 24 are computed water-level hydrographs for BH 8255, 8257 and BH 8274, 8275. Water levels in these boreholes declined by 12 to 13 m after 10 years in this abstraction scenario.

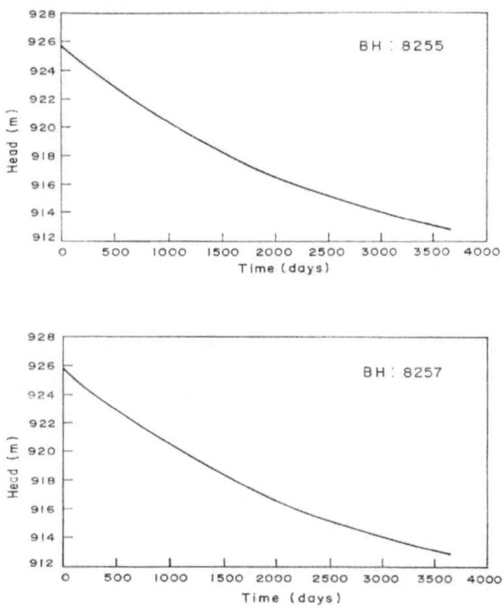

Figure 23. Predicted well hydrograph for boreholes (BH 8255 and BH 8257)

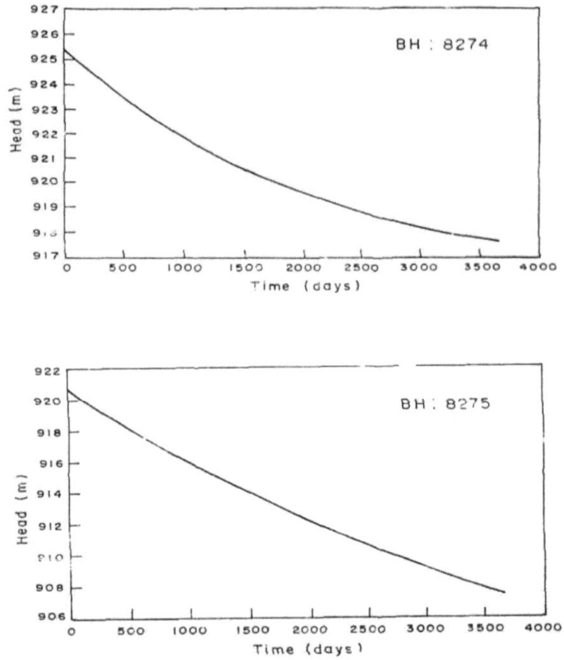

Figure 24. Predicted well hydrograph for boreholes (BH 8274 and BH 8275)

Discussion

The preliminary modeling supports the future consideration of this area for development. It appears that the area can support a considerable amount of freshwater withdrawal without significant water quality changes as a result of upconing. As with most of the explored area, the hydraulic conductivity of the geologic materials that separate the fresh and brackish groundwater is an uncertainty of the model and needs to be investigated in more detail.

Conclusions and Recommendations

The present modeling study is very preliminary and subject to many assumptions and assumed hydraulic parameters. The infiltration due to river flow is not considered in this study, as there was no regular flow for the last three years. The preliminary modeling for this area indicates substantial development potential over a portion of this exploration area. Simulations at pumping 4,800 and 9,600 m^3/day, while producing impacts to the top layer in terms of de-watering, were sustainable if this area receives regular river flow in the future.

REFERENCES

Anderson, T.W. (1968). Electrical analog analysis of groundwater depletion in central Arizona: U.S. Geological Survey, Water Supply Paper 1860.

Anderson, T.W. (1972). Electrical-analog analysis of the hydrologic system, Tucson Basin, Southeastern Arizona. U.S. Geological Survey Water-Supply Paper 1939-C, pp. 34.

Aziz, K. and Settari, A. (1972). A new iterative method for solving reservoir simulation equations. *Jour. Can. Tech.*, **11:** 62-68.

Battermann, G. and Boochs, P.W. (1976). Simulation of groundwater flow with a computer controlled R-C network. Simulation of systems. L. Dekker (Ed.), North Holland Publishing Company, 715-720.

Bear, J. (1972). Dynamics of fluids in porous media. American Elsevier Publishing Company, Inc.

Bear, J. and Schwartz, Y. (1966). "Electric analog for regional groundwater studies," Water Planning for Israel, Ltd., Tel Aviv, Israel, pp. 609.

Bermes, B.J. (1960). An electric analog model for use in quantitative hydrologic studies: U.S. Geological Survey, Mimeographed Report.

Bobba, A.G. and Singh, V.P. (1995). Groundwater contamination modelling. Environmental Hydrology. V.P. Singh (Ed.),. Kluwer Academic Publishers. 225-319.

Bredehoeft, J.B. and Pinder, G.F. (1970). Digital analysis of areal flow in multiaquifer groundwater system. *Water Resources Research*, **6:** 883-888.

Bredehoeft, J.D. and Pinder, G.F. (1973). Mass transport in flowing groundwater. *Water Resources Research*, **9(1):** 194-210.

Brown, R.H. (1962). Theory of Aquifer Tests. Ferris, J.G., Knowles, D.B., Brown, R.H. and Stallman, R.W. (Eds), U.S. Geol. Survey. Water Supply Paper 1536-E, Washington D.C.

Bruch, J.C., Jr. (1970). Two-dimensional dispersion experiments in a porous medium. *Water Resources Research*, **6:** 791-800.

Cahill, J.M. (1973). Hydraulic sand model studies of miscible study flow. U.S. Geological Survey, *Jour. Research*, **6:** 243-250.

Carslaw, H.S. and Jaeger, J.C. (1959). Conduction of Heat in Solids. Clarendon Press, Oxford, England.

Cole, J.A. and Blair, A.H. (1967). Electrical analog techniques in groundwater resource development: Int. Conf. Water Peace, Washington, D.C. **3:** 28-39.

Collins, M.A. et al. (1972). Hele-Shaw model of long Island aquifer system. *Jour. Hydraulics Div., Amer.Soc. Civil. Engrs.*, **98(HY9):** 1701-1714.

Columbus, N. (1966). The design and construction of Hele-Shaw models. *Groundwater*, **4(2):** 16-22.

Dershowitz, W.S., Lee, G. and Geier, J.E. (1991). Fracman: Version B2.3. Interactive Discrete Fracture Data Analysis, Geometric Modeling, and Exploration Simulation. User Documentation. Golder Associates Inc., Redmond, WA., pp. 913-1358.

Emery, P.A. (1966). Use of analog model to predict stream flow depletion. *Groundwater*, **4(4):** 13-19.

Forchheimer, P. (1886). Uber die ergiebigkeit von brunnen, anlagen and sickerschlitzen. *Zeitsch. Archit. Ing. Ver. Hannover*, **32:** 539.

Groundwater Vistas (1998). Groundwater Software by Environmental Simulation International, UK.

Guiguer, N. and Franz, T. (1996). Visual Modflow. Waterloo Hydrogeologic Software, Waterloo, Ontario, Canada.

Gupta, C.P., Thangarajan, M. and Gurunadha Rao, V.V.S. (1979). Electric analog model study of aquifer in Krishni-Hindon interstream region, Uttar Pradesh, India. *Ground Water*, **17(3):** 284-290.

Hartley, L.J. (1998). NAPSAC Technical Summary Document. AEA Technology, Harwell, Didcot, UK, pp. 1-42.

Hele-Shaw, H.S. (1897). Experiments on the nature of the surface resistance in pipes and ships. *Trans. Inst. Naval Architects*, **39:** 145-156.

Javanadal, I. (1984). Methods of Evaluating Vertical Groundwater Movement. Technical report of Lawrence Berkeley Laboratory, University of California, Berkeley, 1-134.

Jorgenson, and Donald, G. (1975). Analog model studies of groundwater hydrology in the Houston District, Texas. Texas Water Development Report 190.

Karplus, W.J. (1958). Analog simulation Solutions of Field Problems. McGraw-Hill Series in Information Processing and Computers, McGraw-Hill, New York.

Kimbler, O.K. (1970). Fluid model studies of the storage of freshwater in saline aquifers. *Water Resources Research*, **6:** 1522-1527.

Konikow, L.F. and Bredehoeft, J.D. (1978). Computer model of two dimensional solute transport and dispersion in groundwater: Techniquies of water-resources investigations of the USGS Chapter C2, Book **7:** 1-79.

Konikow, L.F. and Grove, D.B. (1977). Derivation of equations describing solute transport in groundwater. US Geol. Survey Water Resources Investigations, 77-19, 30.

Kraijenhoff van de Leur (1962). Some effects of the unsaturated zone on non-steady free-surface groundwater flow as studied in a scaled granular model. *Jour. Geophysical Research*, **67:** 4347-4362.

Marino, M.A. (1967). Hele-Shaw model study of the growth and decay of groundwater ridges. *Jour. Geophys. Res.*, **72:** 1195-1205.

Maun Water Supply Hydrogeological Survey (1988). BRGM Technical Report, Vol. 1.

McDonald, M.G. and Harbaugh, A.W. (1988). A modular three-dimensional finite difference groundwater flow model. US Geological Survey Open-file Report 83-875, 528.

McDonald, Michael, G. and Harbaugh, Arlen W. (1988). A modular three-dimensional finite difference groundwater flow model. USGS Technical Report on Modeling Techniques, Book 6.

Moore, J.E. and Wood, L.A. (1967). Data requirements and preliminary results of an analog model evaluation—Arkansas river valley in eastern Colorado. *Groundwater*, **5(1):** 20-23.

Nilson, Guigner and Franz, Thomas (1996). Visual Modflow—Input-Output graphic Software for Modflow. Hydrologic Inc. Waterloo, Ontario, Canada.

Patten, E.P. Jr. (1965). Design, construction and use of electric analog model. *In:* Analog model study of groundwater in Houston District. L.A. Wood and R.K. Gabrysh (Eds), Texas Water commission Bull. No. 65089, 1-103.

Peaceman, D.W. and Rachford Jr., H.H. (1955). Numerical solution of parabolic and elliptic differential equation. *Soc. Indust. Appl. Math. J.*, **3:** 28-41.

Pinder, G.F. and Bredehoeft, J.D. (1968). Application of the digital computer for aquifer evaluation. *Water Resources Research*, **4:** 1069-1093.

Pinder, G.F. and Gray, W.G. (1977). Finite Element simulation in Surface and Subsurface Hydrology. Academic Press, New York, 295 pp.

Prickett, T.A. and Lonnquist, C.G. (1971). Selected digital computer for groundwater resource evaluation. Ill. *State Water Surv. Bull.*, **55:** 62.

Prickett, T.A. (1975). Modeling techniques for groundwater evaluation. *In:* Chow, V.T. (ed.), Advances in hydroscience. Academic Press, **10:** 1-143.

Reddel, D.I. and Sunada, D.K. (1970). Numerical simulation of dispersion in groundwater aquifers. *Hydrology Paper*, **41:** Colorado State University, Fort Collins.

Reed, J.E. (1972). Analog simulation of water level declines in the Sparta Sand, Mississippi embayment. U.S. Geological Survey, Hydrol. Invest., Atlas HA-434.

Remson, I., Hornberger, G.M. and Molz, F.J. (1971). Numerical Methods in Subsurface Hydrology, John Wiley and Sons, New York.

Rumbaugh, J.O. (1991). Quickflow Analytical 2D Groundwater Flow Model, Version 1.0. Geraghty and Miller Inc., Reston, Virginia, USA.

Rushton, K.R. and Ash J.E. (1975). Modeling of aquifer behaviour for long time periods using interactive analogue-digital computers. International conference on mathematical models for environmental problems, Southampton.

Rushton, K.R., Wedderburn, L.A. (1973). Starting conditions for aquifer simulations. *Ground Water* **11(1):** pp. 37-42.

Rushton, K.R. and Redshaw, S.C. (1979). Seepage and Groundwater Flow. John Wiley and Sons Ltd., 1-330.

Segerlind, L.J. (1976). Applied Finite Element Analysis. John Wiley and Sons Inc., 422.

Shamir, U. and Dagan, G. (1971). Motion of seawater interface in coastal aquifers—A numerical solution. *Water Resources Research*, **7:** 644-658.

Shashe Well Field Conceptual Model, Interim Report (Appendix G) 1996, by Eastend Investments (Pty.) Ltd., Gaborone, Botswana.

Skibitzke, H.E. (1960). The use of analog computers for studies in groundwater hydrology. *J. Inst. Water Eng.*, **17**.

Stallman, R.W. (1963). Computation of Groundwater velocity from temperature data. U.S. Geological Survey Water-Supply Paper 1544-H, 36-46.

Stallman, R.W. (1963). Electric analog of three-dimensional flow to wells and its application to unconfined aquifers. U.S. Geological Survey Water-Supply Paper 1536-H, 205-242.

Stone, H.L. (1968). Interactive solution of implicit approximation of multidimensional partial differential equation. *SIAM Jour. of Numerical Analysis*, **5.3:** 530-558.

Strack, O.D.L. (1989). Groundwater Mechanics Computer methods for transient analysis of water table aquifers. *Water Resources Research*, **5:** 144-152.

Thangarajan, M. (1975). R-C Analog Modeling Technique for Groundwater Management. NGRI, Technical Report, No. 4938/G, 1-19.

Thangarajan, M. (1983). Electric Analog Model Study of Some Representative Groundwater Basins in India, Ph.D. Dissertation, Indian School of Mines, Dhanbad, India, 198.

Thangarajan, M. (1999). Numerical simulation of groundwater flow regime in a weathered hard rock aquifer. *Journal of Geological Society of India*, **53(5):** 561-570.

Thangarajan, M. (1999). Modeling multi-layer aquifer system to evolve pre-development management schemes. *Environmental Geology*, **38(4):** 285-295.

Thangarajan, M. (2004). Regional Groundwater Modeling. Capital Publishing Company, New Delhi, 336 pp.

Todd, D.K. (1954). Unsteady flow in porous media by means of a Hele-Shaw viscous fluid model. *Trans. Amer. Geophysical Union*, **35:** 905-916.

Todd, D.K. (1980). Groundwater Hydrology, Second Edition. John Wilely and Sons, (Printed in Singapore, 2001), 1-535.

Toth, J. (1962). A theory of groundwater motion in small drainage basins in Central Alberta, Canada. *Jour. Geophysical Research*, **67:** 4375-4387.

Toth, J. (1963). A theoretical analysis of groundwater flow in small drainage basins. *Jour. Geophysical Research*, **68:** 4795-4812.

Trescott, P.C. and S.P. Larson (1977). Comparison of iterative methods of solving two-dimensional groundwater flow equations. *Wat. Resour. Res.*, **13(1):** 125-136.

Trescott, P.C. (1976). Documentation of Finite-Difference Model for Groundwater flow. U.S. Geological Survey Open file Report 75-438, 1-32.

UNESCO (1999). Water Resources of hard rock aquifers in arid and semi-arid zone. J.W. Lloyd (ed.).

Voss, C.I. (1984). SUTRA—A finite element simulation model for saturated-unsaturated fluid density dependent groundwater flow with energy transport or chemically reactive single species solute transport. U.S. Geol. Surv., Water Resources Invest. Rep., 84-4239.

Walton, W.C. and Prickett, T.A. (1963). Hydrogeologic Electric Analog Computers. Proceedings of American Society of Civil Engineers, 67-91.

Watts, J.W. (1971). An interactive matrix solution method suitable for anisotropic problems. *Soc. Petrol. Engg. Jour.*, **11:** 47-51.

Watts, J.W. (1973). A method for improving line successive over-relaxation in anisotropic problems: A theoretical analysis. *Soc. Petrol. Engg. Jour.* **13:** 105-118.

Weinstein, H.G., Stone, H.L. and Kwan, I.V. (1970). Simultaneous solution of multiphase reservoir flow equations. *Soc. Petrol. Engg. Jour.*, **10:** 99-110.

Willis, R.W. and Yeh, W.W.G. (1987). Groundwater systems planning and Management. Prentice Hall, 416 pp.

Yeh, G.T. and Huff, D.D. (1983). FEW: A finite element model of water flow through aquifers. Oak Ridge. Nat. Lab. Envir. Sci. Div. Pub. 2240, Oak Ridge, Tennessee, USA.

Yen, B.C. and Haise, C.H. (1972). Viscous flow model for groundwater movement. *Water Resources Research*, **8:** 1299-1306.

Zheng, C. (1995). Analysis of particle tracking errors associated with spatial discretization. *Ground Water*, **32(5):** 821-828.

Zienkiewicz, O.C. (1971). The Finite Element Method in Engineering Science. McGraw-Hill, London, 521 pp.

9

Model Calibration and Issues Related to Validation, Sensitivity Analysis, Post-audit, Uncertainty Evaluation and Assessment of Prediction Data Needs

Claire R. Tiedeman and Mary C. Hill[1]

U.S. Geological Survey, Menlo Park, California, USA
[1]U.S. Geological Survey, Boulder, Colorado, USA

INTRODUCTION

When simulating natural and engineered groundwater flow and transport systems, one objective is to produce a model that accurately represents important aspects of the true system. However, using direct measurements of system characteristics, such as hydraulic conductivity, to construct a model often produces simulated values that poorly match observations of the system state, such as hydraulic heads, flows and concentrations (for example, Barth et al., 2001). This occurs because of inaccuracies in the direct measurements and because the measurements commonly characterize system properties at different scales from that of the model aspect to which they are applied. In these circumstances, the conservation of mass equations represented by flow and transport models can be used to test the applicability of the direct measurements, such as by comparing model simulated values to the system state observations. This comparison leads to calibrating the model, by adjusting the model construction and the system properties as represented by model parameter values, so that the model produces simulated values that reasonably match the observations.

This article presents ideas and methods for developing and calibrating groundwater flow and transport models, and for evaluating the calibrated model, its predictions and prediction uncertainty. The methods are intended to help modelers effectively use data on the system characteristics and system state together with conceptual and numerical models, to produce the best possible model of the simulated system. The methods also address diagnosing and reducing model error, which in flow and transport models stems from many sources. These include data error and deficiencies, errors in system conceptual models, limited capabilities of numerical models, and errors that arise when solving these numerical models.

Model calibration methods presented in this article include (1) parameter estimation by nonlinear regression, (2) sensitivity analysis for evaluating the information content of data and identifying existing measurements that dominate model development, (3) model development and calibration strategies for transport models, and (4) model evaluations for diagnosing model error and quantifying parameter uncertainty. Predictive analysis methods are also presented in this article, which focus on assessment of data needs for improving the reliability of model predictions. Three groundwater transport models were constructed to represent a laboratory-scale, a site-scale, and a regional-scale problem to demonstrate the selected methods.

Each of the methods discussed can be placed in the context of the groundwater flow and transport modeling process as shown in Fig. 1. This figure shows how the model, which is characterized by defined parameters, quantitatively connects the system information and the observations to the predictions and their uncertainty. The entities **observations**, **parameters** and **predictions** are in bold type because the quantitative links between these entities are more direct than the link to system information. These entities are directly used by or produced by the model, whereas the system information is used in model development more indirectly. Many of the methods presented take advantage of the links shown in Fig. 1. The objective of the modeling process is to use the system information and observations as effectively as possible to obtain a model that is as constrained as possible. Though not guaranteed, this process is likely to produce predictions that are as accurate as possible.

The methods and ideas presented have broad applicability to many types of scientific and engineering models. The examples cited throughout this article and the studies described subsequently in more detail, focus on applications of inverse and associated methods to models of saturated groundwater flow and transport. For applications of inverse methods to models of flow and transport in the unsaturated zone, see, for example, Zou et al. (2001), Jacques et al. (2002), Kosugi and Inoue (2002), Šimùnek et al. (2002), Olyphant (2003), and references cited therein.

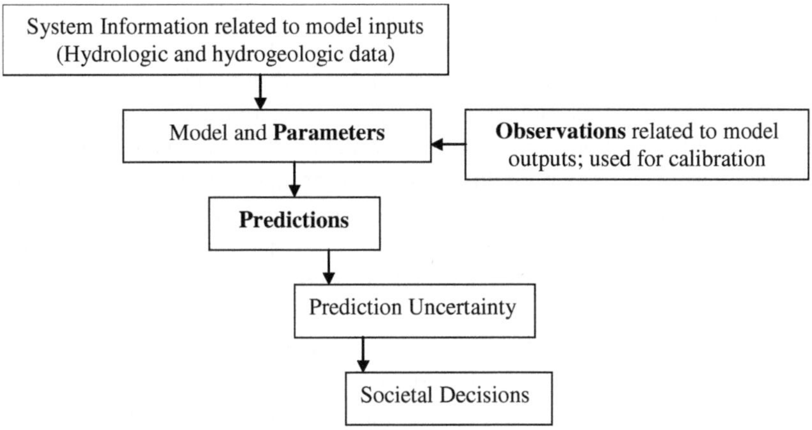

Figure 1. Groundwater flow and transport modeling process, showing how the model directly connects the observations, parameters and predictions.

Many of the methods described in this article have been implemented in the well-documented, generally applicable, public domain computer programmes MODFLOW-2000 (Harbaugh et al., 2000; Hill et al., 2000), UCODE (Poeter and Hill, 1998) and PEST (Doherty, 1994, 2003). Each of these codes performs inverse modeling, posed as a parameter-estimation problem, by calculating parameter values that minimize a weighted least-squares objective function using nonlinear regression. MODFLOW-2000 is an inverse code applicable only to three-dimensional transient groundwater flow and steady-state advective transport simulated using the finite-difference numerical method. The software and its documentation can be downloaded from the following web sites:

<*http://water.usgs.gov/nrp/gwsoftware/modflow2000/modflow2000.html*>
and

<*http://water.usgs.gov/pubs/twri/twri6a1*>

UCODE and PEST are universal inverse codes with broad applicability, and can be used with any simulation model that has ASCII input and output files. The codes and their documentation can be downloaded from the following web sites:

UCODE: <*http://water.usgs.gov/software/ucode.html*>
PEST: <*http://www.sspa.com/PEST/download.html*>

Because they are model-independent, UCODE and PEST can be used to calibrate any type of groundwater flow or transport model. Recent applications of UCODE to transport models include studies by Yager (2002), Ghandi et al. (2002a, 2002b), and Essaid et al. (2003).

MODEL CALIBRATION USING INVERSE MODELING

Inverse modeling refers to formal methods that attempt to estimate parameter values of a mathematical model of a system, given a set of relevant observations of the system state and characteristics. In the context of Fig. 1, the parameter-estimation procedure connects the model and its parameters with the observations used for calibration. The benefits of inverse modeling and associated methods include (1) clear determination of parameter values that produce the best possible fit to the available observations; (2) diagnostic statistics that quantify the quality of calibration and data shortcomings and needs; (3) inferential statistics that quantify reliability of parameter estimates and predictions; and (4) identification of issues such as model limitations that are easily overlooked when calibration is conducted using trial and error methods alone (Poeter and Hill, 1997).

For a given inverse model application, the underlying assumption is that the model correctly represents all system aspects that are not being changed using estimated parameter values. The validity of the model depends on how well this assumption is satisfied. Commonly, many system processes and characteristics that are important to the observed and predicted quantities cannot be determined accurately by system information. It is important to define model parameters to represent as many of these processes and characteristics as possible, and the utility of the methods presented in this article is enhanced by doing so. However, it is difficult in practice to represent some system aspects, such as hydrogeologic unit geometries, with model parameters. Thus, as part of the model calibration procedure, it is important to use information about the model fit to observations and other types of information to guide manual modifications to these system characteristics, in addition to using formal inverse modeling procedures to estimate parameter values.

Parameter Estimation by Nonlinear Regression

One common method for estimating model parameter values is nonlinear regression. Regression and associated methods have been used successfully to calibrate numerous groundwater flow and transport models over the past two decades. In regression, an objective function is used to compare the model with two types of data: (1) observations that are equivalent to model simulated values such as hydraulic heads and concentrations, and (2) measurements of system characteristics (called prior information) that are represented by model parameter values. Parameter values that produce the best model fit are defined as those which produce the smallest value of the objective function.

The weighted least-squares objective function is commonly used and can be expressed as:

$$S(\boldsymbol{b}) = \sum_{i=1}^{ND+NPR} \omega_i \left[y_i - y_i'(\boldsymbol{b}) \right]^2 \qquad (1)$$

where ND is the number of observations, NPR is the number of prior information equations, y_i is the ith observation or prior information value being matched by the regression; $y_i'(\boldsymbol{b})$ is the simulated value that corresponds to y_i and which is a function of $\boldsymbol{b}$, a vector containing values of each of the *NP* parameters being estimated; and ω_i is the weight for the ith contribution to the objective function. Examples of observed quantities y_i in groundwater flow and transport systems are hydraulic heads, flows to or from boundaries such as rivers and lakes, advective transport distances and travel times, concentrations of aqueous and sorbed species such as organic compounds, and prior information on model parameters such as field estimates of hydraulic conductivity from an aquifer test or dispersivity from a tracer test.

The difference $y_i - y_i'(\boldsymbol{b})$ in equation 1 is the residual for the ith observed or prior value, and $\omega_i^{1/2}[y_i - y_i'(\boldsymbol{b})]$ is the corresponding weighted residual. Weights are commonly calculated as $1/\sigma_i^2$, where σ_i is the standard deviation of the error associated with observation i. This weighting strategy achieves two goals: (1) it produces terms of the objective function that are dimensionless and can thus be summed for different observation types, and (2) it reduces the influence of less accurate observations and increases the influence of more accurate observations. For a discussion of issues related to weighting observations in nonlinear regression, see Hill and Tiedeman (2003).

A modified Gauss-Newton method is commonly used for solving the nonlinear regression to estimate the parameters $\boldsymbol{b}$ that minimize the objective function in equation 1. This method performs well relative to alternatives (Cooley, 1985; Hill, 1990 and Cooley and Hill, 1992) in that it is relatively robust and computationally efficient. To estimate the optimal parameter values it uses observation sensitivities, $\partial y_i'/\partial b_j$, which equal the derivative of a simulated equivalent of an observation, y_i', with respect to a model parameter value, b_j. These sensitivities are used to calculate the gradient of the objective-function surface, which the modified Gauss-Newton method uses to determine how to proceed toward the minimum objective-function value. In addition to being used to solve the inverse problem, the observation sensitivities can be used to evaluate the support provided by the observations for the estimation of the parameter values. When combined with prediction sensitivities, which equal the derivative of the simulated predictions with respect to the parameters, observation sensitivities also can be used to evaluate the importance of observations and defined parameters to model predictions.

Sensitivity Analysis for Assessing the Information that Observations Provide about Parameter Values

In models with many defined parameters, the observations may not provide enough information to estimate all parameters by regression. Observation sensitivities, defined above, are powerful measures for determining which model parameters can be estimated, and quantitatively link the observations and parameters shown in Fig. 1. To effectively compare these sensitivities among different observation types and different parameters, scaling is needed. One useful scaling produces dimensionless scaled sensitivities (*dss*), which for a diagonal weight matrix are calculated as (Hill, 1998):

$$dss_{ij} = \left(\frac{\partial y_i'}{\partial b_j} \right) \Bigg|_{\mathbf{b}} |b_j| \, \omega_{ii}^{1/2} \tag{2}$$

where dss_{ij} is the dimensionless scaled sensitivity of observation y_i with respect to parameter b_j. These *dss* can be used in two important ways. First, they can be used to compare the importance of different observations to the estimation of a single parameter b_j. Observations with large dss_{ij} are likely to provide more information about parameter b_j compared to observations associated with small dss_{ij} (large and small in absolute value). Second, the *dss* can be used to compare the importance of different parameters to the calculation of a single simulated value y_i'. Parameters that are more important to the simulated value have dss_{ij} that are larger in absolute value.

The total amount of information provided by the observations for estimation of a particular parameter can be calculated from the *dss* for that parameter. This measure denotes the composite scaled sensitivity (*css*) (Hill, 1998), and is calculated as:

$$css_j = \left[\sum_{i=1}^{ND} (dss_{ij})^2 \Big|_{\mathbf{b}} \Big/ ND \right]^{1/2} \tag{3}$$

Composite scaled sensitivities are often used in a comparative manner, whereby larger values indicate more important parameters, or parameters for which the observations provide more information. If there are *css* that are more than two orders of magnitude less than the largest value, regression often will have trouble converging. In this situation, parameters with small *css* may need to be specified, rather than estimated by the regression.

In addition to their utility for evaluating observations that dominate model calibration, the *dss* and *css* are also used to assess data needs for improving model predictions.

Strategies for Calibrating Transport Models

Groundwater systems are commonly characterized by complex geologic and hydrologic conditions, but data on these conditions and on the system state are typically scarce because of the inaccessibility of the subsurface. As shown in Fig. 1, these system data are the entire basis of the conceptual and numerical models that are developed. Because of the scarcity of these data, many aspects of the developed model are typically uncertain, and there are commonly relatively few observation data available for calibration. These circumstances cause calibration of groundwater models to be difficult and often to be plagued by problems of insensitivity, nonuniqueness and instability. Insensitivity occurs when the observations do not contain enough information to support estimation of the parameters, and can be diagnosed using the *dss* and *css*. Nonuniqueness occurs when different combinations of parameter values match the observations equally well, and can be diagnosed using parameter correlation coefficients. Instability occurs when slight changes in quantities such as parameter values or observations radically change inverse model results, and can be diagnosed during application of the regression method. Hill (1998) presents additional methods and guidelines that can be used to facilitate groundwater model calibration and evaluation, and to help diagnose and alleviate these types of problems.

Application of inverse methods to groundwater transport models can be more difficult than to groundwater flow models because: (1) transport model dependent variables such as concentrations are typically more difficult to simulate than flow model dependent variables such as hydraulic heads, and inaccurate numerical solutions of transport can adversely affect parameter estimation and, to a lesser extent, statistics calculated using simulated sensitivities (Mehl and Hill, 2001); (2) parameters added to represent porosity, dispersivity and, in some cases, more detail in the hydraulic conductivity field can produce greater problems with parameter insensitivity and nonuniqueness; (3) differences between the true and simulated systems can cause much greater differences between simulated and observed transport than between simulated and observed hydraulic heads and flows, and (4) simulation of contaminant transport can be complicated by uncertainty about the source location and release history of the contaminants. These special circumstances need to be recognized and carefully addressed when calibrating groundwater transport models by inverse methods. Strategies for overcoming these difficulties are presented below.

Model Selection and Development

The transport of constituents in natural and engineered systems is the result of many processes, generally including advection, dispersion, retardation, chemical reaction with surrounding material (for example, rocks and sediments in groundwater systems) and other transported constituents, and so on. The first issue that needs to be considered when observations or predictions of transport are involved is which of these processes should be included in the simulation model. In groundwater systems, the options for simulating transport are generally as follows:

Option 1: Use advection to simulate the mean transport travel time and direction;

Option 2: Use advection and dispersion to simulate the arrival of low concentrations at the front of the plume or enduring low concentrations at the tail of the plume; and

Option 3: Use advection, dispersion, reactions and other mechanisms to account for additional processes that can affect arrival times and duration of low concentrations.

Decisions about which transport processes to simulate need to be based on considerations such as the conceptual model of transport; the objective of the transport modeling; the level of detail with which site geologic and hydrologic data are represented in the model; the scale of the simulations, including whether the model is one-, two-, or three-dimensional; the size of the model cells or elements; and model execution times. For example, two of the field applications presented in this article as case studies illustrate transport simulations in models with very different scales and modeling objectives. In a site-scale model of the Grindsted landfill in Denmark with grid spacing as small as one metre, advection and dispersion were chosen to simulate transport observations. In a regional model of the Death Valley region with 1500-metre grid spacing, advective transport was chosen for simulating transport predictions.

Development of models for simulating transport observations or predictions usually requires additional model input, compared to that required for flow models. If only advection is considered, porosity is the sole additional model parameter required. If dispersive processes are included, dispersivity in up to three spatial directions is needed. If multicomponent reactive transport is considered, there are potentially a large number of model inputs (for example, see Prommer et al., 2003). Problems with parameter insensitivity and correlation can be exacerbated in transport models because of the increased number of model inputs for which parameters can be defined. Multicomponent reactive transport models can be especially problematic, because each component is potentially characterized by several separate transport properties.

A number of recent applications of nonlinear regression to transport models have addressed problems with parameter insensitivity and correlation by carefully selecting which parameters to estimate, or by simplifying the simulated processes appropriately. To facilitate calibrating a model of vapour phase hydrocarbon transport, Gaganis et al. (2002) grouped sets of individual hydrocarbon constituents with similar thermodynamic properties into composite constituents. They defined thermodynamic parameters associated with these composite constituents, and thus significantly reduced the total number of model parameters. These transport parameters were estimated using concentrations of the composite constituents as the calibration observations. Ghandi et al. (2002b) calibrated a model of cometabolic trichloroethylene biodegradation by retaining the full model complexity and large number of parameters, and fixing several insensitive parameters at values obtained from laboratory experiments or previous modeling studies. Essaid et al. (2003) made two simplifications when applying regression to a model of hydrocarbon dissolution and biodegradation, because of high parameter correlations. They represented biodegradation by using first-order reactions rather than Monod kinetics, and they defined a single dissolution rate parameter for all hydrocarbon components, which was supported by independent experiments. Even with these simplifications, the model retained a considerable amount of complexity, and the observation data supported estimation of a large number of parameters including individual first-order anaerobic biodegradation rates for all of the hydrocarbon components.

Differences between the model and the true system—that would be inconsequential when simulating flow—can be important when simulating transport. This is partly because transport is sensitive to variations in system characteristics over a wider range of scales, which can cause calibration of transport models to be more difficult than calibration of flow models. The importance of small scale features to transport has been demonstrated for groundwater systems by a number of authors, including Zheng and Gorelick (2003). For example, omitting geologic features smaller than a certain scale might be acceptable when calibrating a flow model because of insensitivity of hydraulic heads to these features, but transport might be highly sensitive to the small-scale geologic detail. Large-scale differences between simulated and true conditions can also be more problematic for transport calibration. For example, inaccurate location of the boundary between two hydrogeologic units with very different hydraulic conductivities might have little effect on simulated heads at moderate distances from the boundary. However, travel times of constituents across the boundary could be very sensitive to the boundary location. Calibration of transport models with these errors can result in problems such as unreasonable parameter estimates, a poor fit to transport observations, or difficulty in converging to optimal parameter estimates.

Numerical Issues: Model Execution Time and Accuracy

The three options for simulating transport that are defined in the previous section each have issues related to model execution time and accuracy of the numerical simulation.

Advective transport (Option 1) is commonly simulated by particle tracking, a method that generally requires only slightly more execution time and computer resources than does simulating groundwater flow without transport. However, to achieve accurate results with this method, a more finely discretized model grid may be needed than for flow modeling alone. For inverse modeling applications with advective-transport observations, inaccurate particle positions can result in meaningless sensitivities and parameter estimates. Zheng (1994) discusses two modeling situations that can cause inaccurate particle tracking results when the model grid discretization is too coarse: the presence of a weak sink or source, and a vertically distorted grid. He presents computational methods for minimizing the particle tracking errors that arise from these problems, which can be used when it is impractical to discretize the grid more finely.

An additional problem with applying regression to models with advective transport observations occurs if a particle exits the grid when the model is solved for a particular set of parameters. In this case, advective transport sensitivities cannot be calculated, and thus the regression lacks the information needed to obtain a more accurate match to observed transport. Anderman and Hill (2001) address this problem by calculating a projected position for any particle that leaves the grid, using the particle velocity at the point where it exits the model. Sensitivities are then calculated using the projected position.

Simulating advection and dispersion (Option 2) requires substantially more execution time than does simulating advection only, and including reactions and other transport mechanisms (Option 3) generally requires even greater execution times. For both of these options, numerical accuracy issues are typically more problematic than for simulation of advective transport. These accuracy problems are inherent in the solution of the advective-dispersive equation in many scientific fields. From the perspective of sensitivity analysis and calibration, one concern is that slight changes in the time-step size can cause slight changes in the concentration simulated at a particular location and time. These slight changes can cause large changes in sensitivities calculated by perturbation methods. When the transport-step size is automatically calculated by the modeling software to satisfy, for example, the Courant number criterion (e.g. Zheng and Bennett, 2002), the step size generally will be different for the simulations needed to compute the perturbation sensitivities for a single parameter (two simulations are needed for forward- or backward-difference sensitivities; three are needed for central-difference sensitivities). This problem can be eliminated by (1) executing the transport code for the base set of parameters, (2) determining a suitably small transport-step size, and (3) imposing this step size rather than allowing the programme to calculate it. In the case of parameter estimation, the parameter values and simulated flow field will change from one regression iteration to the next, and thus the suitable transport-step size will also change. Thus, it is important to ensure that the imposed transport-step size remains valid. This generally can be accomplished by defining a conservative Courant number for the base run.

When simulating advection and dispersion, or when including other mechanisms, similar issues can arise if sensitivities are calculated with different transport solvers or with different grid discretizations. Although it is unlikely that the solver or discretization will change during a single simulation, they may change over the course of a modeling project. It is important that all model sensitivities be recalculated if a different solver is used or if the grid is modified, because the solution method and spatial discretization can have a significant effect on transport model output.

Additional problems with the calculation of concentration sensitivities can arise when using Options 2 or 3 with numerical methods that use particles to represent some or all of the transport processes (e.g. method of characteristics, random walk). By these methods, the concentrations in a model cell (or element) are calculated from masses or concentrations associated with the particles present in that cell and, generally, the concentration in the cell stays the same if no particles leave or enter the cell. Over a period of time the simulated concentration at any cell tends to be accurate on average, but is prone to oscillate as particles enter and leave. Use of perturbation methods to calculate sensitivities can result in artificially small concentration sensitivities. For example, if a small perturbation in a parameter value causes simulated particles to move but not leave or enter a given cell, then the concentration will not change and the calculated sensitivity of zero will be too small. Alternatively, artificially large sensitivities can result when a small perturbation in a parameter value causes one or more particles to move from the edge of one cell into another cell. In these cases, sensitivities may not vary smoothly from one time step to the next, or from one parameter-estimation iteration to the next, which can cause significant problems with convergence of the nonlinear regression procedure. Refining the model grid so that cell or element sizes are smaller can help resolve this problem, but is not always a practical solution. Sonnenborg et al. (1996) addressed this problem by calculating the sensitivities using concentrations averaged over time periods that were longer than the model time steps. This resulted in more accurate sensitivities that changed more smoothly over time, as would be expected.

Numerical dispersion is a common problem when using Options 2 or 3 to simulate transport and its presence can have a significant effect on transport model parameters that are estimated by regression. Mehl and Hill (2001) illustrated this by simulating a two-dimensional laboratory experiment constructed of discrete, randomly distributed, homogeneous blocks of sands. These simulations used the computer programme MT3DMS (Zheng and Wang, 1998), with an added predictor-corrector (P-C) method. Mehl and Hill (2001) first demonstrated that when laboratory measurements of dispersivity and hydraulic-conductivity values are used directly in the transport model, a poor fit to the measured breakthrough curve (BTC) is achieved (Fig. 2A). Results of these simulations also clearly show that the finite-difference (FD) and the modified methods of characteristics (MMOC) exhibit more numerical dispersion than the method of characteristics (MOC), total variation diminishing (TVD), and predictor-corrector (P-C) methods. These results are typical of the methods listed except that other implementations of MMOC can have less numerical dispersion (Russell, 2002).

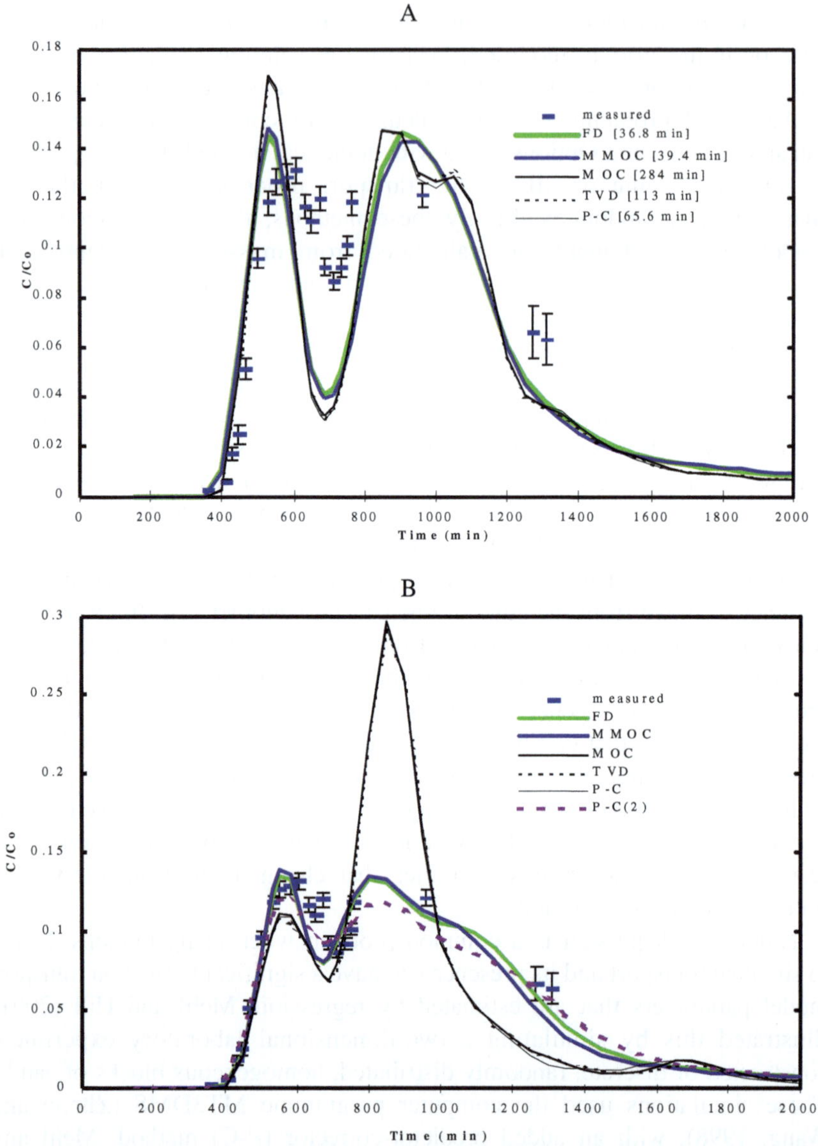

Figure 2. Simulated breakthrough curves (BTC's) from Mehl and Hill (2001). For the measured concentration values, 95% confidence intervals are shown and reflect expected measurement error. (A) BTC's using measured hydraulic conductivities and dispersivities. Computation times are listed in brackets and are from a Linux workstation, Pentium II-333, 64Mb RAM. (B) BTC's using optimized hydraulic conductivities and measured dispersivities. The solution labeled P-C(2) uses dispersivity values increased to approximate the numerical dispersion common to the FD and MMOC methods of MT3DMS.

Mehl and Hill (2001) then applied sensitivity analysis and regression to the models. They found that sensitivities were fairly robust, but estimates of dispersivity and hydraulic conductivity parameters can be strongly affected by the presence of numerical dispersion. When dispersivity parameters were set to the laboratory values and hydraulic conductivities were estimated by regression, the FD and MMOC solution methods produced significantly different optimized values of hydraulic conductivity than did the MOC, TVD and P-C methods. Slightly better fits to measured BTC's were achieved for the methods with more numerical dispersion (Fig. 2B). This result along with the very small measured dispersivity values suggested that the measured dispersivities were consistently too small, and that the estimated hydraulic conductivities were compensating for this bias. When both dispersivities and hydraulic conductivities were estimated, similar hydraulic-conductivity estimates and a similar fit were attained for all solution methods. Estimated dispersivity values were larger for methods not plagued by numerical dispersion.

Transport Models Developed without Concentration Observations

Commonly in groundwater systems, the problem of concern requires an evaluation of transport, but transport observations such as concentrations are not available. This is always a difficult situation, because the other types of observations typical of groundwater systems, heads and flows, are not as sensitive to system features and characteristics that can be very important to transport. Head data used alone is very prone to this deficiency. Barlebo et al. (1998) present an example of using only hydraulic head observations to calibrate a model that is then used to simulate transport. Their study, presented in this article, shows that the plume simulated with the model calibrated using only hydraulic head observations is nearly twice as long as the plume simulated with a model calibrated using heads and concentrations.

Effective Use of Transport Observations

Concentration observations generally provide important information about the estimation of both flow and transport parameters. However, the first attempts to include concentration observations accommodated limited computer capabilities by using sequential parameter-estimation approaches. In these aproaches, head and flow observations were used to estimate flow model parameter values, which were then fixed while concentration data were used to estimate transport parameters (for example, Strecker and Chu, 1986). Yet, concentration observations provide substantial information towards estimating flow model parameters because transport depends strongly on the rate and direction of flow. Wagner and Gorelick (1987) were the first to develop a coupled, or simultaneous, estimation methodology, which they applied to a synthetic example. Several subsequent studies showed that, compared to sequential estimation, simultaneous estimation of flow and transport parameters produced estimates that were more reasonable, in that they were more consistent with independent data, and had reduced uncertainty (for example, Gailey et al., 1991; Sonnenborg et al., 1996; Anderman and Hill, 1999). In some cases, a sequential estimation strategy might produce the same results as those from a coupled inverse procedure (e.g. Jacques et al., 2002), but there is no guarantee of this. When feasible, it is recommended that flow and transport parameters be estimated simultaneously using all available observations.

When applying parameter estimation methods to transport models, a numerical problem can occur when the range of concentration observations spans more than about four orders of magnitude. This range is possible for many constituents, such as dissolved aqueous species or pathogens. The difficulty arises because the observation uncertainty used to determine the weights ω_i for observations often is thought to be proportional to the measurement, and use of a constant proportionality coefficient to calculate weights for all observations can result in enormous weighted residuals (see equation 1) for small concentrations. One solution is to place an upper bound on the weights used for the concentrations (for example, Keidser and Rosbjerg, 1991). It can be important to approach this upper bound gradually.

Often when calibrating transport models, use of point concentrations in the objective function (equation 1) as measures of goodness of fit can be problematic, for reasons related to (1) the potentially large range in the magnitudes of concentration observations, (2) scarcity of concentration observations, and (3) the very large dependence of simulated concentrations on the particular representation of hydrogeologic heterogeneity in the model. Thus, alternative measures of goodness of fit might be preferable when calibrating transport models. For example, for calibrating a model of natural-gradient tracer transport in an extremely heterogeneous aquifer, Feehley et al. (2000) divided the model domain into six zones along the flow direction, and compared simulated and observed masses within each zone. For calibrating a model of a different natural-gradient tracer test in the same aquifer, Julian et al. (2001) compared the maximum simulated concentration from all model layers at a given areal location with the maximum observed value from all vertical sampling points at that location.

Simultaneous Estimation of Model Parameters and Contaminant Source Characteristics

When simulating the transport of contaminants from a disposal source, the location at which the disposal occurred and the release history of contamination might not be accurately known. In this case, if the source location and history are assumed and specified in the model, problems may arise with estimating the parameters of the transport model using observed concentrations. For example, unrealistic parameter estimates might be needed to provide the best match between simulated and observed concentrations downgradient of the assumed source.

An alternative to specifying the source characteristics is to include them with the other parameters that are optimized during model calibration (for example, Wagner, 1992; Mahar and Datta, 2001). This approach can potentially produce a model that more realistically represents the true site conditions. Applications to synthetic examples have shown that the approach has promise, but that nonuniqueness of solutions can be problematic. For example, to facilitate estimation of unique values, Mahar and Datta (2001) needed to place constraints on the parameter values and the source characteristics. When the source location is well known, but the source history is not, nonuniqueness can be less problematic. For field applications in which the known source location was specified in the model, Sonnenborg et al. (1996) and Medina and Carrera (1996) successfully estimated source concentrations along with flow and transport parameters.

Using the Calibrated Model to Diagnose Error and Quantify Uncertainty

Following estimation of model parameters by nonlinear regression, it is important to evaluate (1) the model fit to the calibration observations and (2) the optimal parameter estimates and their uncertainty. These analyses are powerful tools for diagnosing potential errors with the conceptual and numerical models and with the observation data. Three of the most important analyses investigate overall model fit, randomness of weighted residuals, and reasonableness of parameter estimates and confidence intervals. The analyses of model fit and weighted residuals provide an additional quantitative link between the observations and the model shown in Fig. 1. All three of the analyses also provide an indirect link to the system information, as the results are used to help identify model error stemming from incorrect representation of the true system features.

Measures of overall model fit are calculated from the objective function $S(\boldsymbol{b})$ (equation 1), comprising the sum of squared weighted residuals. Commonly used dimensionless indicators of the overall magnitude of the weighted residuals are the calculated error variance, s^2, which equals $S(\boldsymbol{b})/(ND + NPR - NP)$, and its square root, s, which denotes the standard error of the regression. Smaller values of s^2 and s indicate a closer fit to the observations. These measures can also be used to compare regression results among alternative models in which the observation weighting is the same. To obtain dimensional measures of model fit, fitted error statistics can be used, as described by Hill (1998).

While the measures of overall model fit can provide a general indication of how well the model represents the true system, graphical analyses of weighted residuals provide more detail about model fit, and more effectively diagnose potential model error. These analyses involve graphing weighted residuals versus weighted simulated values, as shown in Fig. 3, and versus independent variables (usually space and time), such as on maps of the model domain. Ideally, weighted residuals are randomly distributed on these types of graphs. For example, on graphs of weighted residuals versus weighted simulated values, the weighted residuals should be scattered evenly about 0.0 for the entire range of weighted simulated values, as shown in Fig. 3A. Existence of nonrandom patterns in the weighted residuals relative to weighted simulated values generally indicates model error, or bias. This is illustrated in Fig. 3B, where the presence of weighted residuals that are significantly larger in magnitude for larger simulated values is a strong indication of model error, even though the mean weighted residual is close to zero.

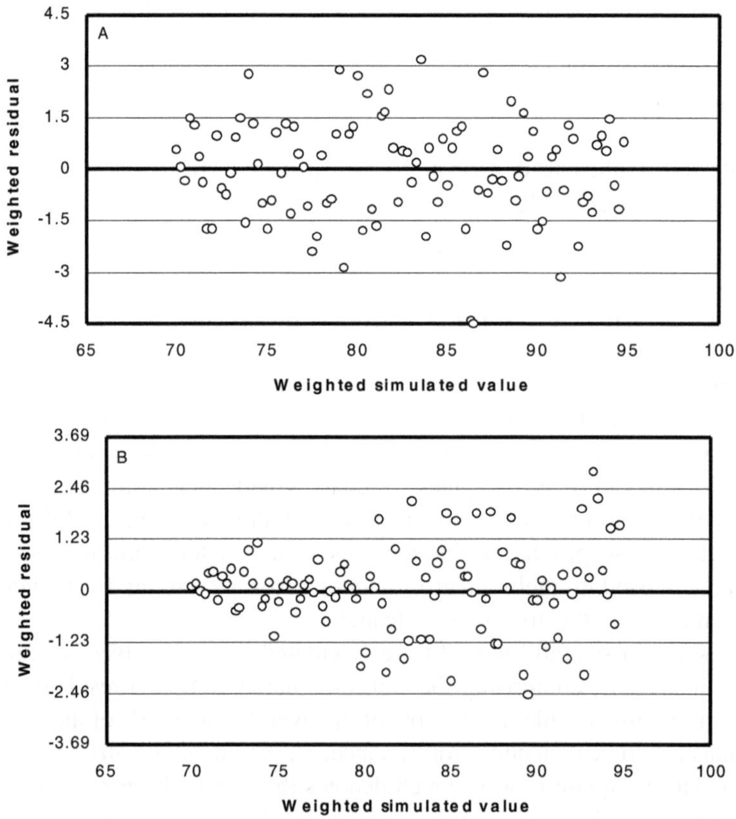

Figure 3. Graphs of weighted residuals versus weighted simulated values showing (A) no model bias and (B) evidence of model bias.

Evaluation of the optimal parameter estimates and their uncertainty can also facilitate diagnosing potential errors in model construction and in interpretation of the observation data. The uncertainty of parameter b_j can be expressed by the individual linear confidence interval for the true, unknown parameter value, calculated as (Hill, 1998; Draper and Smith, 1998):

$$\text{Confidence interval} = b_j - t\left(n, 1.0 - \frac{\alpha}{2}\right)s_{b_j} \;;\; b_j + t\left(n, 1.0 - \frac{\alpha}{2}\right)s_{b_j} \qquad (4)$$

where $t(n,\ 1.0 - \alpha/2)$ is the Student-t statistic for n degrees of freedom (here equal to $ND + NPR - NP$) and a significance level of α (Draper and Smith, 1998); and s_{b_j} is the standard deviation of parameter b_j, calculated as the square root of the parameter variance. This variance is the jth diagonal of the variance-covariance matrix, $\boldsymbol{V}(\boldsymbol{b}')$, calculated as:

$$\boldsymbol{V}(\boldsymbol{b}') = s^2(\boldsymbol{X}^{\mathrm{T}}\omega\boldsymbol{X})^{-1} \qquad (5)$$

where s^2 is the calculated error variance from the regression; $\boldsymbol{X}$ is the sensitivity matrix calculated using the optimal parameter values, and contains observation sensitivities $\partial y_i'/\partial b_j$ as well as entries related to prior information on parameters; ω is the weight matrix containing weights for calibration observations and prior information, and T denotes the transpose operator. (See Hill, 1998, Appendix B, for details on entries in $\boldsymbol{X}$ and ω related to prior.)

The width of the parameter confidence interval in equation 4 is a measure of the likely precision of the estimate, and narrower intervals indicate greater precision. If the model correctly represents the system, the interval also can be thought of as a measure of the likely accuracy of the estimate.

The optimal parameter values and confidence intervals can be used to diagnose model and data error by comparing them to reasonable ranges of parameter values. These ranges are constructed using independent data about the parameters. For example, reasonable ranges of a hydraulic conductivity parameter might be determined from aquifer test data and geologic characterization of the rock types associated with the parameter. Reasonable ranges for groundwater recharge might be determined from field data on precipitation and evapotranspiration.

Ideally, the optimal parameter estimate and a relatively small confidence interval both lie mostly within the reasonable range of values, indicating that the estimate is consistent with the independent information about the parameter. Alternatively, if the estimate and a small confidence interval lie entirely outside the reasonable range, this clearly shows that the estimate is inconsistent with the independent information, in that there is significant model error and the observation data are sufficient to detect it. In this case, both the observation data and the model construction need to be carefully re-evaluated. Finally, if the parameter estimate lies outside the reasonable range but the confidence interval is large and overlaps the reasonable range, this indicates that the data are insufficient to detect whether or not there is significant model error. In this case, the modeler needs to consider (1) the possibility of model error, and (2) additional data that could provide information towards estimating the parameter value more precisely.

If model error is detected by the analyses of weighted residuals and parameter estimates and confidence intervals, then it is important to diagnose the sources of this error by considering the model construction in the areas where error is detected. For example, the results shown in Fig. 3B point to the need to scrutinize the system representation in areas of large weighted simulated values, to investigate why the model fit is preferentially poorer in this area than in areas of small weighted simulated values. Once potential error is identified, strategies for reducing the error include collecting additional information on system characteristics and using these data to modify the model construction, or collecting additional observation data that provide more information towards estimating model parameters in areas where model error is apparent.

One response to poor model fit or unreasonable parameter estimates is to define additional model parameters, which can help address both these problems. However, caution is necessary when adding model parameters. Whereas a closer fit between simulated and observed values often indicates that the model more accurately represents a system, model predictive capability can be degraded by fitting measurements too closely. For example, consider the situation shown in Fig. 4A and B, where the true model is linear. A more complicated model (Fig. 4B) clearly produces a better fit to the observations, but the improved fit is achieved by matching the observation error rather than the system processes, and the predictions are less accurate than in the simpler model. Figure 4C displays the general situation, in which there is a tradeoff between model fit and prediction accuracy with respect to the number of parameters. Thus, if attempts to reduce model error involve defining additional model parameters, it is critical that a thorough evaluation of model and data errors and the possibility of overfitting be conducted as part of the model refinement.

ASSESSING DATA NEEDS FOR IMPROVING MODEL PREDICTIONS

Commonly, models are developed and calibrated for the primary purpose of making predictions related to societal needs. For example, a groundwater transport model might be calibrated using data on an existing contaminant plume, then used to predict the migration time to a boundary of concern or the plume behaviour under various remediation strategies. Once a reasonably accurate simulation of a system has been achieved through careful model development, calibration and error evaluation, the simulation itself becomes a valuable tool not only for making the predictions, but also for other types of predictive analyses.

One important analysis involves identifying potential data that would help improve the predictions and reduce their uncertainty. The expense of data collection for groundwater flow and transport models typically limits the amount of information that realistically can be obtained about the

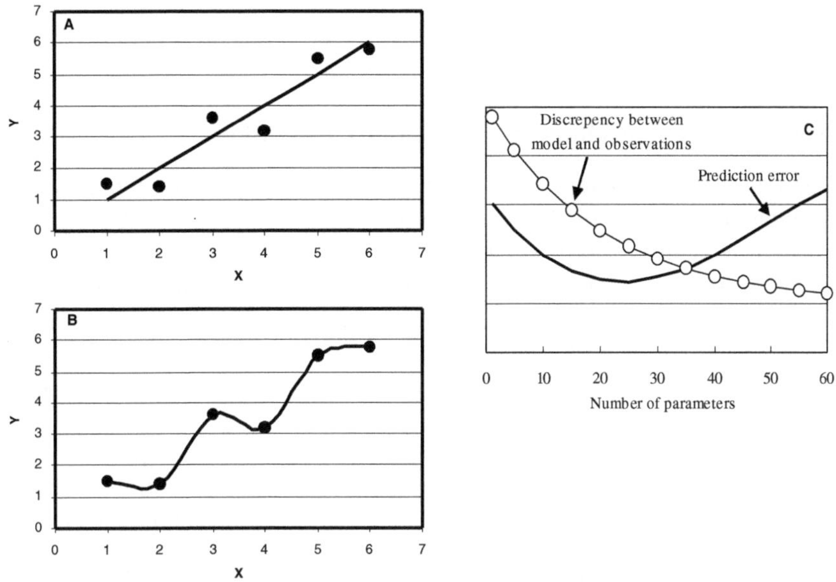

Figure 4. (A) Data with a true linear model. (B) The same data with an overly complex model with little predictive capability. (C) Schematic diagram showing the tradeoff between model fit to observations and prediction accuracy with an increasing number of parameters.

properties and state of a simulated system. Therefore, it is important to design data collection strategies that provide as much information as possible about aspects of the model that are important to the predictions.

Two distinctly different types of data can be collected. The first type includes data that directly characterize the properties of the simulated system, such as field data on stratigraphy, geologic structure, hydraulic conductivity, dispersivity, porosity, and areal recharge. Identifying the model parameters that are most important to the predictions can help guide collection of this type of data. The second type includes system state data that can be used as observations in model calibration, such as hydraulic heads, streamflow gains and losses, and concentrations. Identifying existing and potential system state observations that are most important to the predictions can help guide collection of this type of data. In the context of Fig. 1, methods for identifying these observations provide a quantitative connection between the observations and predictions.

For studies in which enough time has elapsed so that the actual system response can be compared to the model predictions, post-audits can be conducted to evaluate the accuracy of the predictions. These post-audits are additional tools for assessing data needs related to the predictions.

The following sections present methods for evaluating the importance of parameters and observations to model predictions, and a discussion of the utility of post-audits. Application of several of the methods to a regional groundwater flow model with advective transport predictions is illustrated in the forthcoming case studies.

Prediction Sensitivities and Standard Deviations

When model parameters have been estimated by nonlinear regression, quantitative measures of the parameter uncertainty, together with prediction sensitivities, can be used to calculate prediction standard deviations. The prediction standard deviation is a measure of prediction uncertainty, but is not often used by itself. It is used here as the basis for calculating statistics that identify parameters and existing or potential observations that are important to the predictions, which as noted above can be used to help guide future data collection efforts.

The prediction sensitivities, $\partial z'_\ell/\partial b_j$, are the derivatives of prediction z'_ℓ with respect to parameter b_j, and are analogous to the observation sensitivities introduced in this article while discussing 'Parameter estimation by nonlinear regression' and 'Sensitivity analysis for assessing the information that observations provide about parameter values'. The standard deviation of prediction z'_ℓ is denoted $s_{z'_\ell}$, and is calculated as:

$$s_{z'_\ell} = \left[\sum_{i=1}^{NP} \sum_{j=1}^{NP} \frac{\partial z'_\ell}{\partial b_j} \boldsymbol{V}(\boldsymbol{b}') \frac{\partial z'_\ell}{\partial b_i} \right]^{1/2} \tag{6}$$

where $\boldsymbol{V}(\boldsymbol{b}')$ is the parameter variance-covariance matrix for all parameters (equation 5).

It is important to include all defined parameters (those that were and were not estimated by regression) when calculating $\boldsymbol{V}(\boldsymbol{b}')$ that is used in equation 6. In this calculation, the prior weights in matrix ω generally equal zero for parameters for which the calibration observations supply abundant information. Typically, these parameters are estimated by the regression. Parameters that could not be estimated by regression because of insensitivity or correlation, as indicated by small composite scaled sensitivities (equation 3) or extreme parameter correlation coefficients, can be important to predictions. These parameters are likely to be better supported by independent information than by the calibration observations. For such parameters, it is very important that non-zero prior weights be specified in ω when calculating $\boldsymbol{V}(\boldsymbol{b}')$. This allows for a realistic degree of uncertainty in these parameters to be reflected in the analyses of prediction uncertainty.

Assessing System Property Data Needs by Identifying Parameters Important to Predictions

To determine collection strategies for data that directly characterize properties of the simulated system, the methods presented here identify the parameters that are most important to the predictions. Data collection can then focus either on improved field estimation of the values of these parameters, or on improved field characterization of aspects of the model that are associated with the identified parameters, such as the boundaries and geometry of a hydrogeologic unit. In nonlinear regression the predictions of interest may be distinctly different kinds of quantities from the observations. For example, in groundwater models, the observations might be heads and flows, while the predictions might be solute concentrations. For this general situation, classical statistics lack important characteristics, so we have developed alternative methods for identifying parameters important to predictions.

These methods include (1) prediction scaled sensitivities (*pss*), which measure the importance of individual parameters to predictions; (2) combined use of prediction, composite and dimensionless scaled sensitivities (*pss*, *css* and *dss*) and parameter correlation coefficients (*pcc*), to identify parameters important to predictions that are not well supported by the observations and to evaluate whether parameters that are highly correlated in the calibrated model are individually important to the predictions; and (3) the value of improved information (*voii*) statistic, which assesses parameter importance to predictions in a manner that includes the effects of parameter uncertainty, parameter correlation, and prediction sensitivities. The *pss*, *css*, *dss* and *pcc* can be used to help reveal why different parameters are important; the *voii* statistic does the best job of identifying parameters for which additional data would be most advantageous. The broad field of model sensitivity analysis offers many additional measures for evaluating the importance of model inputs to model predictions (e.g. Saltelli et al., 2000, 2004), which range from relatively simple measures similar to the *pss* described below, to more complex and computationally intensive measures that account for model nonlinearity.

Prediction Scaled Sensitivities (pss)

Prediction sensitivities can be scaled to produce prediction scaled sensitivities (*pss*), which are one method of comparing the relative importance of different parameters to the predictions. Prediction scaled sensitivities are generally calculated in one of the following ways:

Sensitivity Scaling

$$pss_{\ell j} = (\partial z'_{\ell}/\partial b_j) \quad (b_j/100) \quad (100/a'_{\ell}) \tag{7a}$$

$$pss_{\ell j} = (\partial z'_{\ell}/\partial b_j) \quad (s_{b_j}/100) \quad (100/a'_{\ell}) \tag{7b}$$

where $pss_{\ell j}$ is the scaled sensitivity of prediction z'_{ℓ} to parameter b_j, s_{b_j} is the standard deviation of parameter b_j calculated from $\boldsymbol{V}(\boldsymbol{b}')$ (equation 5), and a'_{ℓ} is a quantity defined by the modeler. Commonly, a'_{ℓ} equals the predicted value z'_{ℓ}, but could also be set to a regulatory limit, or another value relevant to the particular predictive situation. Larger absolute values of $pss_{\ell j}$ indicate parameters more important to prediction z'_ℓ. The multiplication by $b_j/100$ or $s_{b_j}/100$ in the scalings of equations 7a and 7b results in a $pss_{\ell j}$ that measures the change in the prediction, expressed as a percentage of a'_ℓ, caused by a one-percent change in the parameter value b_j or the parameter standard deviation s_{b_j}, respectively.

The scaling used in equation 7b has three advantages: (1) s_{b_j} almost never equals zero, whereas b_j can equal zero; (2) it is valid for parameters that are affected by the datum of the model, such as groundwater model parameters representing the head at constant-head boundaries; and (3) information about parameter uncertainty is directly included in the calculation. In the context of using the *pss* to guide data collection efforts, the third advantage means that when determining parameter importance, this scaling automatically accounts for the information that existing observations provide about parameters, because s_{b_j} is a function of the observation sensitivities (see equation 5). Thus, given equal prediction sensitivities for two parameters, the $pss_{\ell j}$ calculated by equation 7b will be larger for the parameter that has less support from the observation data.

A disadvantage of the scaling of equation 7b is that it is dependent on the model fit, because of the presence of s^2 in the calculation of $\boldsymbol{V}(\boldsymbol{b}')$. However, s^2 will have the same effect on all $pss_{\ell j}$ calculated using results from a single regression run, so the relative importance of parameters in a single run can be evaluated using this scaling.

Combined Analysis of Prediction, Composite and Dimensionless Scaled Sensitivities (pss, css and dss) and Parameter Correlation Coefficients (pcc)

The scaling in equation 7b for calculating $pss_{\ell j}$ automatically accounts for the information that the observations provide about the parameters, as noted above. However, in some cases, the parameters most important to the predictions can be more clearly determined by separately analyzing a measure of prediction sensitivity and a measure of the information that the observations provide about the parameters. By this analysis, $pss_{\ell j}$ is calculated using the scaling of equation 7a, and these values are compared to *css* calculated for the same set of parameters. If a parameter has a relatively large $pss_{\ell j}$ but a small css_j, this implies that the parameter is important to the predictions but is not well supported by the observations. This result suggests that prediction reliability would be improved by focusing data collection efforts on this parameter, and then incorporating this information into the model.

The *dss* (equation 2) also need to be considered during this type of analysis. If a parameter with large *pss* also has large *css*, indicating that the observation data provide ample support for a parameter important to the predictions, it is important to carefully evaluate the *dss* for that parameter. If the *dss* are relatively large for several observations, this indicates broad support among the observations for the parameter. However, if the *dss* are large for only a very small number of observations, this indicates that the estimated parameter value is largely controlled by these few observations. In this case, the modeler needs to closely examine the interpretation and uncertainty of these observations, and how the model construction affects the simulated equivalents of the observations. If problems with the observations are suspected or if the limited support is of concern, then the parameter might still rank highly as a candidate for future focused data collection, despite its large *css*.

For example, consider a groundwater system in which a pumping well extracts water from a localized zone of high hydraulic conductivity, but in a regional model of the system, the cell containing the well has a much lower hydraulic conductivity. The presence of the well will cause the simulated hydraulic head in this cell to have a relatively large sensitivity to the low conductivity of the cell. In this case, the large *dss* is misleading, because simulated conditions in the cell do not accurately reflect the true conditions. To address this problem, the modeler might modify the model construction by, for example, refining the model grid in the vicinity of the well. However, this may not be practical in a large regional model, so it might be necessary to simply recognize that the support the observation provides toward the low hydraulic conductivity value may not actually be as great as the *dss* indicates, and use this awareness when using the *css*, *pss* and *dss* to determine which parameters are most important to better characterize.

It is also important to determine whether parameters that are highly correlated for the calibrated model are individually important to predictions of interest. This analysis can be achieved by analyzing parameter correlation coefficents (*pcc*). These coefficients are calculated for pairs of model parameters using terms of $\boldsymbol{V}(\boldsymbol{b}')$ (equation 5), and indicate whether the two parameters can be estimated uniquely by regression, given the observation and prior information provided (Hill, 1998). If the correlation coefficient for a pair of different parameters is equal or very close to –1.00 or 1.00, the two parameters generally cannot be estimated uniquely by regression. By the analysis suggested here, two different sets of *pcc* are compared: those calculated using the parameter variance-covariance matrix as described for equation 5, and those calculated with predictions as well.

The *pcc* calculated with the predictions can be produced by augmenting matrices $\boldsymbol{X}$ and ω in equation 5 to include information related to the predictions. The value specified for the prediction as the 'observed value' does not affect the calculation. Values for the weights that augment ω can be established on the basis of one of the three following strategies. The weights can be (1) estimated on the basis of expected measurement error, as is done for observations; (2) established using a statistic that reflects an acceptable range of uncertainty in the prediction (see Sun and Yeh, 1990; Sun, 1994);

and (3) set to values larger than suggested by strategies (1) and (2), so that the effect of the predictive quantity is increased, ensuring that these *pcc* will reveal individual parameter values that are important to predictions.

The *pcc* calculated with only the observations and prior information indicate the uniqueness of parameters estimated by the regression. The *pcc* calculated with the predictions as well indicate whether unique parameter estimates are important to the predictions. If comparison of the two types of *pcc* shows that adding the predictions causes some highly correlated parameter pairs to become much less correlated, this indicates that the predictions are likely to depend on parameter values that the regression could not estimate uniquely. This identifies a weakness in the calibrated model.

To illustrate the utility of the *pcc* with predictions, consider a groundwater flow model calibrated with observations of hydraulic head and streamflow gain or loss. The calibrated model is developed to predict hydraulic head at a location where no measurement can be obtained and advective transport from the site of a contaminant spill. The *pcc* are first calculated using the calibrated model, all observations, and all defined parameters. Two sets of *pcc* with predictions are then calculated, by first including the predicted hydraulic head and then including the predicted advective transport. For a situation similar to this example, Anderman et al. (1996) showed that (1) some parameters of the calibrated model are highly correlated, (2) the hydraulic head prediction did not require uncorrelated parameter estimates, and (3) the advective transport prediction did require uncorrelated estimates. Thus, this analysis concluded that hydraulic head could be predicted with some confidence, but that the transport prediction using the estimated parameters is highly suspect.

A classification system summarizing the analysis of *css*, *pss* and *pcc* is shown in Fig. 5. The upper and lower portions of this figure classify, respectively, the precision and uniqueness of parameter estimates in the context of parameter importance to the predictions. If the analysis of a parameter indicates a classification in box IV of Fig. 5A or B, this means that improved estimation of this parameter and improved representation of the system features with which it is associated are likely to improve prediction accuracy. If the analysis indicates a classification in boxes I, II or III, the term 'acceptable' means that a parameter is estimated well, is unimportant to the predictions of interest, or both. Improved estimation of the parameter and improved representation of the system features with which it is associated are likely to be less beneficial to improving prediction accuracy than for parameters that are classified in box IV. This method of using *css*, *pss* and *pcc* can be revealing, but is awkward. Fortunately, the *voii* statistic described next incorporates the effects of parameter correlation as well as observation and prediction sensitivity.

<table>
<tr><td colspan="2" rowspan="2">A</td><td colspan="2">Precision of the parameter estimate</td></tr>
<tr><td>Imprecise: Composite scaled sensitivity >1.0 and not small relative to other parameters[1]; large confidence interval</td><td>Precise: Composite scaled sensitivity <1.0 or small relative to other parameters[1]; small confidence interval</td></tr>
<tr><td rowspan="2">Importance of the parameter to predictions of interest</td><td>Not important: Small prediction scaled sensitivity</td><td>I. Acceptable</td><td>II. Acceptable</td></tr>
<tr><td>Important: Large prediction scaled sensitivity</td><td>IV. Improve estimation of this parameter and representation of associated system features.</td><td>III. Acceptable</td></tr>
</table>

<table>
<tr><td colspan="2" rowspan="2">B</td><td colspan="2">Uniqueness of the estimates for a parameter pair</td></tr>
<tr><td>Nonunique: Parameter correlation coefficient[2] is near |1.00|</td><td>Unique: Parameter correlation coefficient[2] is less than about |0.95|</td></tr>
<tr><td rowspan="2">Importance of unique parameter estimates to predictions of interest</td><td>Not important: Parameter correlation coefficient with predictions[3] is near |1.00|</td><td>I. Acceptable</td><td>II. Acceptable</td></tr>
<tr><td>Important: Parameter correlation coefficient with predictions[3] is less than about |0.95|</td><td>IV. Improve estimation of one or both of the parameters and improve the representation of associated system features</td><td>III. Acceptable</td></tr>
</table>

1. A composite scaled sensitivity (*css*) is small relative to other parameters if it is less than about one percent of the largest *css*. This is a rule of thumb but is sufficient for identifying potentially problematic parameters.
2. Calculated using all parameters, and including only the observations and prior information used in the calibration.
3. Calculated as in footnote 2, but also including the predictions.

Figure 5. Classification of the need for improved estimation of a parameter and, perhaps, associated system features. The classification is based on statistics that indicate (A) the precision and importance to predictions of a single parameter and (B) the uniqueness and importance to predictions of a pair of parameters. See text for additional explanation.

The Value of Improved Information (voii) Statistic

The value of improved information (*voii*) method assesses the importance of parameters to predictions in a way that accounts for parameter correlations (Tiedeman et al., 2003). In the context of Fig. 1, this statistic links the parameters and predictions. The equation and procedure for obtaining the *voii* statistic is more complicated than for *pss*, which can cause interpretation of the results to be more difficult. Evaluating the *css* and *pss* as described above can facilitate this interpretation.

The *voii* approach takes advantage of the connection between parameter uncertainty, parameter correlation, and prediction uncertainty provided by equation 6 for the prediction standard deviation $s_{z'_\ell}$. The method calculates the reduction in $s_{z'_\ell}$ produced by a reduction in the uncertainty of one or more parameters. First, $s_{z'_\ell}$ is computed with equation 6 using a calibrated model and existing independent information about parameter values, then is recomputed under the assumption of improved information about one or more parameter values. The resulting reduction in $s_{z'_\ell}$ is used to calculate the *voii* statistic.

The prediction standard deviation produced with improved information on parameter b_j is calculated using a modified form of equation 6 for $s_{z'_\ell}$. In the modified equation, a modified matrix ω, denoted $\omega_{(j)}$, is used to calculate a modified $\boldsymbol{V}(\boldsymbol{b}')$, denoted $\boldsymbol{V}(\boldsymbol{b}')_{(j)}$, which is then used to calculate a modified $s_{z'_\ell}$, denoted $s_{z'_\ell}(j)$. Improved information on parameter b_j is implemented in matrix $\omega_{(j)}$ by specifying the weight on prior information for this parameter to be larger than its weight in ω. Conceptually, this larger weight represents the increased certainty in the prior value that would result from collection of additional field data; that is, from improved information about the parameter.

Because of the increased weight in $\omega_{(j)}$, the variance of parameter b_j will be smaller in $\boldsymbol{V}(\boldsymbol{b}')_{(j)}$ than in $\boldsymbol{V}(\boldsymbol{b}')$. Parameters that do not have improved information, but that are correlated with parameter b_j, will also tend to have smaller variances in $\boldsymbol{V}(\boldsymbol{b}')_{(j)}$ compared to those in $\boldsymbol{V}(\boldsymbol{b}')$. Primarily because of these reductions in parameter variances, the prediction standard deviation calculated with improved information ($s_{z'_\ell(j)}$) will generally be smaller than the prediction standard deviation calculated without improved information ($s_{z'_\ell}$).

The scaled difference between $s_{z'_\ell(j)}$ and $s_{z'_\ell}$ measures the value of the improved information on parameter b_j with respect to prediction z'_ℓ, and is calculated as:

$$voii_{\ell(j)} = 100 \times \left(\frac{s_{z_\ell} - s_{z'_\ell(j)}}{s_{z_\ell}} \right) = 100 \times \left(1 - \frac{s_{z'_\ell(j)}}{s_{z_\ell}} \right) \tag{8}$$

where $voii_{\ell(j)}$ is the value of improved information statistic, and equals the percent reduction in the standard deviation of prediction z'_ℓ that results from improved information on parameter b_j.

To rank the importance of individual parameters to prediction z'_ℓ, $voii_{\ell(j)}$ is calculated once for each parameter b_j. The parameter associated with the largest value of $voii_{\ell(j)}$ ranks as most important to prediction z'_ℓ. To implement improved information on each model parameter in a consistent manner, Tiedeman et al. (2003) suggest individually increasing the weight of the prior information on each parameter so that the standard deviation of the parameter estimate is decreased by a specified percent.

The *voii* methodology presented above is easily extended to the case of evaluating improved information on more than one parameter (Tiedeman et al., 2003). When evaluating multiple parameters, the effect of parameter correlations can strongly influence which parameters are important to a prediction. This effect can produce situations in which the set of parameters with the highest individual $voii_{\ell(j)}$ values is not identical to the set of parameters that are most important when improved information on multiple parameters is considered.

The *voii* method assumes that the model is linear with respect to the parameter values. However, the method has been found to be fairly robust for a mildly nonlinear groundwater model (Tiedeman et al., 2003). Methods for assessing parameter importance to predictions that account for model nonlinearity are presented by Sulieman et al. (2001) for models calibrated by regression, and by Saltelli et al. (2000, 2004) for the general case in which the model may not have been calibrated by regression. These methods are more complex and computationally intensive than the *voii* method.

Assessing System State Data Needs by Identifying Existing and Potential Observations Important to Predictions

Using *pss*, *css* and *dss* together, as already discussed, provides one method for identifying observations important to the predictions. However, this procedure can require numerous graphs, and it can be difficult to summarize these graphs and extract the key results. An alternative is the observation-prediction (*opr*) statistic. This statistic essentially integrates the information contained in the *css*, *dss*, *pcc* and *pss*, and it often is useful to investigate those measures to better understand the *opr* results.

Observation-Prediction (opr) Statistic

The observation-prediction (*opr*) statistic assesses the effect on prediction uncertainty of either removing existing observations or adding potential new observations. In the context of Fig. 1, this statistic provides a quantitative link between the observations and predictions. The *opr* statistic can be used to address issues related to monitoring the state of the simulated system, such as: 'What observations in an existing monitoring network might be removed, with little impact on the predictions?' and 'What potential new observations are most important to the predictions?'

The *opr* method does not involve recalibrating the model with these observations added or removed. Thus, the leverage of the observations, rather than their influence (Helsel and Hirsch, 2002), is determined. Observations have large leverage if their omission or addition could have a substantial effect on the predictions, and observations have large influence if their omission actually has a substantial effect on the predictions. In general, influential observations are a subset of those with significant leverage, and therefore the observations identified using the *opr* statistic are likely to include those with high influence. An advantage of the *opr* statistic is that it requires trivial computational effort. In contrast, identifying existing observations that are influential with respect to the predictions requires jackknifing, bootstrapping, or similar methods, which repeat the nonlinear regression with one or more observations omitted (Efron, 1982; Davison and Hinckly, 1997). Influence cannot be determined for potential observations because assessing influence requires the observed value in addition to other information associated with the potential observation, such as its type, location and time.

As for the *voii* method, a modified version of equation 6 for the prediction standard deviation $s_{z'_\ell}$ is used to evaluate the effect on prediction uncertainty of omitting or adding one observation y_i (Hill et al., 2001; Ely et al., in review). In this modified equation, modified matrices $\boldsymbol{X}$ and ω, respectively denoted $\boldsymbol{X}_{(\pm i)}$ and $\omega_{(\pm i)}$, are used to calculate a modified $\boldsymbol{V}(\boldsymbol{b}')$, denoted $\boldsymbol{V}(\boldsymbol{b}')_{(\pm i)}$, which is then used to calculate a modified $s_{z'_\ell}$, denoted $s_{z'_\ell}(\pm i)$. The matrix $\boldsymbol{X}_{(\pm i)}$ is formed by either adding (+*i*) or removing (–*i*) the sensitivities of the simulated equivalent of y_i, and the matrix $\omega\,(\pm i)$ is formed by either adding (+*i*) or removing (–*i*) the weight associated with y_i.

In practice, observation y_i is removed by setting its weight equal to zero and leaving the sensitivity matrix $\boldsymbol{X}$ unchanged. An observation is added by calculating the related sensitivities and assigning weights on the basis of an analysis of errors that would be expected for the potential observed values. The value of the potential observation does not affect the *opr* statistic because s^2 from the regression is used in the calculation of $\boldsymbol{V}(\boldsymbol{b}')_{(\pm i)}$, and also cancels out of the *opr* equation, as discussed below.

The percent change in prediction standard deviaion that results from removing or adding observation y_i is used as the measure of its importance to prediction z'_ℓ:

$$opr_{\ell(\pm i)} = 100 \times \left| \frac{s_{z'_\ell} - s_{z'_\ell(\pm i)}}{s_{z'_\ell}} \right| = 100 \times \left| 1 - \frac{s_{z'_\ell(\pm i)}}{s_{z'_\ell}} \right| \tag{9}$$

where $opr_{\ell(\pm i)}$ is the observation-prediction statistic, and the vertical brackets denote absolute value. Hill et al. (2001) and Ely et al. (in review) provide a more thorough presentation of the *opr* method, including its extension to accommodate evaluation of adding or omitting any combination of existing or potential observations. In addition, they show that the method has been found robust for a mildly nonlinear model, even though the *opr* statistic assumes that the model is linear with respect to the model parameters.

The crux of the *opr* statistic is the use of equation 6 for prediction standard deviation as the basis for evaluating the importance of existing or potential observations to model predictions. This concept has also been used in the design of groundwater observation networks for improving model predictions by, for example, Sun and Yeh (1990) and Wagner (1995, 1999).

It is important to understand why the *opr* statistic indicates that particular observations are important to the model predictions. Equation 6, from which the *opr* statistic is derived, includes the prediction sensitivities and the parameter variance-covariance matrix (equation 5). The prediction sensitivities can be investigated using the *pss* of equation 7. The variance-covariance matrix can be investigated by first noting that in equation 9, the s^2 term of equation 5 cancels out, causing the *opr* statistic to be fit-independent. Thus, only the term $(\boldsymbol{X}^T\omega\,\boldsymbol{X})^{-1}$ from equation 5 remains. The contribution of this term to the *opr* statistic can be investigated by considering *dss* (equation 2) and *pcc*. If the observation y_i provides substantial information about parameter b_j (dss_{ij} is large) and this parameter, or a parameter b_k with which b_j is highly correlated (pcc_{kj} is large), is important to prediction z'_ℓ ($pss_{\ell j}$ or $pss_{\ell k}$ is large), then it is likely that $opr_{\ell(\pm i)}$ will be large.

It is important to recognize that in some situations, values of *dss* can be large because of simplifications made during model construction rather than because of actual hydrogeologic conditions. Additional insight into why certain observations can rank as important is provided by the example application of the *opr* statistic to the DVRFS model.

Using Post-Audits to Evaluate Prediction Accuracy and Assess Data Needs

Post-audits are designed to evaluate the accuracy of model predictions, by comparing predicted values to the true system responses. Post-audits performed several years after the end of the model calibration period generally serve as more rigorous tests than those conducted at earlier times, because at later times, system conditions have typically changed more substantially from the calibration conditions (Anderson and Woessner, 1992). In the context of assessing data needs for improving the model predictions, the results of a post-audit can be valuable for guiding data collection and refinements to the model so that it better represents the true system.

Most published post-audits of regional-scale groundwater flow and transport models have found that actual system responses differ from responses predicted by the model. Examples include post-audits by Konikow and Person

(1985), Alley and Emery (1986), Konikow (1986), Reichard and Meadows (1992), Hanson (1996) and Stewart and Langevin (1999), which involved regional-scale models with prediction times several years after the model calibration period. In most of these studies, there were significant differences between the actual and predicted responses. The primary reasons for these large discrepancies were model errors and incorrect projected stress conditions. The post-audits cited above identified the following types of model errors as contributing to inaccurate model predictions: (1) incorrect model parameter values that reasonably calibrate the model but that produce inaccurate predictions; (2) errors in the system conceptualization such as incorrect locations of system boundaries and inappropriate representation of flow and transport processes; and (3) incorrect assumptions about the model calibration period being representative of the predictive period.

In contrast to the post-audits cited above with predictions at long times after the calibration period, Andersen and Lu (2003) conducted a post-audit of a local-scale flow model used for plume containment design with a predictive period only one year after the calibration period. Heads in the true system with the containment design implemented differed from predicted heads, and model error was identified as the cause of these inaccuracies. However, Andersen and Lu (2003) concluded that the prediction errors were small enough and that the model had been a valuable tool for design of the containment system.

In each of the cited post-audits, the results were used to gain considerable insight into the simulated groundwater systems, even though the predictions were incorrect to some degree. This insight was used to help detect model error, and to identify data needs and changes to the conceptual model that could help reduce this model error. Thus, the post-audits served as valuable tools for guiding data collection and ultimately improving the models.

APPLICATIONS

Selected aspects of three model applications are presented to demonstrate some of the ideas and methods presented in this article. The applications involve models of a theoretical study of virus transport; contaminant transport underlying the Grindsted Landfill, Denmark; and the regional groundwater flow system surrounding Death Valley, California and Nevada, USA.

Assessing the Information that Observations Provide About Parameters of a Synthetic Virus Transport Model

Sensitivity analyses methods presented earlier in this article were used by Barth and Hill (in review-a, b) to determine which observations are most important to estimating flow and transport parameters of a synthetic virus-transport model, and which parameters are most important to simulating virus concentrations at different times and distances from the source. The simulation mimics conditions of field experiments conducted by Schijven et al. (1999) includes observations of hydraulic head (which have small sensitivities because the system is homogeneous and constant-head and head-dependent boundaries are imposed), total bulk flow through the system, normalized first temporal moments of conservative-transport concentrations, and virus concentrations.

The composite scaled sensitivities (*css*) are shown in Fig. 6. The numerical model transport step size (TSS) is included in this figure, to allow evaluation of whether the information provided by the observations is sufficient to overcome typical numerical inaccuracies (Barth and Hill, in review-a).

Results suggest that most observation types provide substantial information about hydraulic conductivity and porosity, as illustrated by the large *css* for these parameters (Fig. 6). Also, although not depicted in Fig. 6, only virus-concentration observations provide information about sorption, represented by parameter Kd. This result suggests that when virus-concentration observations are not available, as occurs for many field problems, estimation of important sorption parameters is likely to be problematic. Finally, even with virus concentration observations, the *css* for one of the virus transport parameters, $\lambda 1$, is smaller than that for TSS, indicating that changes in TSS affect model simulated values more than do changes in $\lambda 1$. This indicates that sensitivities to $\lambda 1$ are strongly affected by the size of TSS, which suggests that estimation of $\lambda 1$ is likely to be affected by numerical inaccuracies related to variability in TSS. This problem was addressed by fixing TSS at a value smaller than the default calculated by the transport simulation code, which increased the accuracy of the sensitivities for $\lambda 1$, and caused the estimation of this parameter to be less affected by variations in TSS.

Figure 6 shows that the *css* values for all parameters are within two orders of magnitude, suggesting the likelihood that all parameters can be estimated by regression. However, the observations were not sufficient to estimate the parameters uniquely. Even when using all observation types, there was extreme parameter correlation between porosity and hydraulic conductivity and between sorption and in-solution inactivation. Parameter estimation was accomplished by fixing values of porosity and in-solution inactivation.

Development and Calibration of a Groundwater Flow and Transport Model for the Grindsted Landfill, Denmark

An existing landfill in Grindsted, Denmark (Fig. 7) that has contaminated the underlying groundwater system was the focus of a study by Barlebo et al. (1998). They investigated some of the transport model development

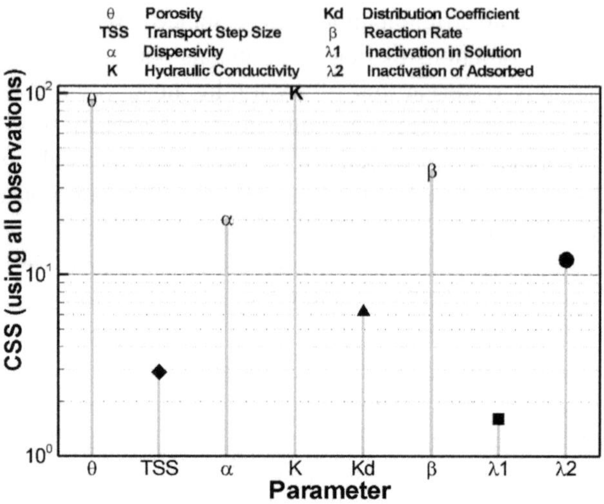

Figure 6. Composite-scaled sensitivities (*css*) of seven system parameters and the simulation transport step size, TSS. Observations include hydraulic heads, moments of conservative transport, and reactive transport concentrations. (From Barth and Hill, in review-b).

and calibration issues, discussed earlier in this article, including: (1) the accuracy of simulated concentrations using a transport model calibrated with hydraulic-head observations and no concentration observations, and (2) the advantages and disadvantages of using a two-dimensional cross-sectional model instead of a three-dimensional model. Observation data included 139 hydraulic-head and 210 concentration observations from about 120 areal locations, and the inverse groundwater flow and transport model was developed using an early version of the finite-element model WATFLOW/WTC (Molson and Frind, 2002).

A map and cross-section of the system are shown in Figs 7 and 8, and the model domain is shown in Fig. 9. The layers shown in Fig. 8 persist over most of the simulated area and were used to represent the system in all of the models developed. Most heads and all concentrations were measured using wells screened only above the silt layer, and the concentrations were measured along the four transects labeled A, B, C and D (Fig. 9).

Investigation of the two issues of interest proceeded as follows. First, a three-dimensional model was developed and calibrated using the head and concentration observations. This was expected to be the most accurate model, and the plume simulated using this model is shown in Fig. 10. Then, the significance of concentration data for calibrating a model used to predict transport was investigated by estimating hydraulic parameters for the three-dimensional model using only the head observations. The parameters needed only for simulating transport were assigned values equal to those estimated

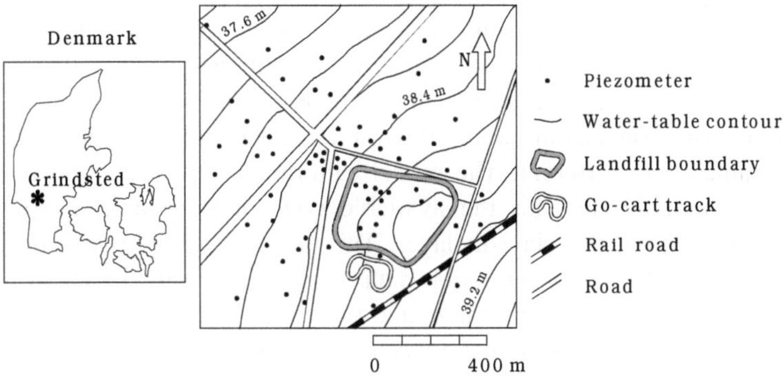

Figure 7. The Grindsted landfill, with local water-table contours and piezometer locations. (From Barlebo et al., 1998).

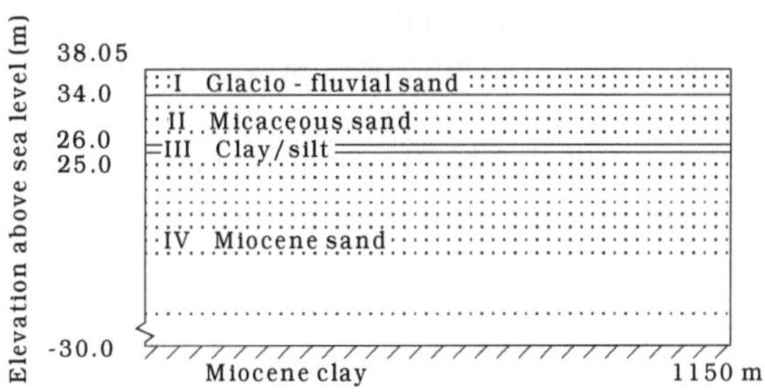

Figure 8. Northwest-southeast cross-section of modeled area (dots represent nodes in the model). (From Barlebo et al., 1998).

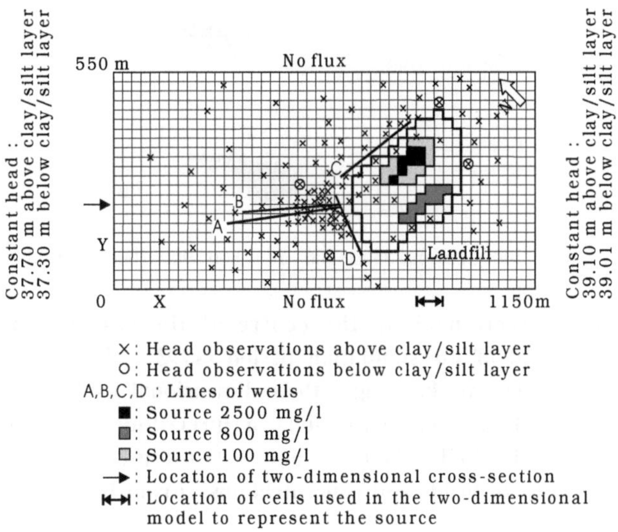

Figure 9. Plan view of modeled area in the vicinity of the Grindsted landfill. (From Barlebo et al., 1998).

for the first model. The resulting plume is shown in Fig. 10. Finally, use of a two dimensional cross-sectional model was calibrated using both observation types, and produced the plume shown in Fig. 10.

Two main conclusions were drawn. First, the simulated plumes in Fig. 10 are dramatically different. The model calibrated with heads and concentrations reasonably matches the available concentration data and simulates transport through the clay/silt layer. The model calibrated with only heads does not match the concentration data as well, simulates transport to much greater distances from the source, and the plume remains above the clay-silt layer. This demonstrates that inaccurate transport behaviour is predicted using a model calibrated only with hydraulic head data. Second, the major benefit of two-dimensional cross-sectional modeling was that execution times were reduced by a factor of seven. As shown in Fig. 10, the plume simulated by this model is similar to that simulated by the three-dimensional model calibrated with heads and concentrations; however, the three-dimensional plume matches the measured concentrations more closely. Advantages of the three-dimensional simulation were primarily that accurate representation of the source was possible and that all hydraulic head and concentration observations could be used.

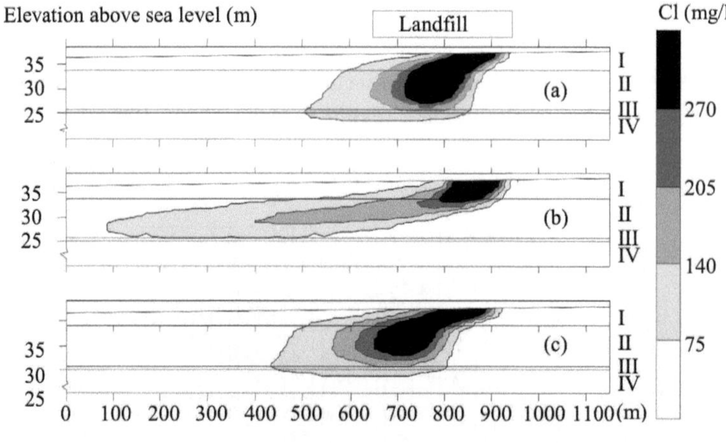

Figure 10. Cross-section along the centre of the chloride (Cl) plume simulated at the end of the simulation period (1993) using a model of the Grindsted Landfill, Denmark, using a three-dimensional model that produces the best fit to heads and concentrations, a three-dimensional model that produces the best fit to heads only, and a two-dimensional model that produces the best fit to heads and concentrations. (From Barlebo et al., 1998)

During calibration of the three-dimensional model, estimated parameter values were compared to reasonable ranges as described earlier in this article. Figure 11 demonstrates why these comparisons also need to consider confidence intervals on the estimated parameters. An unreasonable parameter value for the horizontal hydraulic conductivity in geologic layer II is estimated when only head observations are used, but because of its large confidence interval, it is not clear if the unreasonable estimate indicates errors in model construction. When concentration observations are included in the regression, the reasonable estimate and small confidence interval indicate clearly that the problem was not model error. For the estimate obtained when only heads were used, the information provided by the large confidence interval helped prevent an unnecessary search for model errors.

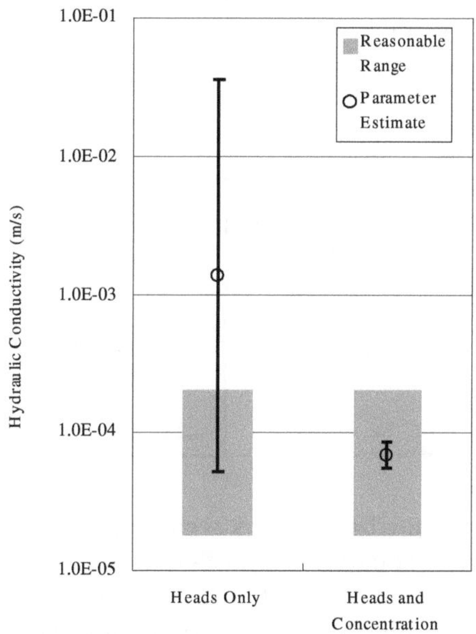

Figure 11. Estimates of parameter khII (horizontal hydraulic conductivity of micaceous sand [see Fig. 8]), in metres per second, from a groundwater model of the Grindsted Landfill calibrated with only head observations and with head and concentration observations. Black bars show the linear individual 95-percent confidence intervals. (From data reported by Barlebo et al., 1998)

Evaluating Parameters and Observations Important to Advective-Transport Predictions in a Model of the Death Valley Regional Groundwater Flow System

Several of the methods for evaluating model predictions have been applied to a model of the Death Valley regional groundwater flow system (DVRFS), located in southern California and Nevada, USA (Fig. 12A). The groundwater system has been studied extensively because of the potential effects of activities at the Nevada Test Site, where underground nuclear testing has been conducted, and Yucca Mountain, where high-level radioactive waste is scheduled to be stored in thick unsaturated volcanic deposits located above the water table. In addition, local pumping has affected the groundwater system. Concerns related to these activities include: transport paths and times of potential contaminants from the Nevada Test Site and Yucca Mountain, and the effects of these potential contaminants at possible receptors such as Death Valley National Park and water supply wells.

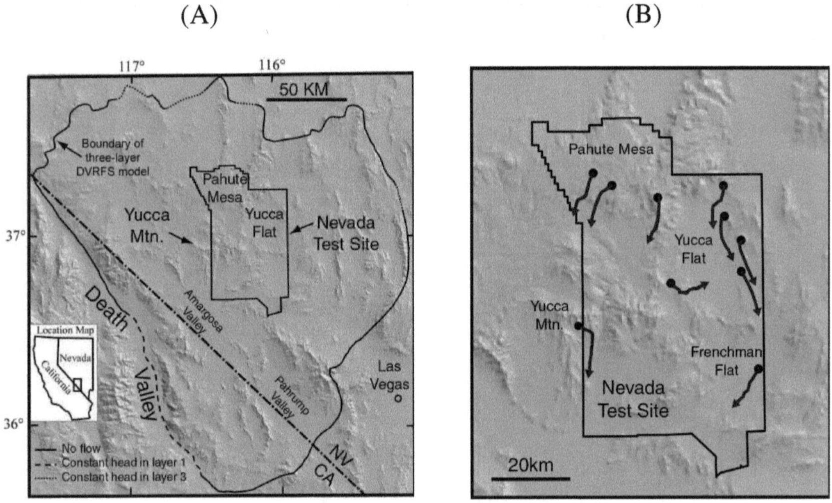

Figure 12. (A) Location of the Death Valley regional groundwater flow system (DVRFS) and the boundary of the three-layer flow model. (B) Selected predicted advective transport paths from Yucca Mountain and Underground Test Area (UGTA) sites simulated using the three-layer regional saturated model of D'Agnese et al., 1997. The paths shown start slightly below the water table. The vertical component of advective transport is small compared to the lateral component, and is not shown.

Groundwater Flow Model and Advective-Transport Predictions

A steady-state, three-layer model of the flow system (D'Agnese et al., 1997, 1999) is used to illustrate the methods for evaluating model predictions. More detailed 15-layer models of the flow system have been subsequently developed (D'Agnese et al., 2002; Faunt et al., in review). In the DVRFS, the predictions of interest involve potential transport of contaminants from beneath Yucca Mountain and Underground Test Area (UGTA) sites on the Nevada Test Site (Fig. 12). Accurate simulation of this transport is plagued by a number of problems, including the fractured nature of the subsurface rocks and the regional scope of the model. In a regional model, it is impossible to represent accurately small features or processes such as dispersion and retardation that can be important to transport. A useful approach is to consider only the transport processes appropriate to the scale of the model. Thus, only advective transport is considered, which is the transport that would occur if the solute did not spread and encountered no reactions with the surrounding rocks. It can be considered the first building block of transport, upon which other complexities are added. Calculation of advective transport over large distances and times is consistent with the scale of a regional model because regional conditions generally influence this component of transport more than other transport processes.

Advective transport was simulated using the Advective-Transport Observation (ADV) Package of MODFLOW-2000 (Anderman and Hill, 2001). The ADV Package uses particle-tracking methods nearly identical to those in MODPATH (Pollock, 1989) to determine advective-transport paths. To compute the particle trajectory, particle displacement is decomposed into displacements in the three spatial dimensions of the DVRFS model: north-south (N-S), east-west (E-W), and vertical. This analysis of the directional components of transport allows parameter and observation importance to be evaluated with respect to each direction of transport. The predicted transport paths are shown in plan view in Fig. 12B.

Combined Analysis of Composite (css), Prediction (pss) and Dimensionless (dss) Scaled Sensitivities to Identify Parameters Important to Predictions

Figure 13 presents graphs of *css*, *pss* and *dss* for selected DVRFS model parameters and for predicted advective transport from one UGTA site. Figure 13B shows the mean and range of the *pss* for five particles that originate in the single model cell containing the UGTA site. Here, the *pss* values are defined using the scaling in equation 7a, with a'_ℓ equal to the predicted value, which is the simulated advective transport distance in the north or south coordinate direction. Thus, the *pss* equal the percent change in the advective transport distance that is caused by a one-percent change in parameter value.

In Fig. 13B, the *pss* show that HK2, HK3 and POROS (the hydraulic conductivity of two rock types and porosity) are the most important parameters to the determination of advective transport. In Fig. 13A, the *css* show that the observations used in the regression provide ample information for parameters HK2 and HK3, but provide no information for POROS. This is problematic, because a parameter very important to the predictions is not supported by the observation data. In Fig. 13C, the *dss* show that although the *css* for parameter HK4 is large, the support primarily comes from just four observations. HK4 is not highly important to the predictions, but if it were, the *dss* suggest that these four observations should be closely investigated, because they largely control the estimate of HK4. This type of combined analysis of *pss*, *css* and *dss* can be used to understand and communicate model strengths and weaknesses and to justify and plan additional model development and data collection efforts.

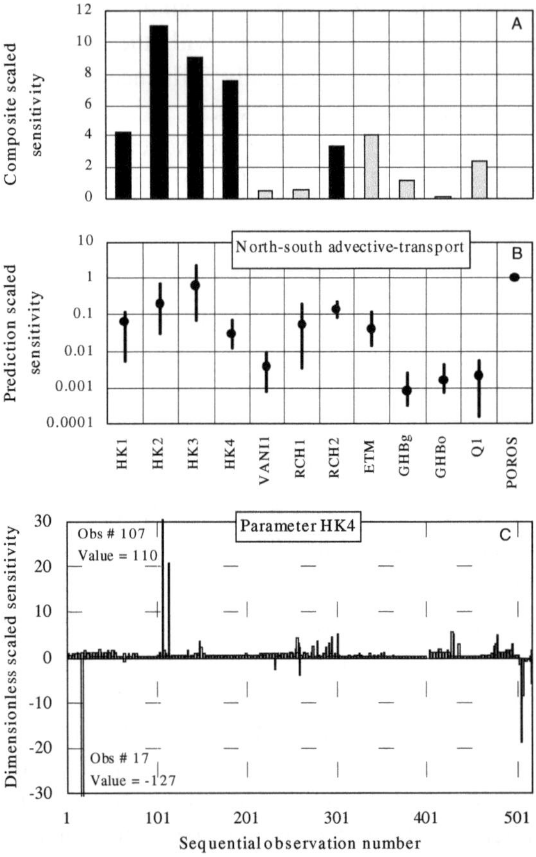

Figure 13. (A) Composite scaled sensitivities for selected DVRFS model parameters estimated (black bars) and not estimated (gray bars) by the regression. (B) Prediction scaled sensitivities (calculated using equation 7a) for the north-south component of predicted advective transport. (C) Dimensionless scaled sensitivities showing the support provided by the observations for parameter HK4.

Using the Value of Improved Information (voii) Statistic to Identify Parameters Important to the Predictions

The *voii* statistic was used to evaluate the importance of all 23 parameters of the three-layer DVRFS model to an advective-transport path in Yucca Flat (Fig. 12) (Tiedeman et al., 2003). The *voii* results are compared to the *pss* calculated using the scaling in equation 7b; by this scaling, the *pss* include information on the support provided by the observations for the parameters. Comparison of these *voii* and *pss* results illustrates how the *voii* method accounts for parameter correlations.

The *voii* results show that HK1 and RCH3 are the two most important parameters to advective transport in the E-W and N-S directions (Fig. 14A). Reducing the standard deviation of HK1 by 10 percent reduces the uncertainty of both the E-W and the N-S transport components by about 3.5 percent. The *pss* show that parameters HK5 and HK1 rank as the two most important parameters to these components of advective transport (Fig. 14B). The results indicate that increasing the value of HK5 by one percent of its standard deviation would change the distance travelled in the E-W direction by about 0.5 percent and the distance travelled in the N-S direction by about 0.7 percent.

Figure 14 shows that some of the same parameters are identified by both the *pss* and *voii* methods as important to the prediction of interest, but also shows that the parameters identified as important can differ among the two methods. For example, the $pss_{\ell j}$ indicate that predicted advective transport in the E-W and N-S directions is relatively insensitive to RCH3. However, the $voii_{\ell(j)}$ indicate that RCH3 is the second-most important parameter to advective transport in these directions. These different rankings by the two methods are an example of how parameter correlations affect the *voii* results. The difference in the ranking of RCH3 by the *pss* and *voii* methods can be explained by the large correlations between RCH3 and HK1 (0.96) and between RCH3 and HK3 (0.94), in combination with the relatively large sensitivities of advective transport in the E-W and N-S directions to HK1 and HK3 (Fig. 14B). Because of these correlations and sensitivities, specifying improved information on RCH3 will reduce the prediction uncertainty (and thus produce a relatively large value of $voii_{\ell(j)}$) even though the prediction is not highly sensitive to RCH3.

These effects of parameter correlations on the *voii* results have an important consequence for the cost-effectiveness of data collection. For example, to improve predicted advective transport at this site, field data could be collected about RCH3 instead of about HK1 or HK3. Collecting data about a recharge rate or the geographic extent of a recharge zone is likely to be less expensive than collecting subsurface information about a hydraulic-conductivity value or hydrogeologic unit.

Using the Observation-Prediction (opr) Statistic to Identify Existing and Potential Observations Important to the Predictions

The *opr* statistic was used to identify the importance of observations to the advective-transport predictions. An averaged measure of the statistic was

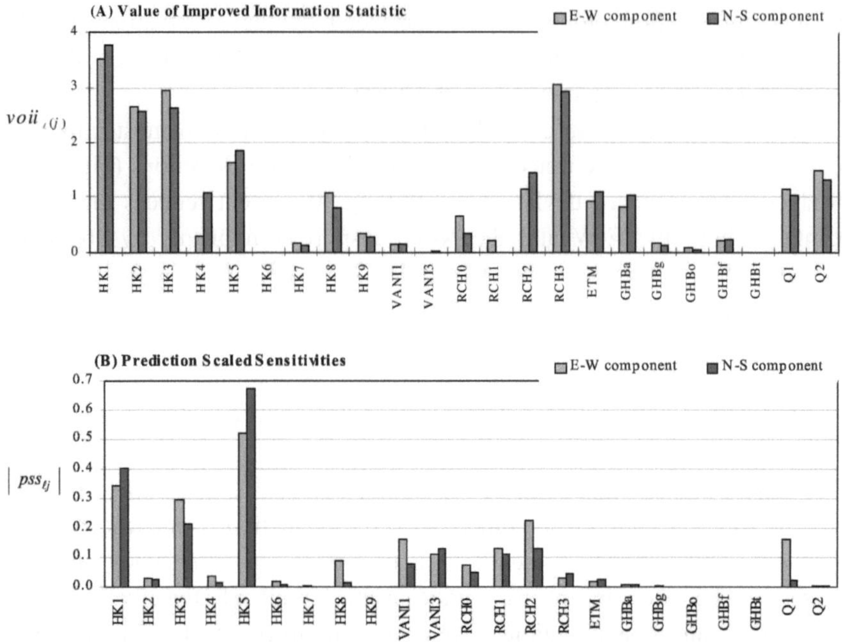

Figure 14. Results of evaluating the importance of all 23 parameters of the Death Valley regional flow model to a predicted advective transport path. (A) The value of improved information (*voii*) statistic, calculated by specifying that the standard deviation of each parameter is reduced by 10 percent. (B) Absolute value of *pss*, defined as the percent change in predicted value produced by a one percent change in the standard deviation of a parameter. (From Tiedeman et al., 2003)

used to gauge the importance of individual observations in an existing network of 501 hydraulic-head monitoring locations to all advective transport paths shown in Fig. 12. Figure 15A shows the results of calculating the averaged *opr* statistic for each of the existing observations. This figure indicates the sets of 100 observations that are most and least important to the advective-transport predictions. The results clearly show that important observations are not necessarily located near the predictions of interest on or near the Nevada Test Site.

Figure 15B shows the results of using the averaged *opr* statistic to assess the importance of potential new hydraulic-head observations in the uppermost layer of the DVRFS model, in the context of the advective-transport predictions. Areas with large values (shown in darker shades of gray) on this map are locations where adding a hydraulic-head observation would produce a relatively large percent decrease in the calculated prediction standard deviation. As for the analysis of the existing monitoring network at this site, areas that rank as important can be distant from the locations of the advective transport predictions.

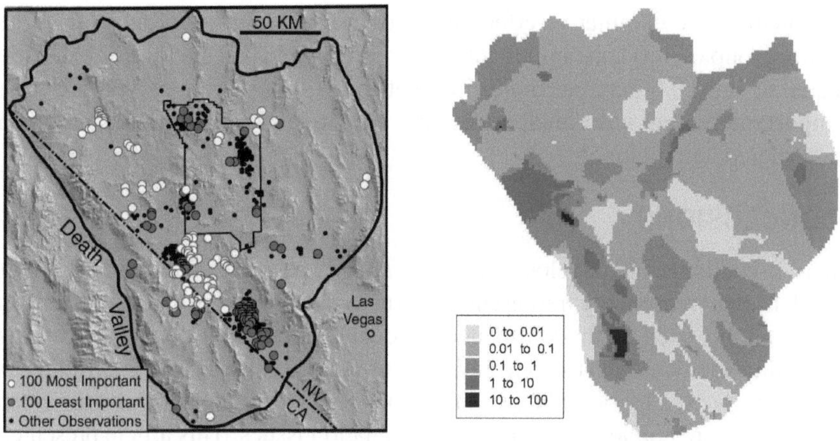

(A) (B)

Figure 15. Results of determining observation importance to advective transport predictions in the DVRFS model. An averaged value of the $opr_{\ell(-i)}$ statistic was used to rank (A) the 501 existing head observation locations (modified from Hill et al., 2001) and (B) potential new observation locations in each cell of model layer 1. (Modified from J. Fenelon and M.C. Hill, U.S. Geological Survey, written communication, 2001) (Colour reproduction on Plate 4)

There are two reasons why important existing or potential observations can be located relatively far from the predictions of interest, as shown in Fig. 15. First, the hydrogeologic units and patterns of areal recharge and boundary flows in this system control the shape of the water table, which varies considerably in gradient. These variations in regional gradients can influence the hydraulic gradients in the vicinity of the advective transport paths of interest, even though they are some distance away. Second, existing or potential observations that rank as important to the predictions are generally in locations where hydraulic head is sensitive to one or more parameters to which the predictions are also sensitive. In this model, a single hydraulic conductivity zone typically represents several different rock types located in different geographic areas that are thought to have similar hydraulic conductivity values. Thus, observations far from the predictions can lie in a hydraulic-conductivity zone that is also near the predicted paths. If the rocks at the different locations are actually similar hydraulically, it allows the model to gain information about the hydraulic conductivity of the rocks near the predictions from a hydraulic head at a distant location. However, problems may occur if the hydraulic conductivity of a distant rock type is actually different from that of a rock type in the same hydraulic conductivity zone near an advective path.

The results shown in Fig. 15A were obtained by evaluating each observation individually. Because these results might be used to remove several of the less important observations from the network, it is important to calculate the cumulative effect of removing such a group. Removal of the entire group of 100 least important observations yields an average *opr* statistic equal to only 0.6 percent, indicating that the 100 observations that are least important on an individual basis are also relatively unimportant when considered cumulatively.

CONCLUSIONS

It is a challenge to develop and calibrate groundwater flow and transport models that reasonably represent the true system and produce predictions that can be used with confidence in decision making. Reasons for this difficulty include inaccessibility of the subsurface, data scarcity, differences between the scales of field measurements and the model characteristics to which they are applied, and numerical simulation issues. This article presented inverse and associated methods for addressing some of these modeling challenges. These methods provide strategies for (1) using existing data about the system state and system characteristics as effectively as possible, (2) addressing special issues related to transport model calibration, (3) diagnosing model error, and (4) identifying potential data that would improve prediction reliability. Use of these strategies can help modelers gain considerable insight into the system being analyzed, identify the most appropriate simulation approaches, and achieve a model that is the best possible representation of the true system.

REFERENCES

Alley, W.M. and Emery, P.A. (1986). Groundwater model of the Blue River basin, Nebraska—Twenty years later. *J. Hydrol.*, **85:** 225-249.

Anderman, E.R., Hill, M.C. and Poeter, E.P. (1996). Two-dimensional advective transport in groundwater flow parameter estimation. *Ground Water*, **34(6):** 1001-1009.

Anderman, E.R. and Hill, M.C. (1999). A new multi-stage groundwater transport inverse method: Presentation, evaluation, and implications. *Water Resour. Res.*, **35(4):** 1053-1063.

Anderman, E.R. and Hill, M.C. (2001). MODFLOW-2000, the U.S. Geological Survey modular groundwater model—Documentation of the Advective-transport observation (ADV2) Package, version 2. *U.S Geol. Surv. Open-File Rep.* 01-54, 69 pp.

Andersen, P.F. and Lu, S. (2003). A post-audit of a model-designed groundwater extraction system. *Ground Water,* **41(2):** 212-218.

Anderson, M.P. and Woessner, W.W. (1992). *Applied Groundwater Modeling.* Academic Press, San Diego, 381 pp.

Barlebo, H.C., Hill, M.C., Rosbjerg, D. and Jensen, K.H. (1998). Concentration data and dimensionality in groundwater models. Evaluation using inverse modeling. *Nordic Hydrology*, **29:** 149-178.

Barth, G.R. and Hill, M.C. (2005). Numerical methods for improving sensitivity analysis and parameter estimation of virus transport simulated using sorptive-reactive processes. *Journal of Contaminant Hydrology. 76*: 251-277.

Barth, G.R. and Hill, M.C. (in press). Parameter and observation importance in modeling virus transport in saturated systems - Investigations in a homogenous system. *Journal of Contaminant Hydrology.*

Barth, G.R., Hill, M.C., Illangasekare, T.H. and Rajaram Harihar (2001). Predictive modeling of flow and transport in a two-dimensional intermediate-scale, heterogeneous porous media. *Water Resour. Res.*, **37(10):** 2503-2512.

Cooley, R.L. (1985). A comparison of several methods of solving nonlinear regression groundwater flow problems. *Water Resour. Res.*, **21(10):** 1525-1538.

Cooley, R.L. and Hill, M.C. (1992). A comparison of three Newton-like nonlinear least-squares methods for estimating parameters of groundwater flow models. *In:* Russell, T.F., Ewing, R.E., Brebbia, C.A., Gray, W.G., and G.F. Pinder (eds), Computational Methods in Water Resources 9th, vol. 1: Numerical methods in water resources. Elsevier, 379-386.

D'Agnese, F.A., Faunt, C.C., Turner, A.K. and Hill, M.C. (1997). Hydrogeologic evaluation and numerical simulation of the Death Valley Regional groundwater flow system, Nevada and California. *U.S. Geol. Survey Water-Resources Inv. Rept.* 96-4300, 124 pp.

D'Agnese, F.A., Faunt, C.C., Hill, M.C. and Turner, A.K. (1999). Death Valley regional groundwater flow model calibration using optimal parameter estimation methods and geoscientific information systems. *Adv. Water Resour.*, **22(8):** 777-790.

D'Agnese, G.M., O'Brien, F.A., Faunt, C.C., Belcher, W.R. and Carma San Juan (2002). A three-dimensional numerical model of predevelopment conditions in the Death Valley Regional groundwater flow system, Nevada and California. *U.S. Geol. Survey Water-Resources Inv. Rept.* 02-4102, 114 pp.

Davison, A.C. and Hinckley, D.V. (1997). *Bootstrap methods and their application.* New York: Cambridge University Press, 582 pp.

Doherty, J. (1994). PEST: Corinda, Australia. Watermark Computing, 122 pp.

Doherty, J. (2003). PEST-2000: Corinda, Australia. Watermark Computing, *<http://www.sspa.com/PEST/index.html>*

Draper, N.R. and Smith, H. (1998). *Applied regression analysis* (3nd ed.), New York, John Wiley & Sons, 706 pp.

Efron, B. (1982). *The jackknife, the bootstrap, and other resampling plans.* Philadelphia, Society of Industrial and Applied Mathematics, 92 pp.

Ely, D.M., Tiedeman, C.R., Hill, M.C. and O'Brien, G.M. In review, A method for evaluating the importance of observations to model predictions, with application to the Death Valley Regional Groundwater Flow System, journal article.

Essaid, H.I., Cozzarelli, I.M., Eganhouse, R.P., Herkelrath, W.N., Bekins, B.A. and Delin, G.N. (2003). Inverse modeling of BTEX dissolution and biodegradation at the Bemidji, MN crude-oil spill site. *J. Contam. Hydrol.*, **67:** 269-299.

Faunt, C.C., Blainey, J.B., Hill, M.C., D'Agnese, F.A. and O'Brien, G.M. In review, Chapter F, Transient numerical model of groundwater flow, in Belcher, Wayne, ed., Evaluation of the Death Valley regional groundwater flow system, Nevada and California. U.S. Geological Survey.

Feehley, C.E., Zheng, C. and Molz, F.J. (2000). A dual-domain mass transfer approach for modeling solute transport in heterogeneous porous media, application to the MADE site. *Water Resour. Res.*, **36(9):** 2501-2515.

Gaganis, P., Karapanagioti, H.K. and Burganos, V.N. (2002). Modeling multi-component NAPL transport in the unsaturated zone with the constituent averaging technique. *Adv. Water Resour.*, **25:** 723-732.

Gailey, R.M., Gorelick, S.M. and Crowe, A.S. (1991). Coupled process parameter estimation and prediction uncertainty using hydraulic head and concentration data. *Adv. Water Resour.*, **14k(5):** 301-314.

Ghandi, R.K., Hopkins, G.D., Goltz, M.N., Gorelick, S.M. and McCarty, P.L. (2002a). Full-scale demonstration of in situ cometabolic biodegradation of trichloroethylene in groundwater, 1. Dynamics of a recirculating well system. *Water Resour. Res.*, **38(4):** 10.1029/2001WR000379.

Ghandi, R.K., Hopkins, G.D., Goltz, M.N., Gorelick, S.M. and McCarty, P.L. (2002b). Full-scale demonstration of in situ cometabolic biodegradation of trichloroethylene in groundwater, 2. Comprehensive analysis of field data using reactive transport modeling. *Water Resour. Res.*, **38(4):** 10.1029/2001WR000380.

Hanson, R.T. (1996). Post-audit of head and transmissivity estimates and groundwater flow models of Avra Valley, Arizona. *U.S. Geol. Survey Water-Resources Inv. Rept.* 96-4045.

Harbaugh, A. W., Banta, E. R., Hill, M.C. and McDonald, M.G. (2000), MODFLOW-2000. The U.S. Geological Survey modular groundwater model-Users guide to modularization concepts and the groundwater flow process. *U.S. Geol. Surv. Open-File Rep.* 00-92, 121 pp.

Helsel, D.R. and Hirsch, R.M. (2002). Statistical methods in water resources. *U.S. Geol. Survey Techniques in Water Resources*, Book 4, Chapter A3, 510 pp, *<http://pubs.water.usgs.gov/twri4a3>*.

Hill, M.C. (1990). Relative efficiency of four parameter-estimation methods in steady-state and transient groundwater flow models. *In:* Gambolati, G., Rinaldo, A., Brebbia, C.A., Gray, W.G. and Pinder, G.F. (eds.), Computational Methods in Subsurface Hydrology, International.

Hill, M.C. (1998). Methods and guidelines for effective model calibration. *U.S. Geol. Survey Water-Resources Inv. Rept.* 98-4005, 90p. Accessed 21 February 2004 at *<http://pubs.water.usgs.gov/wri984005/>*

Hill, M.C. and Tiedeman, C.R. (2003). Weighting observations in the context of calibrating groundwater models. *In:* Kovar, K. and Hrkal, Z. (eds.), Calibration and Reliability in Groundwater Modeling: A Few Steps Closer to Reality (ModelCARE 2002). IAHS Publication 277, 196-203.

Hill, M.C., Banta, E.R., Harbaugh, A.W. and Anderman, E.R. (2000). MODFLOW 2000. The U.S. Geological Survey modular groundwater model, User's guide to the observation, sensitivity, and parameter-estimation processes. *U.S. Geol. Surv. Open-File Rep.* 00-184, 209 pp.

Hill, M.C., Ely, M.D., Tiedeman, C.R., D'Agnese, F.A. Faunt, C.C. and O'Brien, G.A. (2001). Preliminary evaluation of the importance of existing hydraulic-head observation locations to advective-transport predictions, Death Valley regional flow system, California and Nevada. *U.S. Geol. Survey Water-Resources Inv. Rept.* 00-4282, 82p, Accessed 21 February 2004 at *<http://water.usgs.gov/pubs/wri/wri004282/>* .

Jacques, D., Šimùnek, J., Timmerman, A. and Feyen, J. (2002). Calibration of Richards' and convection-dispersion equations to field-scale water flow and solute transport under rainfall conditions. *J. Hydrol.*, **259:** 15-31.

Julian, H.E., Boggs, J.M., Zheng, C. and Feehley, C.E. (2001). Numerical simulation of a natural gradient tracer experiment for the Natural Attenuation Study: Flow and physical transport. *Ground Water*, **39(4):** 534-545.

Keidser, A. and Rosbjerg, D. (1991). A comparison of four inverse approaches to groundwater flow and transport parameter identification. *Water Resour. Res.*, **27(9):** 2219-2232.

Konikow, L.F. (1986). Predictive accuracy of a groundwater model-Lessons from a post-audit. *Ground Water*, **24(2):** 173-184.

Konikow, L.F. and Person, M.A. (1985). Assessment of long-term salinity changes in an irrigated stream-aquifer system. *Water Resour. Res.*, **21:** 1611-1624.

Kosugi, K. and Inoue, M. (2002). Estimation of hydraulic properties of vertically heterogeneous forest soil from transient matric pressure data. *Water Resour. Res.*, **38(12):** 1322, doi:10.1029/2002WR001546.

Mahar, P.S. and Datta, Bithin (2001). Optimal identification of groundwater pollution sources and parameter estimation, *J. Water Resour. Plann. Manage.*, **127(1):** 20-29.

Medina, A. and Carrera, J. (1996). Coupled estimation of flow and transport parameters. *Water Resour. Res.*, **32(10):** 3063-3076.

Mehl, S.W. and Hill, M.C. (2001). A comparison of solute-transport solution techniques and their effect on sensitivity analysis and inverse modeling results. *Ground Water*, **39(2):** 300-307.

Molson, J.W. and Frind, E.O. (2002). WATFLOW/WTC User Guide and Documentation, Version 3: Waterloo, Ontario, Canada. Department of Earth Sciences, University of Waterloo, 75 pp.

Olyphant, G.A. (2003). Temporal and spatial (down profile) variability of unsaturated soil hydraulic properties determined from a combination of repeated field experiments and inverse modeling. *J. Hydrol*, **281:** 23-35.

Poeter, E.P. and Hill, M.C. (1997). Inverse modeling: A necessary next step in groundwater modeling: *Ground Water*, **35(2):** 250-260.

Poeter, E.P. and Hill, M.C. (1998). Documentation of UCODE: A computer code for universal inverse modeling. U.S. Geol. Survey Water-Resources Inv. Rept. 98-4080. 116 pp.

Pollock, D.W. (1989). Documentation of computer programs to compute and display pathlines using results from the U.S. Geological Survey modular three-dimensional finite-difference groundwater flow model. U.S. Geol. Surv. Open-File Rep., 89-381, 188 pp.

Prommer, H., Barry, D.A. and Zheng, C. (2003). MODFLOW/MT3DMS-based multicomponent transport modeling. *Ground Water*, **41(2):** 247-257.

Reichard, E.G. and Meadows, J.K. (1992). Evaluation of a groundwater flow and transport model of the upper Coachella Valley, California. U.S. Geol. Survey Water-Resources Inv. Rept. 91-4142.

Russell, T.F. (2002). Numerical dispersion in Eulerian-Lagrangian methods. *In:* S.M. Hassanizadeh et al. (eds), Proceedings of the XIVth International Conference on Computational Methods in Water Resources. Elsevier, Amsterdam, 963-970. Accessed 18 May 2004 at *<http://www-math.cudenver.edu/ccm/reports/rep182.ps.gz>*

Saltelli, A., Chan, K. and Scott, E.M. (2000). *Sensitivity Analysis*. John Wiley & Sons, NY, 475 pp.

Saltelli, A., Tarantola, S., Campolongo, F. and Ratto, M. (2004). Sensitivity Analysis in Practice: A Guide to Assessing Scientific Models. John Wiley & Sons, NY, 232 pp.

Schijven, J.F., Hoogenboezem, W., Hassanizadeh, S.M. and Peters, J.H. (1999). Modeling removal of bacteriophages MS2 and PRD1 by dune recharge at Castricum, Netherlands. *Water Resour. Res.*, **35(4):** 1101-1111.

Šimůnek, J., van Genuchten, M.Th., Jacques, D., Hopmans, J.W., Inoue, M. and Flury, M. (2002). Solute transport during variably saturated flow: Inverse methods. *In:* Dane, J.H. and G.C. Topp (eds.), Methods of Soil Analysis. Part 4. Physical Methods. SSSA Book Series 5, Soil Science Society of America, Madison, WI, p. 1435-1449.

Sonnenborg, T.O., Engesgaard, P. and Rosbjerg, D. (1996). Contaminant transport at a waste residue deposit: 1. Inverse flow and nonreactive transport modeling. *Water Resour. Res.*, **32(4):** 925-938.

Stewart, M. and Langevin, C. (1999). Post-audit of a numerical prediction of wellfield drawdown in a semiconfined aquifer system. *Ground Water*, **37(2):** 245-252.

Strecker, E.W. and Chu, W. (1986). Parameter identification of a groundwater contaminant transport model. *Ground Water*, **24(1):** 56-62.

Sulieman, H., McLellan, P.J. and Bacon, D.W. (2001). A profile-based approach to parameteric sensitivity analysis of nonlinear regression models. *Technometrics*, **43(4):** 425-433.

Sun, N.-Z. (1994). *Inverse problems in groundwater modeling*. Boston, Kluwer Academic Publishers, 337 pp.

Sun, N.-Z. and Yeh, W.-G. (1990). Coupled inverse problems in groundwater modeling, 1, Sensitivity analysis and parameter identification. *Water Resour. Res.*, **26(10):** 2507-2525.

Tiedeman, C.R., Hill, M.C., D'Agnese, F.A. and Faunt, C.C. (2003). Methods for using groundwater model predictions to guide hydrogeologic data collection, with application to the Death Valley regional groundwater flow system. *Water Resour. Res.*, **39(1):** 1010, doi:10.1029/2001WR001255.

Wagner, B.J. (1992). Simultaneous parameter estimation and contaminant source characterization for coupled groundwater flow and contaminant transport modeling. *J. Hydrol.*, **135:** 275-303.

Wagner, B.J. (1995). Sampling design methods for groundwater modeling under uncertainty. *Water Resour. Res.*, **31(10):** 2581-2591.

Wagner, B.J. (1999). Evaluating data worth for groundwater management under uncertainty. *J. Water Resour. Plann. Manage.*, **125(5):** 281-288.

Wagner, B.J. and Gorelick, S.M. (1987). Optimal groundwater quality management under parameter uncertainty. *Water Resour. Res.*, **23(7):** 1162-1174.

Yager, R.M. (2002). Simulated Transport and Biodegradation of Chlorinated Ethenes in a Fractured Dolomite Aquifer Near Niagara Falls, New York. U.S. Geol. Survey Water-Resources Inv. Rept. 00-4275, 55 pp. Accessed 14 April 2004 at *http://ny.water.usgs.gov/pubs/wri/wri004275/*

Zheng, C. (1994). Analysis of particle tracking errors associated with spatial discretization. *Ground Water*, **32(5):** 821-828.

Zheng, C. and Wang, P. (1998). MT3DMS-A modular three-dimensional multispecies transport model for simulation of advection, dispersion and chemical reactions of contaminants in groundwater systems. University of Alabama, Tuscaloosa.

Zheng, C. and Bennett, G.D. (2002). Applied contaminant transport modeling, second edition, New York, John Wiley and Sons, 621 pp.

Zheng, C. and Gorelick, S.M. (2003). Analysis of solute transport in flow fields influenced by preferential flowpaths at the decimeter scale. *Ground Water*, **41(2):** 142-155.

Zou, Z.-Y., Young, M.H., Li, Z. and Wierenga, P.J. (2001). Estimation of depth averaged unsaturated soil hydraulic properties from infiltration experiments. *J. Hydrol.*, **242:** 26-42.

10

Groundwater Development and Management of Coastal Aquifers (including Island Aquifers) through Monitoring and Modeling Approaches

A. Ghosh Bobba

National Water Research Institute
Burlington, ON, Canada, L7R 4A6

PREAMBLE

More than half the world's population now lives within 80 kilometres of a shoreline and human population in these regions is growing exponentially, such that it is expected to double or triple in the next two decades in North America alone. Human activities on coastal and island watersheds provide the major sources of nutrients and pollutants entering shallow coastal ecosystems. Contamination loadings from watersheds are the most widespread factor that alters structure and function of receiving aquatic ecosystems. Surface water in the coastal and island watersheds interacts with adjacent subsurface water. Subsurface water discharges are recognized as potentially significant contaminant pathway from the land to the coast by human activities. This interaction affects the water quality and quantity in both surface and subsurface waters. The diversity, in combination with the range of disciplines and of time and space scales involved; complicate the use of data for purposes other than those envisioned by the original investigators. As land and water resource development increases in the coastal watersheds, it is becoming readily apparent that subsurface and surface waters interaction must be considered in establishing water management policies. On a more regional scale, human activity has already demonstrated heavy impacts on coastal zones and further demands are increasingly threatening the integrity and health of biologically rich aquatic ecosystems, including coral reefs, mangroves, sea grass beds, lagoons, wetlands, bays and near shore waters. Describing environmental pollutants and pollution in coastal areas, application of a finite element numerical model to four cases is presented.

INTRODUCTION

We live on a planet covered nearly 70 percent by water. However, almost all of the Earth's water is in the ocean; less than five percent is fresh. Of all the freshwater, just over 30 percent is groundwater. As an example, in the United States, groundwater supplies 50 percent of the people with drinking water.

Ecosystem of the world's coastlines is receiving extraordinary amounts of nutrients as consequences of human activities such as fertilizers, industrial emissions to the atmosphere, and disposal of wastewater in watersheds adjoining coastal waters. The loadings of nitrogen and phosphorous to coastal aquatic environments even exceed those to fertilized agro-ecosystem. Increased nutrient loading from anthropogenic sources is pervasive and function of shallow coastal ecosystems during coming decades.

Although increased nutrient loading by precipitation has been documented, most research has focused on deeper estuaries in which flow from rivers and streams dominates water budgets and contributes the major part of nutrients. Rivers and direct precipitation, however, are not the sole source of freshwater borne nutrients to coastal environments. Even in places without rivers, salinity is often depleted in coastal waters due to groundwater input. Groundwater flow is especially important where underlying coastal sediments are coarse, unconsolidated sands of glacial or marine origin. In such situations flow of groundwater may be the major source of nutrients to coastal waters.

Groundwater and pollution discharge from coastal watersheds to coastal waters and estuarine waters has been a topic of theoretical and practical interest for at least a century (Valiela et al., 1990, 1992; Bobba and Singh, 2003). Most works have stressed controls on saltwater intrusion to freshwater aquifers and it has been only recently that field measurements of groundwater discharge have shown the importance of subsurface water flow on water and nutrient budgets in coast and estuaries. Pulled by gravity, groundwater seeps from the surface slowly downward through aquifers in the earth's subsurface and eventually discharges into lakes, rivers, and the coastal ocean. The discharge of groundwater directly into marine waters is called **S**ubmarine **G**roundwater **D**ischarge (SGD). While it is an unseen phenomenon, the influence of submarine groundwater discharge on the ecology of coastal systems may be more important than once thought, due to the potential impacts resulting from contaminants carried in groundwater. This phenomenon is being studied by scientists to better understand the interaction of groundwater and surface waters along coastlines.

In unconsolidated sediments, groundwater moves through the watershed shoreward in paths that have downward vertical as well as horizontal vector. Downward flow is caused by additional water infiltrating along the path of the water. Freshwater eventually moves close enough to shore to meet the denser saltwater that saturates interstitial space in sediments beneath the sea. The presence of sea water in the pore space acts together with lower head pressures in the near shore zone compared with offshore to deflect the path of fresh groundwater sharply upward (Bobba, 1993a). As a result, most of the groundwater flow occurs very near the shore. The importance of groundwater is not so much because of the magnitude of flow rates, but rather because of the high nutrient concentrations in groundwater compared to those in receiving coastal water. Although highly variable, the nutrient content of groundwater discharging onto coastal water may be up to five orders of magnitude larger than concentrations in receiving seawater (Valiela et al., 1990, 1992).

GROUNDWATER BASICS

Groundwater is one part of a continuous water cycle between land, ocean and atmosphere called the Hydrologic Cycle (Fig. 1). The details of groundwater basics are given in hydrological literature (Bobba and Singh, 1995; Freeze and Cherry, 1975). Surface water, from bodies of water such as oceans and lakes, evaporates from the Earth's surface by energy from the Sun. Water vapour forms clouds and eventually returns to the Earth's surface in the form of rain, snow, or other precipitation. This water re-enters surface waters either directly or through surface runoff.

The remainder seeps underground and flows slowly through geologic formations called aquifers. An aquifer is an underground formation of loose soil and permeable rock that is capable of storing useful quantities of water. Aquifers also transmit water. The rate at which the groundwater travels depends on the pressure from the upgradient direction, the size of the spaces in the soil or rock and how well the spaces are connected, or simply the extent of permeability. There are two primary types of aquifers: (1) unconfined or water table (Fig. 3A) and (2) confined (Fig. 4A).

An unconfined aquifer, or water table aquifer (Fig. 4), has no confining layer between the top of its saturated zone and the earth's surface. The water level, or water table, in this aquifer rises and falls in response to infiltration of rainfall, evapotranspiration, pumpage, and discharge to surface waters. Water table aquifers are typically the uppermost aquifers and are more susceptible to contamination resulting from surface intrusions such as wastewater systems, chemical spills and fertilizers.

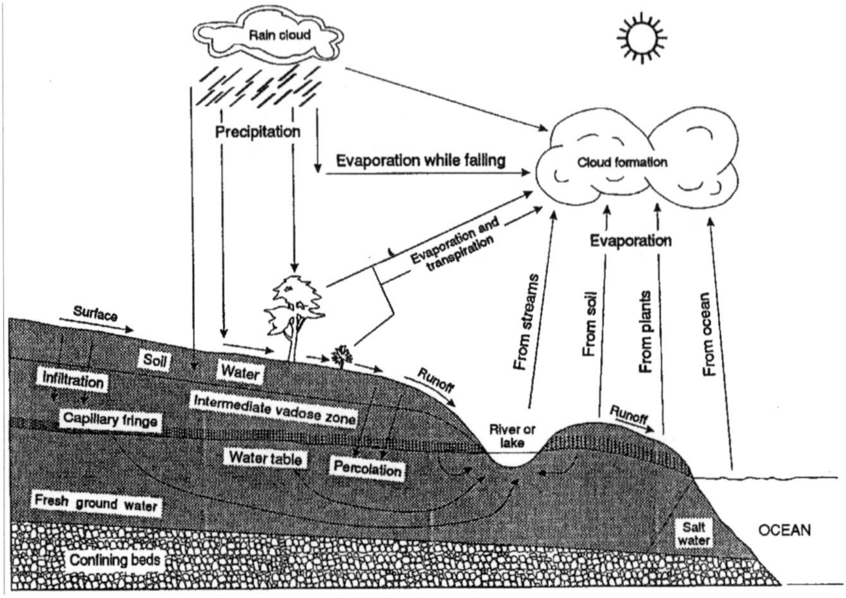

Figure 1. Hydrological Cycle.

A relatively impermeable layer, a confining unit or aquitard, overlies a confined aquifer (Fig. 4). Usually, the groundwater found in these confined aquifers is under extreme pressure. If a well penetrates the aquifer, the water will rise above the top of the aquifer, referred to as the piezometric surface. Water that flows freely from a well drilled into a confined aquifer is called an artesian well. The water moves upward due to pressure within the aquifer.

Groundwater in any aquifer typically flows from areas of high hydraulic head to areas of low hydraulic head under the force of gravity. More simply, groundwater flows downhill. Yet it must follow a tortuous path through small pores in the aquifer material, typically slowing its flow to a crawl. Unlike surface waters that move at noticeably rapid rates, groundwater may only move a few inches in a day. Ultimately, it flows into the coastal ocean through seeps and springs, thus closing the continuous water cycle between land, ocean and atmosphere.

There has been relatively little scientific study of the magnitude and effects of submarine groundwater discharge, although its importance has been recognized for some time (Bobba, 1993a). People have observed springs and seeps of fresh water in the marine environment over the centuries.

Groundwater/Coastal Water Interface

The Badon Ghyben-Herzberg approximation defines the depth to the freshwater/salt-water interface beneath a coastal aquifer under static conditions (Bobba, 1993a). However, groundwater is not static and this simplified relationship is not reliable wherever significant vertical flow occurs. The fresh water and salt water may be separated by a brackish mixing zone in which the interface is defined by the isochlor describing a mixture of 50% sea water and 50% groundwater. The thickness of the mixing zone is a function of the physical agitation of tides and the pumping regime as well as the hydraulic character of the aquifer. Groundwater seeks a discharge area at the coast and flows from beneath the area of highest groundwater head (inland) upwards and outwards along the saltwater interface with the effect that the mixing zone is constantly flushed to brackish coastal and submarine springs. A strong horizontal discharge maintains the freshwater lens and tends to limit the development of the mixing zone, which may be only a few metres thick. In the absence of a significant vertical groundwater flow component, the Badon Ghyben-Herzberg approximation provides a good estimate of the depth to the interface (Fig. 2).

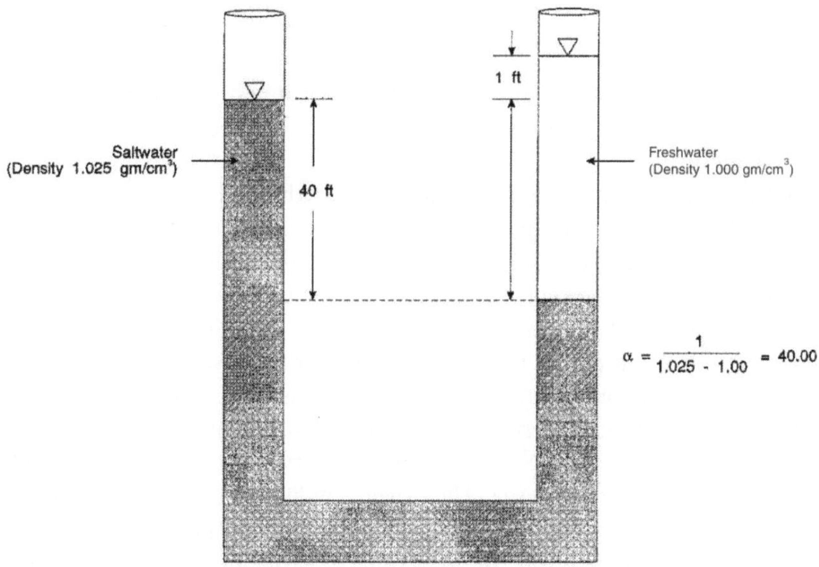

Figure 2. Tube concept illustrating the hydrostatic balance between fresh water and salt water.

Saltwater intrusion of aquifers may originate from various sources, including (1) encroachment of sea water in coastal areas, (2) sea water that entered aquifers during deposition or during a high stand of the sea in past geologic time (connate water), (3) salt in salt domes, thin beds, or disseminated in the geologic formations, (4) slightly saltwater concentrated by evaporation in tidal lagoon, playas, or other enclosed areas, (5) return flows to streams from irrigated lands, and (6) man's saltwater wastes. A major factor in determining the movement of freshwater and saltwater is the relative density between the two. The density of water in natural aquifer systems ranges from 0.9982 kg L^{-1} at 20° C for pure freshwater to greater than the 1.345 kg L^{-1} reported for the Salado brine of New Mexico. The density of "average" surface seawater ranges between 1.022 and 1.028 kg L^{-1}; this value is partly dependent on the temperature as well as the solute concentration. Because freshwater is less dense than saltwater, the former will tend to float above the latter. For the hydrostatic case shown in Fig. 2, the freshwater stands 1 ft higher than the saltwater. Writing an equation for the equality of pressure on each side of the U-tube yields

$$h_s = \frac{\rho_f}{\rho_s - \rho_f} h_f \tag{1}$$

where ρ_f and ρ_s are the freshwater and saltwater densities, respectively, h_f is the freshwater head above the saltwater, and h_s is the distance from the saltwater surface to the interface between the freshwater and saltwater. For seawater with a density of 1.025 g/cm³, Equation (1) reduces to

$$h_s = 40 \times h_f \tag{2}$$

Although not a hydrostatic situation, subsequent investigations have shown that Eq. (2) is a good first approximation to the depth for nearly horizontal flow conditions.

The interface, consisting of the surface between the freshwater and saltwater, has a parabolic form, with the saltwater tending to under ride the less dense freshwater. Figure 3a shows cross sections of an unconfined aquifer for two conditions; Fig. 3a, represents equilibrium between the seaward-flowing fresh water and seawater, whereas Fig. 3b indicates the intrusion of seawater into the aquifer when extractions reduce the freshwater flow. Similar situations for a confined coastal aquifer are shown in Figs. 4a and 4b. Whatever the flow situation, a dynamic equilibrium tends to develop. It can be shown that the length of the intruded wedge of saltwater varies inversely with the magnitude of the freshwater flow to the sea. Thus, even a reduction of freshwater flow is sufficient to cause intrusion. A diagnostic approach should precede the solution of any problem; it is, therefore, of great value to define the modes of saltwater invasion. There are always intricate cases that defy any simple analyses, and for this reason it is usually necessary in applied sciences to idealize the natural conditions in order to have a better comprehension of the problem. Experience and judgment are needed for application of the results of the theoretical ideal cases. In the following discussion, various idealized cases and the availability of their solutions is explained; the practical unknown features are emphasized in the hope of stimulating research workers. It should be noted in this respect that some of these solutions couldn't satisfy practitioners of average training. On the other hand, some of these solutions may include mathematical approximations or certain assumptions that may result in errors of unknown magnitudes.

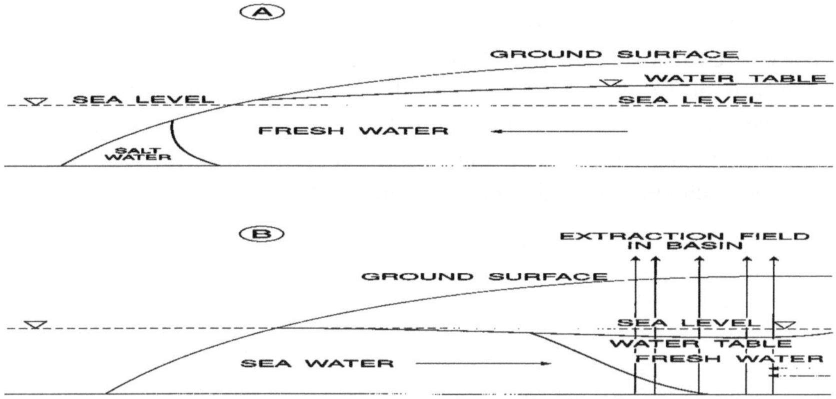

Figure 3. Hydrologic conditions in an unconfined coastal aquifer. (A) Not subject to sea water intrusion. (B) Subject to seawater intrusion.

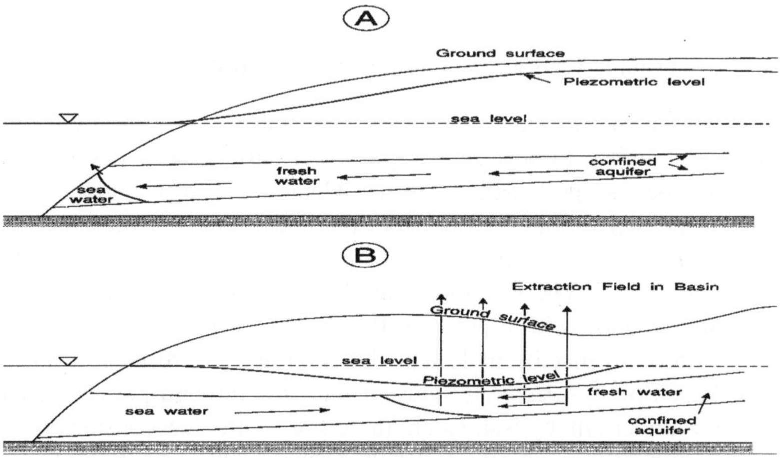

Figure 4. Hydraulic conditions in a confined aquifer in continuity with the seacoast. (A) Not subject to seawater intrusion. (B) Subject to seawater intrusion.

The Transition Zone

The phenomenon of freshwater underlain by saltwater is quite a complex one. In reality the two fluids are miscible and are separated by a transition zone (Figs. 5 and 9) with a continuous upward gradient of salt concentration from saltwater below to uncontaminated above. This results in a slight upward density gradient in the transition zone. The two fluids tend to mix by molecular diffusion and macroscopic dispersion due to bulk flow (Fig. 9). Because saltwater intrusion represents immiscible displacement of liquids in porous media, the processes of hydrodynamic dispersion and molecular diffusion tend to mix the two fluids. As a result, the idealized interfacial surface becomes a transition zone (Fig. 5). The thickness of the zone depends upon the physical structure of the aquifer, the history of extractions from the aquifer, the variability of recharge, and the tides. Steady flows minimize the transition zone thickness, but variable influences such as pumping, recharge, and tides increase the thickness.

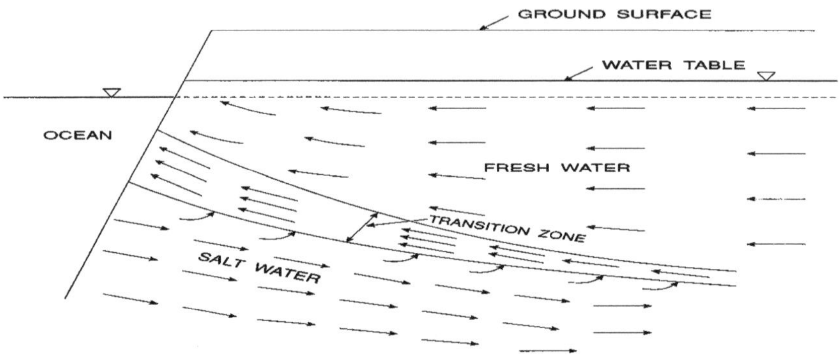

Figure 5. Schematic vertical cross-section showing fresh water and salt water

Flow within the transition zone varies from that of the freshwater body at the upper surface to near zero at the lower surface. The velocity in the transition zone transports salt to the sea. Continuity considerations suggest that the salt discharge must come from the underlying saltwater dispersing upward into the zone. It follows that there must be landward saltwater flow, as shown in Fig. 5. The transition zone becomes wider and more diffused as the interface moves in response to flow in one or both of the fluids. The two fluids tend to mix by molecular diffusion and macroscopic dispersion (due to bulk flow). A rigorous study of the flow through porous media of freshwater underlain by saltwater, therefore, calls for consideration of hydrodynamic dispersion (Bear, 1979; Kashef, 1986, 1977; Reddel and Sunada, 1970) and simultaneous solution of equations of hydrodynamic dispersion and bulk flow with appropriate boundary conditions.

Types of Islands

From a hydrological perspective, a small island can be considered to be one on which water resources are very scarce. Water development measures on small islands are special and usually they are not those normally considered as standard on continents or larger islands. By definition, small islands are those having an area equal or less than 2000 km^2 and the width does not exceed 10 km (Falkland, 1991). The definition also introduces the concept of "very small islands" which includes islands whose surface area does not exceed 100 km^2 or the width is not greater than 3 km and where the water scarcity is even more acute and surface water resources are normally absent.

On small or very small islands groundwater is normally the only naturally occurring water resource, where surface water is very scarce and difficult to catch if any.

A detailed generic classification of small islands has been developed by Falkland (1991), based on the key aspects of geology, climate, freshwater lens geometry and water balance. Islands are often conveniently separated into two broad groups: 'Low' islands and 'High' islands. The high islands are primarily comprised of volcanic rocks, but also often contain substantial quantities of sedimentary materials, including raised coralline limestone and debris. The high islands are older and much larger than the low islands, and generally contain developable quantities of surface water and groundwater. The low islands, which are mainly comprised of coralline atolls and reef material, are very small and usually stand only a few metres above sea level. They have little if any surface water, thus, users must rely on direct rainfall catchment and a thin, often brackish, groundwater lens. This classification can be reduced to six basic island types and classification of an island within these types may help to identify its groundwater flow regime. Further sub-division is always possible.

Each of the six main island types can be analyzed by the adequacy of four key characteristics: (i) recharge, (ii) storage, (iii) yield, and (iv) quality. Examples of islands and their island type are given in Table 1. Adequacy,

Table 1. Hydrological Characters of Islands (Falkland, 1991)

Island Type	*Recharge*	*Storage*	*Yield*	*Water quality*	*Examples*
Geologically young 'high rise' volcanic island	Adequate	Adequate	Adequate	Adequate St. Vincent	Hawaii
Geologically old 'high rise' volcanic island	Adequate	Possibly Adequate	Possibly inadequate	Adequate	St Helena
Low elevation coral limestone island	Possibly Inadequate	Inadequate	Adequate	Possibly inadequate	Caribbean and Pacific Islands e.g. Bermuda
Recent calcareous sedimentary island	Inadequate	Possibly adequate	Adequate	Possibly adequate	Turcs and Calcos, Pacific
Upland limestone island	Adequate	Inadequate	Inadequate	Adequate	Malta
Near continental bedrock island	Possibly adequate	Possibly adequate	Possibly inadequate	Adequate	Jersey

although subjective, is used here as ability to sustain normal demand levels for potable water supply and is partly a function of population density. Of the four characteristics, recharge depends principally on climate, but it also relates to island size, soil cover, vegetation, elevation and topography. Storage depends on the effective porosity of the aquifer, and also on the elevation of the water table and the location of the saltwater interface. Secondary features, such as the development of weathering and active and fossil karstic horizons in limestone aquifers, are also important. Yield, or more particularly sustainable yield, relates to recharge, storage and transmissivity. Higher transmissivity produces higher borehole yields, but also increases the hydraulic gradient, the thickness of the freshwater lens, and the quantity of groundwater discharge to the sea. Groundwater quality is normally related to seawater intrusion and mixing, but may also relate to land-based pollution.

HYDROLOGICAL CHARACTERISTICS

The major influences on the small islands hydrological characteristics are the climate, physiography, geology, hydrogeology, soils, vegetation cover, location, shape and human intervention.

The climate influences the availability of freshwater. The rainfall quantity and its variation with respect to time and space and the evapotranspiration play an important role on the availability of the freshwater resources. The temporal variation is usually high in small islands where the spatial variation is a function of the islands physiography. The physiography plays an important role on the classification of small islands high or low (Table 2). High islands are those that can influence the precipitation patterns and with some runoff and low islands are those without any effect on the precipitation pattern and no significant surface runoff. High islands have steep topography with flashy surface water resources, since they are volcanic islands, with low permeability bedrocks. These islands have small perennial streams at high elevations and the surface water may form an important component of the water balance. Low islands have a flat topography with a minimum of surface runoff and the groundwater is a major component of the water balance. Low islands are also affected by their height above sea level, which determines the risk of overtopping.

The geology and hydrogeology influences the availability, the type and the distribution of the water resources. Groundwater is abundant in islands with soils and rocks of moderate to high permeability, where surface water resources occur on islands with soils and rock with low permeability. Very high permeability causes the mixing of freshwater and seawater resulting to brackish groundwater.

Table 2. Problems of water quantity and quality on small islands

Water Quantity	*Water Quality*
Climatic conditions are generally difficult, such as capricious rainfall or torrential rains	Seawater intrusion is a serious problem, especially where overexploitation occurs due to increasing population, tourists, industrial or agricultural developments.
Permanent rivers and springs occur only where rainfall is relatively high and well distributed over the year	Islands in tropical regions are frequently exposed to torrential rains, which combined with local factors can cause siltation of water storages and adverse effects on water quality in lagoons. High turbidity may cause water supplies to become non-potable.
River basins are often large in number but small in size and with limited regulation capacity.	High population densities increase the risk of pollution and wastewater disposal options are as important as water supply distribution system.
Few islands have topographical and hydrological conditions suited to the construction of significant storages.	Ocean out falls for disposal of domestic and industrial waste waters are frequently used, but sometimes without proper design, operation and site identification.
Turnover time of groundwater systems tends to be short, generally a few years at most.	Water pollution caused by the uncontrolled use of fertilizers herbicides and pesticides are prevalent in some islands.
Many small islands have high population densities, which places great stress on naturally occurring water resources	The tourist industry, a major economic activity in many small islands, demands large volumes of water with high physical, chemical and bacteriological standards. There is also a need for satisfactory lagoon water quality standards.
Geological conditions may be difficult such as extremely permeable rocks or impermeable rocks without significant storage potential.	There is often the lack of skilled local personnel for the operation and maintenance of conventional treatment plants.

Soil types play an important role in the hydrological cycle. Sandy soils found in limestone and coral islands with high permeability do not allow the creation of surface runoff, where clay soils found in volcanic islands have a lower permeability which allow the creation of surface runoff. Soil water retention capacity is an important factor in the evapotranspiration and recharge. The retention capacity is a function of the texture of the soil and the thickness of the soil. Recharge is enhanced by coarse grain soils (sand) and fine grain soils reduce the recharge.

Vegetation in small islands, generally adapted to the local climatic conditions, consists of a variety of trees, bushes and grasses, and normally does not require irrigation. The type and density of the vegetation affects the hydrological cycle, water interception and transpiration reduce recharge, slow surface runoff and reduce erosion on high islands thus increasing infiltration of water into ground. Generally the benefits of vegetation outweigh the negative effects on the water resources.

Climate change and the potential rise in sea level are critical to many low-lying islands (Fig. 6). A 1-m rise in sea level would create an overall increase in the potentiometric head of 1 m throughout an island relative to a fixed datum. However, many islands are low-lying with a shallow water table, which could not universally accommodate a 1-m rise. The overall reduction in the area of a low-lying island will reduce recharge, and consequently also base flow (Fig. 6). According to Darcy's Law, this will, in turn, reduce the hydraulic gradient towards the coast and the reduced head will cause a reduced depth to the saltwater interface. It is also possible that a rise in the water table could saturate any higher permeability near-surface weathered zone which would allow more rapid drainage to the coast, inhibiting the water table from rising by the full 1-m. Therefore a rise in sea level of 1-m would cause a reduction in the head of groundwater above the new sea level by some thickness less than 1-m, and the thickness of the fresh water lens would correspondingly be reduced (Fig. 6). This might destroy many marginal lenses beneath small low elevation islands (Bobba, 1998; Bobba et al., 2000). A small rise in sea level would turn many low-lying islands into potential wetlands. Elevation of the water table into the root zone or to open water would increase evaporation and reduce aquifer storage.

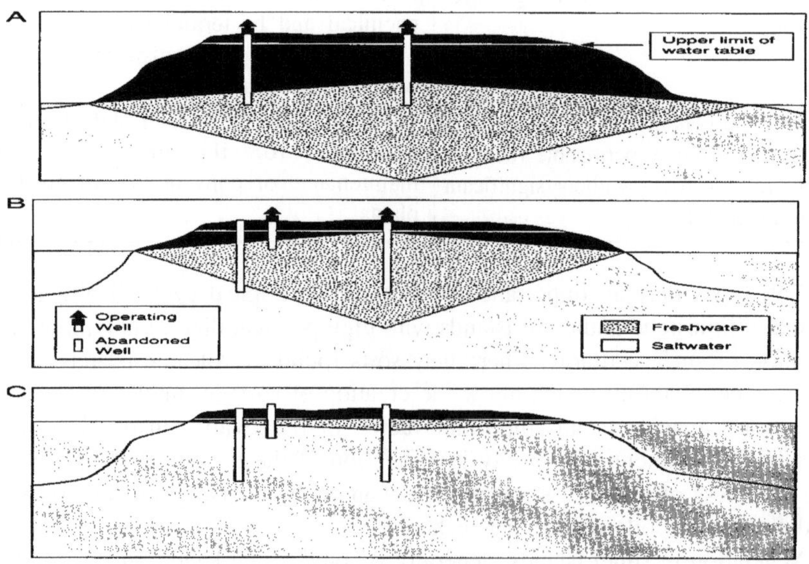

Figure 6. Climate change effects in island

Coastal Watershed Hydrology

A watershed is topographically defined as an area where water enters through precipitation and leaves as evapotranspiration, surface runoff, and subsurface water discharge (Fig. 7). In the case of a coastal watershed, runoff and subsurface water discharge enter the sea (Fig. 7). Rainfall and potential rate of evapotranspiration are determined by climate. Actual evapotranspiration is limited by the climatically controlled potential rate, vegetation and the wetness of the soil. Soil properties, topography, and the history of rainfall and evapotranspiration determine runoff and subsurface water discharge. For the purpose of this discussion, surface runoff includes direct runoff, which occurs as stream flow immediately following a rainstorm, and drainage of subsurface water into creeks, which accounts for stream flow between streams. The amount of direct runoff generated by a storm depends on the amount of rainfall and on the moisture condition of the soil. In general, more runoff occurs when the soil is initially wet. The subsurface water discharge can be calculated if rainfall, evapotranspiration, runoff, and the change in the amount of water stored on the watershed are known.

A link between rising sea level and changes in the water balance is suggested by the general description of the hydraulics of subsurface water discharge at the coast.

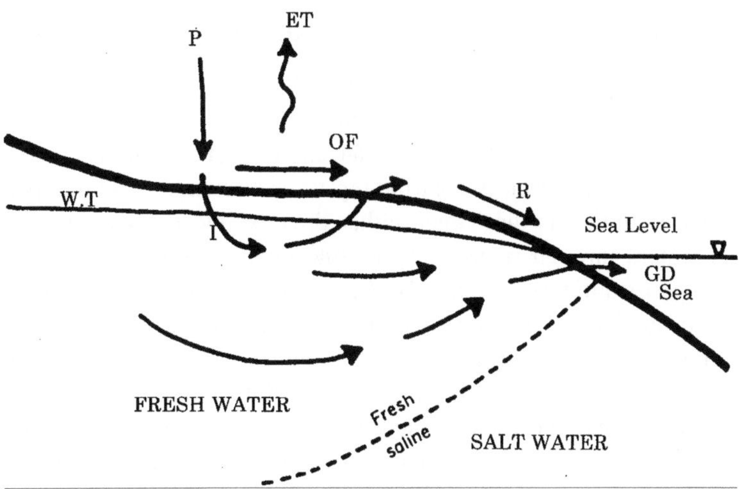

Figure 7. Water balance in coastal watershed

Fresh subsurface water rides up over denser salt water in the subsurface system on its way to the sea (Fig. 7), and subsurface water discharge is focused into a narrow zone that overlaps with inter tidal zone. The width of the zone of subsurface water discharge measured perpendicular to the coast is indirectly proportional to the discharge rate. The shape of the water table and the depth to the fresh/saltwater interface are controlled by the difference in density between freshwater and saltwater, the rate of freshwater discharge and the hydraulic properties of the subsurface system. The elevation of the water table is controlled by mean sea level through hydrostatic equilibrium at the shore. Because pollutants are transported in large part by the bulk motion of subsurface water, the parameters of subsurface water flow are of major importance in the understanding of pollution processes. The various aspects of the subsurface water environments, as well as stratigraphic factors that control or could influence subsurface water motion are also of major consideration. The subsurface hydrology environment is shown schematically in Fig. 7. It consists mainly of saturated and unsaturated zones. The unsaturated zone occurs above the capillary fringe where the soil pores are partially saturated with water. This zone is important in waste management because in most cases, it is the burial zones for wastes. Consequently, a thick unsaturated zone may sometimes be preferred for waste disposal since it would take a much longer time for pollutants to reach the water table. In the saturated zone, the pores are saturated with water. When this zone is capable of transmitting significant quantities of water for economic use it is referred to as an aquifer. In most field situations, two or more aquifers occur, separated by impermeable strata or aquitard. In the situation illustrated in Fig. 7, the upper or unconfined aquifer is much more prone to pollution than the lower confined aquifer.

Environmental Problems in the Coastal and Island Watersheds

The major sources of coastal and marine pollution originating from the land vary from coast to coast. The nature and intensity of development activities, the size of the human population, the state and type of industry and agriculture are but a few of the factors contributing to each country's unique pollution problems. Pollution is discharged either directly into the sea, or enters the coastal waters through rivers and by atmospheric deposition.

In order to mitigate and control the impact of pollution on coastal and marine resources, it is essential that the type and load of pollutants be identified. This involves determination of the sources and their location, and the volume and concentration of the pollutants. Point sources of pollution (Table 3) are sources that can be identified to one location, such as industrial and sewage treatment plants. Point sources, though easy to identify, account only for a fraction of the land-based sources of pollution affecting coastal and marine environments. Non-point sources (Table 4) are harder to identify, and include urban storm water run-off and overflow discharges, as well as runoff from forest and agriculture.

Pollution sources can be located relatively far away from coastal areas and still have an impact. Pollutants from sources and activities within a drainage area can be carried to the coast by rivers. Pollution from distant sources can also enter into the marine environment through atmospheric deposition. Based on current information, the land-based pollutants constituting the greatest threat

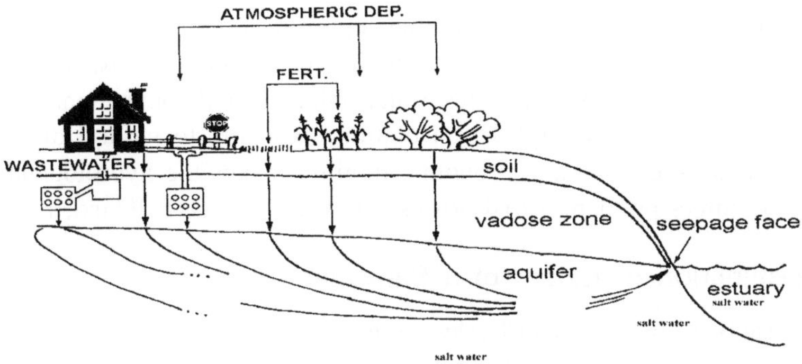

Figure 8. Different sources of pollutants in a coastal watershed

to coastal and marine ecosystems and to public health in the coastal regions are oil hydrocarbons, sediments, nutrients, pesticides, litter and marine debris, and toxic wastes.

Table 3. Point sources of pollution in a coastal watershed

Type of Pollution	*Examples*
Sewage disposal systems	Sewage lagoons, septic systems, cesspools, barnyards/feed lots.
Surface waste disposal sites	Landfills, garbage dumps, surface waste dumps.
Underground waste disposal sites	Storage tanks (low, medium, high level wastes).
Spills, washings, and intrusions	Oil, gas, waste spills: Auto workshop washings, Research laboratory washings, seawater or saltwater intrusions.
Mining sources	Acid mine drainage: Mine waste dumps, seepages gas explosions.
Natural mineral/ore deposits	Saltwater springs, hot spring waters, anhydrite, pyrite deposits etc.

Table 4. Non-point sources of pollution in a coastal watershed

Source	*Examples*
Agriculture	Cropland, irrigated land, woodland, feed lots
Silviculture	Growing stock, logging, road building
Construction	Urban development, highway construction
Mining	Surface, underground
Utility maintenance	Highways, streets and de-icing
Urban runoff	Floods and snowmelt

There is the need for some degree of flexibility in applying the definition of smallness to islands. There is obviously a range of areas, and other characteristics, which together make an island small in the sense that methods, techniques and approaches to hydrology and water resource issues cannot be directly applied from continental situations. Their reduced areas, shortage of natural resources (arable land, freshwater, minerals and conventional energy sources), geological complexity, isolation, exposure to natural disasters, demographic pressures and economic fragility makes the hydrological and water resources problems of islands usually very serious (Falkland, 1991).

Determination of Extent of a Saltwater Body

The location of the sea water/freshwater interface is important to groundwater management in coastal regions. The exact locations of the freshwater/saltwater interface and the transitional zone are difficult to determine, either theoretically or by field-testing. There are many approaches to delineating the depth to, and extent of, a saltwater body. First it is necessary to define what is meant by saltwater. Terms relating to the degree of salinity were suggested by the USGS (Hem, 1970) (Table 5).

Saltwater and freshwater are miscible fluids (Fig. 9). When water bodies with differing salinities are in contact, molecular diffusion causes mixing across the line of contact. Diffusion coefficients are relatively high in unconsolidated deposits, and therefore a wide zone of mixing would be encouraged. The effects of diffusion are exacerbated when groundwater and seawater are in motion, and the intergranular structure of the formation causes dispersion and the development of a transition zone due to groundwater and saltwater movement. External influences such as tides, recharge events and pumping will cause movement of the interface and encourage mixing, thus increasing the interface thickness. Thus the processes of diffusion and dispersion result in a transition zone where the salinity gradually changes from completely fresh to fully saline, the thickness of the zone depending on these two components. In permeable coastal aquifers, which are subjected to heavy abstraction, the zone may attain a thickness of up to 100 m (Bobba, 1993a). As, in reality, there is no sharp interface, a threshold level must be decided, at which water is considered saline. When investigating the use of geophysical and chemical techniques, Tellam et al. (1986) used the horizon at which salinity started to rise near the saltwater body.

Table 5. Terms describing degree of salinity as used by USGS (Hem, 1970)

Description	*TDS (mg/l)*
Fresh	< 1000
Slightly saline	1000 – 3000
Moderately saline	3000 – 10000
Very saline	10000 – 35000
Brine	> 35000

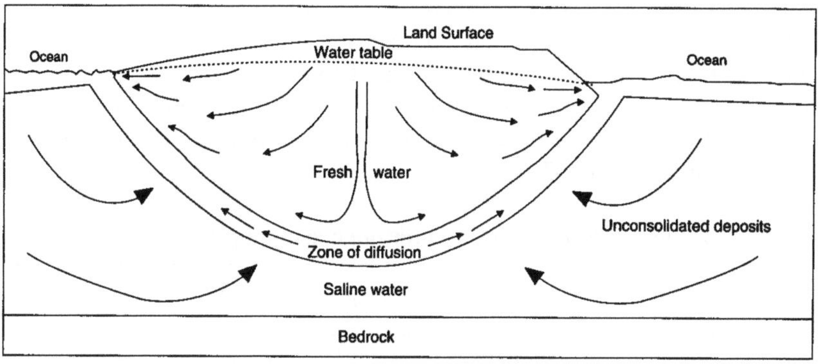

Figure 9. Dynamics of subsurface freshwater and subsurface saltwater in island

There are no significant differences between water table and confined aquifers, and the same principles apply. However, in water table coastal aquifers, fresh water is able to discharge freely to the sea. Under confined conditions, a finite head of fresh water is needed to compensate for the denser seawater column at the submarine outlet (Custodio, 1985). If this situation occurred, no fresh water flow could occur, and the seawater would act as a plug; the mixing zone would therefore expand by diffusion and dispersion, aided by tidal and seasonal fluctuations, and a long, thick mixing zone would develop.

The presence of saltwater in an aquifer is established either by the measurement of conductivity, or from chemical analysis. Historical data may be available in some areas. Such records may include partial chemical analyses or observations of the existence of saltwater, indicating that a saltwater intrusion problem existed in the past, and may still exist. Other indications that a problem existed include records stating that a borehole was partially backfilled to improve quality, or abandoned because of deteriorating water quality.

SAMPLING METHODS

Samples of groundwater may be obtained by various means. In the interpretation of the results of chemical analysis particular attention should be paid to the method of acquisition of the samples. It should be stressed that aquifers are complex, three-dimensional flow systems, and that the placement of boreholes within these systems can complicate the hydraulics even further. Methods and their problems are briefly discussed below and a summary given in Table 6.

Grab samples include bailed samples and depth samples. The former are obtained using a bucket or bottle attached to a line. Although very simple, the method only samples the top of the water column, and the results may be difficult or impossible to interpret. Depth samples are obtained from various depths in the water column using devices with valves that can be remotely closed at a certain depth. These provide some depth discrimination; however, it should be noted that this approach samples the artificial regime in the borehole. Rushton (1980) indicated that, if influenced by abstraction, the apparent interface within the borehole does not necessarily coincide with the actual interface within the formation, i.e. if there is vertical flow within the borehole, the sample will not be representative of the aquifer at that depth. Therefore, boreholes will only reflect the true position of the interface under conditions of horizontal flow.

Table 6. Summary of groundwater sampling techniques

Sample Method	*Advantages*	*Drawbacks*
Bailed	Easy operation, cheap	Only samples top of water column, making interpretation difficult or impossible
Depth Sample	Some depth, discrimination possible	Samples may not be representative of sample depth, may disturb water column
Pumped Sample	Provides composite sample, cheap	Sample may not be an 'average' of the borehole
Pore Water	Detailed, specific samples	Expensive and difficult
Packer	Detailed, specific samples	Good knowledge of vertical profile required for accurate sitting; uncertainty about seal in rough boreholes; relatively expensive and difficult to use.
Geochemical logging	Continuous monitoring of pH, SEC; relatively cheap	

Pumped samples are generally thought to provide an average sample of the water in the borehole (Kashef, 1986). However, the composition of the sample is a function of the heads of the contributing aquifers, which is not therefore a true average of the contributing aquifers. In addition, such samples do not aid in detecting the transition between salt and fresh water. These problems can be overcome by using packers to obtain samples from precise intervals; although care should be taken in screened boreholes where water may bypass the packed interval (Fetter, 1994). Pore water abstracted from cored samples probably provides the most accurate data (Tellam et al., 1986). However, there may be problems in obtaining such samples from unconsolidated formations.

Chemical Analysis for Saltwater

The two most widely used approaches to detect the presence of saltwater are measurement of conductivity and chloride concentration. Tellam et al. (1986) suggested the use of simple measures of conductivity to identify horizons where salinity was increasing. However, natural waters are not simple solutions, but contain a variety of ionic and undissociated species, the amounts and proportions of which may range widely. Therefore conductivity cannot be simply related to ion concentrations or dissolved solids (Hem, 1990). However, in practice, when investigating salinity problems, known to be a result of seawater intrusion, conductivity is probably a reasonable parameter.

Determination of chloride is also widely used. Chloride is a conservative element and is usually only found at low concentrations in groundwater, but is the dominant anion in seawater. In contrast, bicarbonate is the most abundant anion in groundwater but not in saline water. Revelle (1941) proposed that the chloride-bicarbonate ratio could be used to evaluate saltwater intrusion. Although this approach will distinguish saltwater from fresh water, it cannot differentiate between different sources of saltwater.

Cation determination can also provide an indication of the dynamics of saltwater intrusion. Under conditions of saltwater intrusion, Na in seawater replaces Ca adsorbed onto the surface of clays. This results in relative depletion of Na in seawater, and Na concentrations plot below the seawater/freshwater mixing line. When seawater is flushed by groundwater, Ca expels adsorbed Na, causing a relative surplus of Na (Geirnaert and Laeven, 1992). Therefore samples from the mixing zone may provide data to indicate whether the saltwater front is advancing or retreating.

Measuring Submarine Groundwater Discharge

Although springs and seeps funnel large amounts of fresh water and dissolved constituents into a small area, the slow, more diffusive seepage of groundwater that flows out along most shorelines of the world may be volumetrically more important. How much water and dissolved constituents (nutrients, carbon, contaminants) this dispersed source of groundwater delivers to the coastal ocean is unknown because of the difficulty involved in trying to measure it. Three different methods of measurement are typically used to study Submarine Groundwater Discharge (SGD):

- The French engineer, Henry Darcy, first described the movement of water through the ground analytically in 1856. His mathematical expression has been widely used in hydrological studies ever since ($Q = KiA$). He observed that at steady-state, the volumetric flow rate of water (Q) is directly proportional to the local hydraulic gradient, or the change in hydraulic head over some distance (i), the aquifer hydraulic conductivity (K), a measure of the ease that water will pass through an aquifer, and the cross sectional area (A) of the aquifer. This is one of the simplest equations used to describe groundwater flow and acts as the foundation for more complex systems.

- Direct measurements of SGD—A seepage meter provides a simple, inexpensive method to measure the discharge of groundwater into surface waters. The details and development of seepage metre was explained by Bokuniewicz (1980). The basic seepage meter may consist of the top or bottom section of a standard 55-gallon drum with an open port placed near the rim to attach a plastic water collection bag. The volume of water that enters the bag over a known time and area is a direct measure of the seepage rate. More recently, scientists have begun to automate seepage meters by use of thermal and ultrasonic technologies for measuring flow rates.
- Tracer studies—A tracer is a natural or artificial substance dissolved in the fluid so that it can be followed in space and time, providing information on the patterns of events. When using a tracer to study SGD, the tracer would typically occur at high concentrations in the groundwater relative to surface waters. Measuring the groundwater and surface water concentrations and accounting for all other sources and sinks of the tracer can estimate the magnitude of groundwater flux in the system.

THEORETICAL METHODS

The following calculation methods use formulae to describe the interface depth and saltwater wedge penetration under steady state conditions and simple geometric configurations. Various workers have arrived at different solutions based on different assumptions. If the possible influence of dispersion, diffusion, tidal influences, barometric effects, evaporation, leakage from and into confining beds, earthquakes, non-steady flow, etc., are considered, it can be appreciated that any solution will be complex. Generally accuracies are acceptable for most practical purposes (Bobba, 1993a; Bobba and Singh, 1995), although errors may become significant close to the coast and under transient flow conditions when vertical flow components become important.

Modeling

In the past four decades, modeling has become an important and powerful tool in many branches of science. Models allow engineers and scientists a way to test hypotheses in a manner that is removed and nondestructive to the actual problem at hand. In studies involving salt-water intrusion, modeling has been used for many purposes. One common goal of these models is to predict and characterize the movement of the transition zone in the aquifer where fresh water and salt-water meet. Another purpose of modeling is to predict the degree and extent of mixing that occurs in this transition region. In this way, models allow problems to be defined before they actually occur. Bobba (1993a) and Bobba et al. (2000) have reviewed the various model studies carried out in the past. Bobba and Singh (1995) presented a detailed description of available model characteristics and recent trends in subsurface water flow and transport modeling. Mathematical groundwater models can be classified into four distinct categories according to their purposes: (1) prediction models that simulate the behaviour of subsurface water systems and their response to stresses; (2) management models that integrate prediction models and simulation/optimisation models for evolving alternative management decisions; (3) identification models which are used for parameter estimates; and (4) models used for storage, retrieval, manipulation and management of data bases. The development of finite element model is explained in Appendix. It should be noted that the distinction between these categories is artificial and hazy. For example, a prediction model can be used iteratively for estimation of parameters (Bobba and Joshi, 1989).

Flow models are used to determine the quantitative aspects of groundwater motion, such as direction, rate, changes in water table or potentiometric head, stream-aquifer interaction, etc. Transient 2-D models for simulating groundwater flow are widely available. The problem of identifying and estimating parameters associated with subsurface flow and transport can be addressed in two different ways. One way is to develop a model exclusively for this purpose. Another way is to use these groundwater flows and transport simulation models iteratively to derive parameter values such that the predicted output is as close to the observed values as possible.

Saltwater intrusion problems are intrinsically complex due to the number of interacting phenomena. It is therefore difficult to acquire the detailed understanding needed to effectively manage such situations. Several quite common real-world occurrences do not lend themselves to conventional analytical techniques. Numerical models are therefore valuable tools for quantifying the hydraulic parameters and testing conceptual models of the aquifer system.

The dominant mechanism of saltwater intrusion is mass transport: the salt in solution is carried with the water by advection (Bobba, 1993a). In addition to the normal flow equations, saltwater intrusion models usually take account of the effects of density variations within the groundwater due to the mass of salt in solution. These introduce vertical flow to the system and lead to stratification in the eventual salinity distribution. This is important because a number of common conceptual models assume groundwater flow as entirely horizontal. The transport is often regarded as conservative though at a higher level of sophistication solute production or decay and chemical evolution can be included.

The principle distinction between classes of saltwater intrusion model lies in their treatment of the interface between the separate bodies of water within the aquifer. As described earlier, saltwater and fresh water are miscible, and mixing occurs at the interface due to molecular diffusion and mechanical dispersion. The parameters that control diffusion and dispersion are difficult to quantify directly and in any case the width of the transition zone is often small compared to the dimensions of the saltwater body. Consequently many models have included a sharp interface approximation, with the advance of one water body completely displacing the other body by piston flow.

The vertical flow components mentioned above means that saltwater intrusion problems are fully three-dimensional. However often some line of symmetry can be deduced (perpendicular to a straight coastline) so that vertical strips, 2-d models can provide a useful insight to the situation while minimizing the computational complexity. Horizontal 2-d models imply a vertical interface, which can be misleading: wells that appear safely on the fresh side of the interface might be at risk of degradation by up-coning at the toe of the intruding wedge. The following sections described SUTRA model and application to real field.

SUTRA Model

The SUTRA (**S**aturated, **U**nsaturated, **Tra**nsport) model is a finite-element groundwater flow and energy (heat) and solute (contaminant) transport code, which was developed by U.S. Geological survey (Voss, 1984). The SUTRA model employs a two-dimensional, finite-element method to simulate variable density groundwater flow, and transport of the total dissolved solids (TDS) in the groundwater. This model has many of the advantages associated with finite element models. It produces stable solutions with a relatively coarse mesh and introduces small numerical dispersion. The two primary variables upon which the model is based are fluid pressure, and TDS concentration expressed as mass fraction. The solute mass balance per unit aquifer volume at a point (x, z) in an aquifer with variable-density fluid is given by (Voss, 1984):

$$\theta\rho\frac{\partial C}{\partial t} + \theta\rho \boldsymbol{v}.\underline{\Delta}C - \underline{\Delta}.[\theta\rho(D_{\mathrm{m}}I + D).\underline{\Delta}C] = Q_{\mathrm{p}}(C^{*} - C) \qquad (3)$$

where $C(x,z,t)$ is solute concentration as a mass fraction (mass solute/mass fluid) $[M_s/M]$ where $[M_s]$ are units of solute mass and [M] are units of fluid mass; θ is porosity; $q(x,z)$ is aquifer volumetric porosity; $v(x,z,t)$ is fluid velocity (L/T); D_m is molecular diffusivity of solute in pure fluid including aquifer material tortuosity effects $[L^2/T]$; $\boldsymbol{I}$ is the identity tensor; $\boldsymbol{D}(x,z,t)$ is the dispersion tensor $[L^2/T]$; Q_p (x,z,t) is a fluid mass source (mass fluid/ aquifer volume/time) $[M/L^3.T]$; $C^*(x,z,t)$ is concentration of solute as a mass fraction in the source fluid $[M_s/M]$; $r(x,z,t)$ is fluid density $[M/L^3]$ where $[L^3]$ is fluid volume; and density is given as a linear function of concentration:

$$\rho = \rho_0 + \frac{\partial\rho}{\partial C}(C - C_0) \qquad (4)$$

where ρ is fluid density when $C = C_0$; C_0 is a base solute concentration. Darcy's law gives the mass-average fluid velocity as:

$$\mathbf{v} = \left(\frac{k}{\theta\mu}\right)\cdot(\Delta p - \rho g) \qquad (5)$$

where $k\ (x,y)$ is intrinsic permeability tensor $[L^2]$; μ is fluid viscosity [M/ L.T]; **g** is the gravity vector $[L/T^2]$; $p(x,\ y,\ t)$ is the fluid pressure $[M/L.T^2]$. The mass balance of fluid per unit aquifer volume at a point in the aquifer, assuming a rigid solid matrix, may be given by:

$$\rho S_{\mathrm{op}}\frac{\partial p}{\partial t}+\theta\frac{\partial \rho}{\partial t}\frac{\partial C}{\partial t}+\underline{\Delta}.(\theta\rho \boldsymbol{v})=Q_{\mathrm{p}} \tag{6}$$

where Darcy's law, Eq. (2), for mass-average velocity may be substituted in the fluid mass balance Eq.(4). The specific pressure storovity, S_{op}, for a rigid solid aquifer matrix $[M/(L.T^2)]^{-1}$ is given by:

$$S_{\mathrm{op}}=(1-\theta)\alpha+\theta\beta \tag{7}$$

where α is porous matrix compressibility $[M/(L.T^2)]^{-1}$, and β is fluid compressibility $[M/(L.T^2)]^{-1}$.

A general boundary condition for the fluid mass balance that applies at stationary boundaries as well as approximately at a water table (Bear, 1979) is:

$$\frac{S_{\mathrm{y}}}{|g|}\frac{\partial p}{\partial t}+\theta\rho v\cdot n+Q^{*}=0 \tag{8}$$

where $S_{\mathrm{y}}(x,z)$ is the water table specific yield (volume fluid released/aquifer volume) for unit drop in hydraulic head; $|g|$ is the magnitude of gravitational acceleration, $[L/T^2]$; and Q^* is a fluid mass source due to flow across boundaries (mass fluid recharged per unit area of boundary/time), $[M/L^2T]$. This condition is accurate when at a water table $|p| = r|g|$, that is, when pressure gradients are nearly hydrostatic at the water table. This condition is usually satisfied because commonly vertical pressure gradients are often nearly hydrostatic and horizontal gradients are much smaller than vertical gradients.

The mechanical dispersion tensor in two spatial dimensions is given by:

$$D=\begin{vmatrix} D_{\mathrm{xx}} & D_{\mathrm{xz}} \\ D_{\mathrm{zx}} & D_{\mathrm{zz}} \end{vmatrix} \tag{9}$$

$$D_{\mathrm{xx}}=\left(\frac{1}{|v^2|}\right)\left(d_{\mathrm{L}}\,v_{\mathrm{x}}^2+d_{\mathrm{T}}\,v_{\mathrm{z}}^2\right) \tag{10}$$

$$D_{\mathrm{xx}}=\left(\frac{1}{|v^2|}\right)\left(d_{\mathrm{T}}\,v_{\mathrm{x}}^2+d_{\mathrm{L}}\,v_{\mathrm{z}}^2\right) \tag{11}$$

$$D_{\mathrm{xz}}=D_{\mathrm{zx}}=\left(\frac{1}{|v^2|}\right)\left(d_{\mathrm{L}}+d_{\mathrm{T}}\right)\left(v_{\mathrm{x}}\,v_{\mathrm{z}}\right) \tag{12}$$

$$d_L = \alpha_{L_v} \tag{13}$$

$$d_T = \alpha_{T_v} \tag{14}$$

where $|v|$ is the magnitude of velocity and d_L and d_T are longitudinal and transverse coefficients of mechanical dispersion, and $\alpha_L(x, z)$ is longitudinal dispersivity [L] and $\alpha_T(x,z)$ is transverse dispersivity [L].

The intrinsic structure of finite element models provides substantial flexibility in designing the mesh and conforming to the boundary conditions as well as making the placement of material property boundaries such as hydraulic conductivity boundaries more realistic. Although the code does not accommodate triangular elements, it compensates for this by providing pinch nodes. Pinch nodes are nodes that end on an element side. The input also does not allow direct variation of the storage coefficient but allows variability of this coefficient through changes in porosity, saturated thickness, and compressibility. This model has been applied to real field data and observed favourable results (Bush, 1988; Bobba, 1993b; Piggott et al., 1994, 1997).

The tidal variations in sea level were represented by a sine wave with amplitude of 1.0 m for a period of 12 h. The sine wave is expressed as pressure varying with time according to

$$p(x, y, t) = (0.9 \sin 4\pi t + \text{elevation} - y)\rho g \tag{15}$$

Initial conditions for simulation were taken to completely correspond to a saltwater system with pressure everywhere reflecting MSL. This is expressed as

$$p(x_i, y_i, 0) = z_i \, \rho \, g_{salt} \tag{16}$$

With a time step of 1.0 h, it required two simulated days for pressure to reach a stable pattern of hydraulic response.

APPLICATION OF NUMERICAL MODELS TO THE COASTAL WATERSHED AND THE ISLAND

In this article, a finite element model is applied to different cases. This model simulates fluid movement and the transport of either dissolved substances or energy in the subsurface system. The model can be applied areally or in cross section. It uses a two-dimensional, finite-element method to approximate the equations that describe the two interdependent processes being simulated. When used to simulate saltwater movement in the subsurface system in cross section, the two interdependent processes are the density-dependent saturated subsurface-water flow and the transport of dissolved solids in the subsurface water. Either local- or regional-scale sections having dispersed or relatively sharp transition zones between saltwater and freshwater may be simulated. The results of numerical simulation of saltwater movement show distributions of fluid pressures and dissolved solids concentrations as they vary with time and also show the magnitude and direction of fluid velocities as they vary with time. Almost subsurface properties that are entered into the model may vary in value throughout the simulated section. Sources and boundary conditions may vary with time. The finite-element method using quadrilateral elements allows the simulation of irregular areas with irregular mesh spacing. The model has been applied to real field data and observed to give favourable results (Bobba, 1993b, 1998; Bobba et al., 2000).

Case Study 1: Lambton County, Ontario, Canada

Lambton County is located in the province of Ontario, Canada (Fig. 10). The study region occupies most of Lambton County located along the Saint Clair River between Lakes Huron and Saint Clair. The region is bounded in the north by Lake Huron and in the west by the Saint Clair River. Numerous chemical and petrochemical companies operate in this region and the wastes generated by these facilities have historically been managed using near surface burial and deep well injection. Sporadic discharge of formation fluids, and possibly industrial waste, to the surface has caused concern over the contamination of Lake Huron, the Saint Clair River, and the regional aquifer that serves as a water supply for rural Lambton County.

Migration of wastes injected at depth to the near surface via discontinuities in the confining strata and abandoned deep wells is a plausible mechanism for the contamination of surface and subsurface waters. The geology and hydrogeology of the site are described earlier (Bobba, 1993b; Piggott et al., 1997). The groundwater is obtained from sand and gravel deposits at depths below 60 m. The aquifer called as Fresh Water Aquifer is floating on top of high saline water due to density difference. Flow in this aquifer is likely to be strongly influenced by precipitation events and results in either discharge to shallow ditches, creeks and rivers of recharge to the deeper formations.

The details and the application of the numerical models are explained earlier (Bobba, 1993b). Figure 11 shows the computed hydraulic head map which best approximates the observed water table.

The model results indicate that virtually all of the water flowing through the aquifer discharges to the St. Clair River. The water level contours in the bedrock valley indicate a generally westward groundwater flow system. Since hydraulic gradient varies across the bedrock valley, the rates of groundwater flow also vary. Calculations indicate that groundwater flux into and out of the valley are about 0.9 and 0.5 m^3/sec, respectively, for a 100 m strip of aquifer. The quantum, which flows westward out of the 0.5 m^3/sec probably discharges to the St. Clair River. Discharge of groundwater from the Canadian side of the Fresh Water Aquifer is calculated to be between 0.45

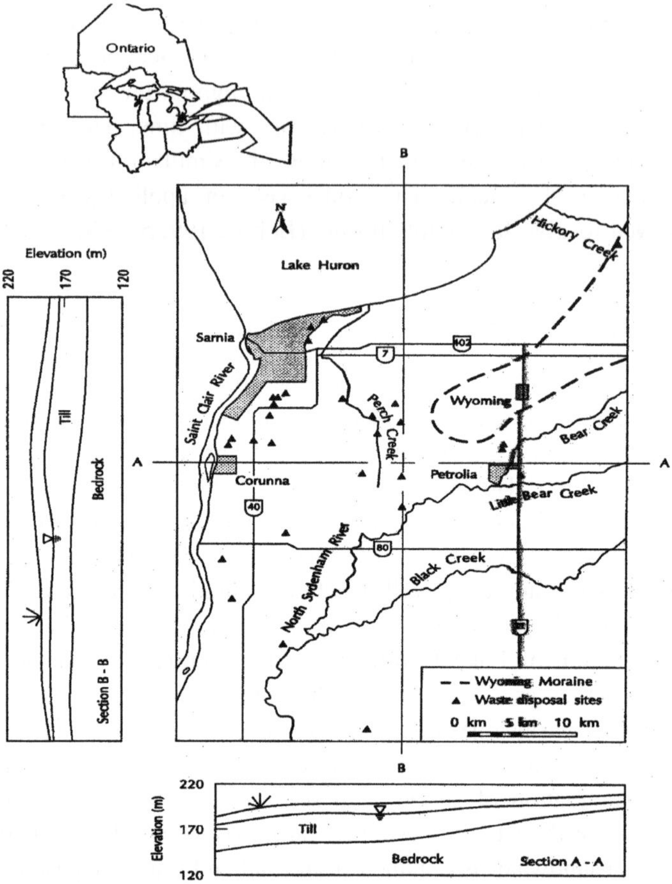

Figure 10. Location of Lambton County where Sections A-A and B-B show the surface geology and hydrogeology of region in East-West and North-South cross sections.

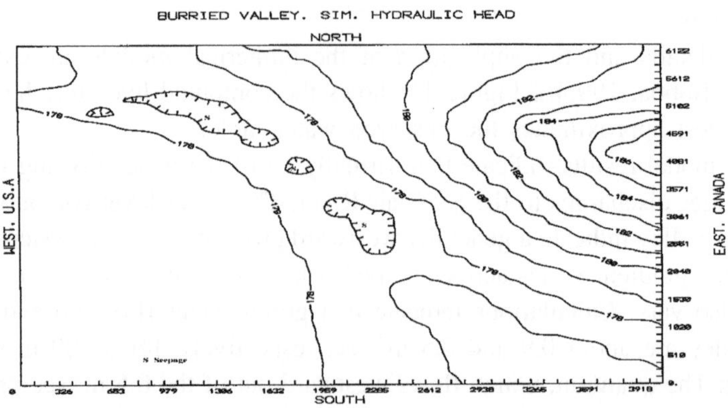

Figure 11. Simulated hydraulic head map of the buried valley, and groundwater seepage locations in Saint Clair River, Canada and USA boundary.

and 0.50 m^3/sec for that portion of the river between Lake Huron and Stag Island.

Case Study 2: Port Granby Radioactive Disposal Site

Canada has a uranium refinery at Port Hope, Ontario. The waste from the refinery is disposed at Port Granby waste management site located on the north shore of Lake Ontario (Fig. 12). In recognition of concern over the possible contamination of surface lake waters, the concentrations of radium and uranium are measured in water samples collected at Lake Ontario coastal zone near waste site. These data show the leachate infiltrating and seeping to coastal zone of Lake Ontario. The plume is moving parallel to the shoreline in the direction of the prevailing wind (Fig. 13). The waste disposal site acts as a source for tracer test to calculate the discharge concentration. The finite element model was applied to calculate hydraulic head and contamination discharge to lakeshore. The predicted Ra-226 contamination concentration from the waste disposal site to the beach is shown in Fig. 13. The details of the application of the model and interpretation of results were presented earlier (Bobba and Joshi, 1988, 1989).

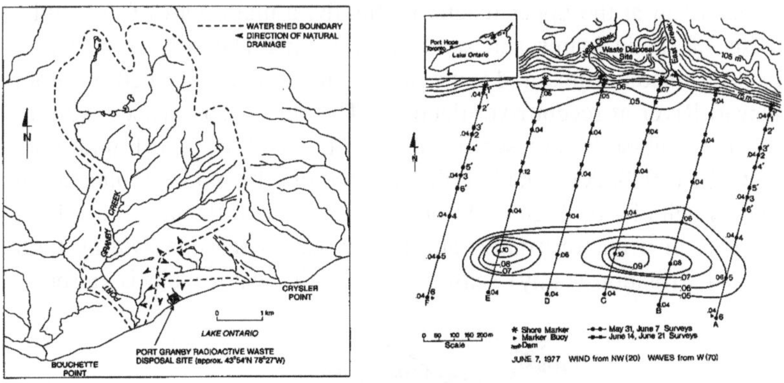

Figure 12. Location of waste disposal site and Ra-226 concentration in Lake Ontario, Canada.

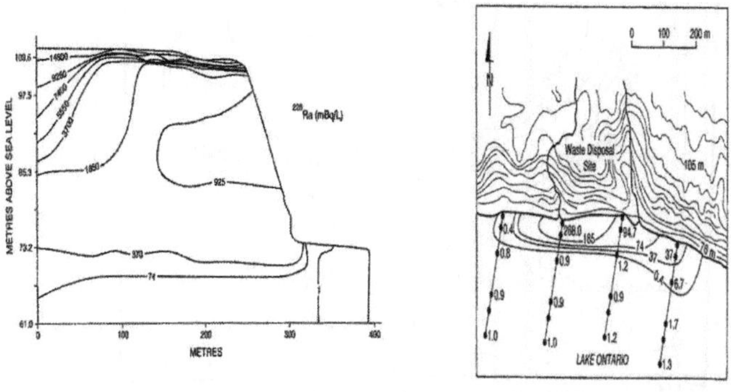

Figure 13. Computed Ra-226 concentration in waste site, beach and observed concentration in coast (Bobba and Joshi, 1988).

The groundwater flow to Lake Ontario from the site is 2.5×10^5 m^3/y. This annual volume of water is about 3-4% of the total solid volume of the site, assuming a depth of about 35 m. If the ^{226}Ra activity in the shore piezometers averages about 100 Bq m^{-3} , about 2.5×10^7 Bq y^{-1} is carried into the lake. But the total amount of ^{226} Ra which has been disposed of at the site is 2.3×10^{13} Bq, so that only about 10^{-6} of this is lost from the site per year. A comparable calculation for uranium indicates that about 25 kg of this element reaches the lake (Bobba and Joshi, 1988, 1989).

Case Study 3: Godavari Delta, India

The Godavari delta is located in East Coast of India (Fig. 14). The details of geology and environmental problems have been explained earlier (Bobba 2000, 2002). The Godavari delta lies between the sea level and 12-m contour. The delta has a projection of about 35 km into the sea from the adjoining coast. The Godavari delta consists of alluvial plain. It has a very gentle land slope of about 1 m per km. The coastal line along the study area measures to about 40 km and the general elevation varies from about 2 m near the sea to about 13 m at the upper reach. Texturally, a major part of the study area consists of sandy loams and sandy clay loams. The silty soils, which are very deep, medium textured with fine loamy soils are located all along the Godavari River as recent river deposits. The very deep, coarse textured soils with sandy sub-soils representing the coastal sand are also found along the sea. Bobba (2000, 2002) had earlier detailed out the model application to the Godavari delta basin. Figures 15 and 16 depict the influence of sea level rise variations, in irrigation (rainy) and non-irrigation seasons. During high tide level and irrigation (rainy) season periods, the water table is raised.

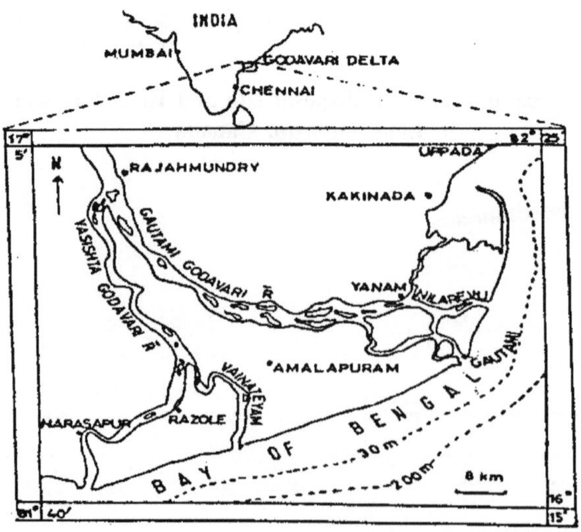

Figure 14. Location of the Godavari delta.

The distance between surface soil and water table in the coastal area is very small, and the material is generally composed of sands, which do not retain significant amount of moisture under unsaturated conditions. Hence, the water that overflows the soil directly recharges the groundwater. The distance between the water table and surface soil is at a minimum in the central portion of the delta. It has been observed that areas of minimum depth from the ground level to the water table have high freshwater potential whereas lowering of the water table from the ground surface reduces the freshwater potential substantially. The water table elevation varies from 0.5 m to 1 m from MSL and decreases gradually towards the coastal side. Patches of freshwater zones are also present along coastal areas (Figs. 15b and 16b). The areas of coastal aquifer contaminated by salt water are delineated in Fig. 15b. The sea level raise (SLR) and non-irrigation season may cause an upward movement of saline water in coastal aquifers. The freshwater potential in the aquifer is depicted in Figs. 15b and 16b. The aquifer likely to be saline is more along the eastern side than the southeastern side. Saline water contamination due to SLR and non-irrigation may be critical to the southern tapering segment of the island.

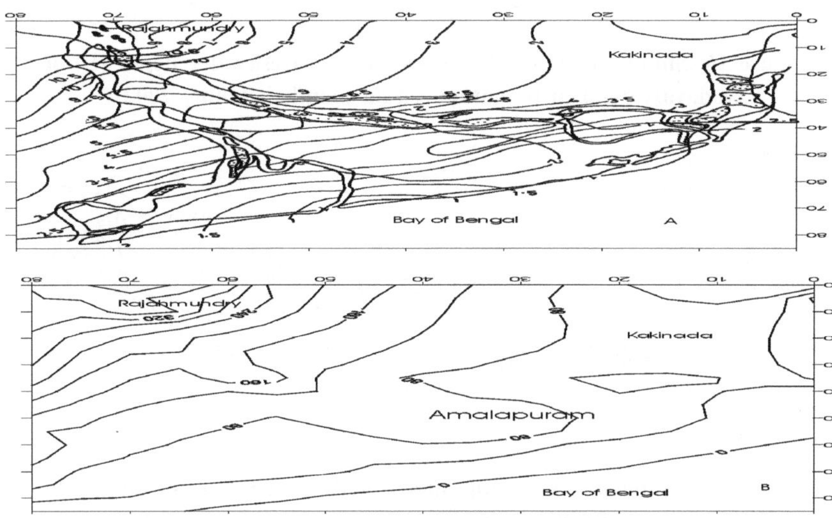

Figure 15. Simulated hydraulic heads (A) and freshwater depth (B) of Godavari delta in non-irrigation months (Bobba, 2002).

Figure 17 shows the results in different environmental conditions due to heavy and long rainy season and drought conditions due to high temperature and less rainfall (climate change). Higher water table conditions are observed due to more rain and irrigated water is recharged to the aquifer. The salt water was flushed out or stopped seawater intrusion to the aquifer. However, if the severe drought conditions (higher temperature, lesser rainfall) occur in the delta, the water table is reduced due to higher evapotranspiration and

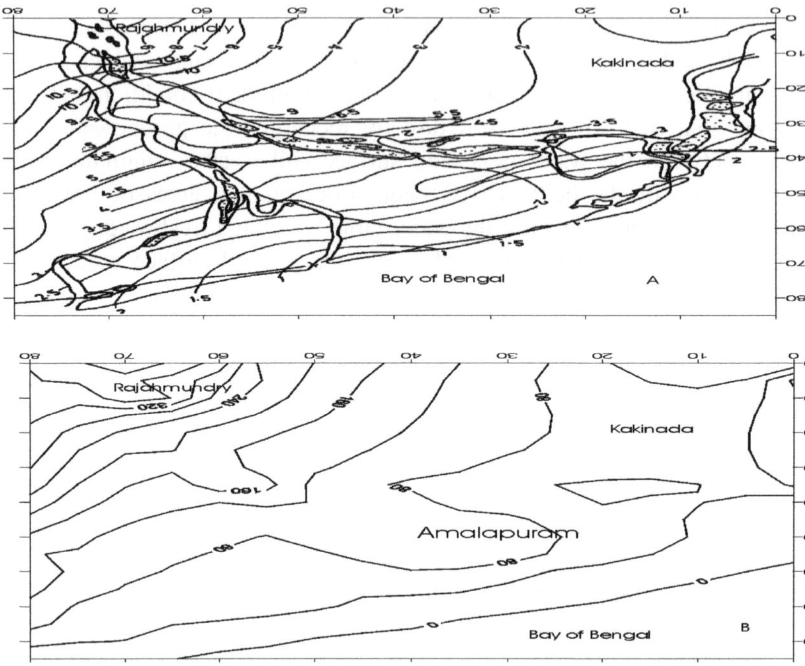

Figure 16. Simulated hydraulic heads (A) and freshwater depth (B) of Godavari delta in irrigation season months (Bobba, 2002).

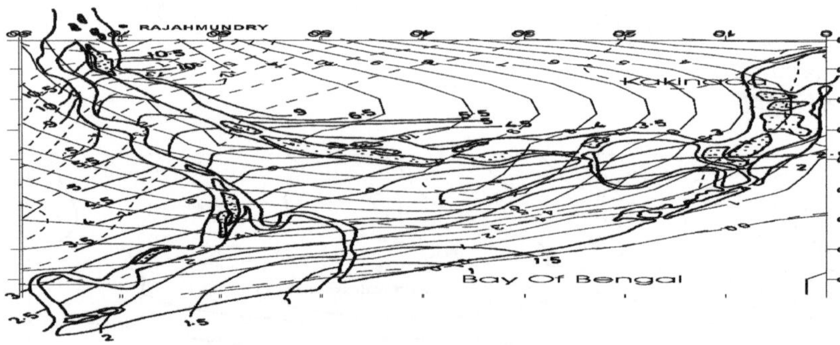

Figure 17. Simulated hydraulic heads of Godavari delta in different seasons (solid line in heavy rainy season, ------ drought conditions) (Bobba, 2002).

over pumping the groundwater for irrigation and domestic purposes. The salt water intruded to the aquifer and freshwater thickness was reduced in the delta.

Case Study 4: Agati Island, India

The Agati island is part of Laccadive Islands located in the Arabian Sea off the coast of Kerala (India), some 200 to 300 km away consisting of a group of 36 Islands reposing in Arabian sea. The Laccadive group comprises a number of coral atolls enclosing lagoons and submerged reefs and banks, lying in the Arabian Sea between latitudes 8 and 14° N and longitudes 74°41' and 74°10'E, 200-300 km off the west coast of India. Ten islands are inhabited. The main islands are Kavaratti, Agati, Minicoy and Amini. The capital is Kavaratti located in island Kavaratti.

The Laccadive Islands have an average elevation of 3-5 m with an area of 108.78 km^2. Only about 25% of the area is inhabited. The total population is about 51,000. Hills and streams are conspicuous by their absence. In general, the lagoons are on the western (windward) side and relatively steep slopes predominate along the eastern margin except in the Andrott island which extends east-west. In the islands, the climatic conditions do not show much variation. As a matter of fact, the region remains muggy throughout the year, as the temperature never falls below 23.8°C. The mean annual total rainfall recorded at Amini is 255 mm and at Minicoy 220 mm. Only one-fourth (28.49 km^2) of the total area (108.7 km^2) of the islands is inhabited with a total population of about 51,000 (1990). The details of geology, hydrogeology and application of model to Agati Island were explained earlier (Bobba, 1998; Bobba et al., 2000).

Figures 18, 19 and 20 illustrate the contour maps of simulation results on two horizontal dimensions for the Agati island aquifer. In the contour maps representing the interface, the area occupied by the interface of the freshwater and saltwater thickness is shown. The boundaries of the inland area allocated to the interface on the map represent the zero saltwater thickness (interface toe); beyond these limits, the contour lines represent the thickness of fresh water only. The sector occupied by seawater intrusion has similarly been illustrated when its presence is apparent in the simulation results. In this regard, seawater intrusion is considered to exist where the freshwater thickness is nil and where the position of the interface coincides with that of the water table level. For the purpose of obtaining a Ghyben-Herzberg theoretical approximation and a steady-state solution, the sector occupied by seawater intrusion naturally coincides with that occupied by the phreatic levels below sea level. The effect of various management alternatives for the reservoir on water table levels, thickness of fresh water and the position of the interface were considered.

These balances are mainly a consequence of natural recharge, pumping exploitation and flow regulations. The contour maps illustrate the possible evolution on the successive states reached by the aquifer in a stationary flow regime. It can be seen, for example, that when the water balance is negative, the aquifer shows clear signs of seawater intrusion. Figures 18, 19 and 20 depict the influence of sea level rise variations. During low tide level periods the water table is raised. The distance between surface soil and water table in this atoll is very small, and the material is generally composed of coral sands, which do not retain significant amount of moisture under unsaturated conditions. Hence, the water that overflows the soil directly recharges the groundwater. The distance between the water table and surface soil is at a minimum in the central portion of the island. It has been observed that areas of minimum depth from ground level to water table have high freshwater potential whereas lowering of the water table from the ground surface reduces the freshwater potential substantially.

The water table elevation varies from 60 cm to 395 cm from MSL and decreases gradually towards the coastal side. Patches of freshwater zones are also present on this island (Figs. 18 and 19).

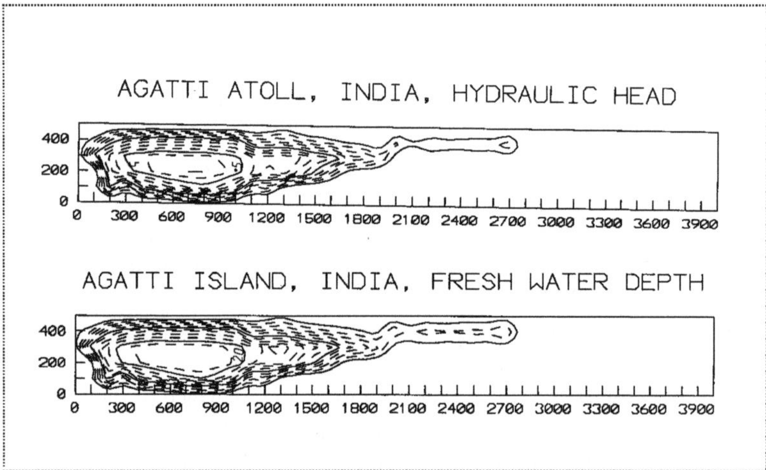

Figure 18a and 18b: Simulated hydraulic head and freshwater depth of Agati Island without climate change effect in two dimensions (Bobba et al., 2000).

The areas of coastal aquifer contaminated by salt water are delineated in Fig. 20 due to tidal effect. Salt water is present at the southern end up to a distance of 0.1 km from the southern tip (Fig. 21). However, during the low tide period, the saline wedge is limited to a distance of only 0.2 km from the tip. In the tapering southern region, with a width less than 20 m, the stormy beaches, due to reduced lagoonal effects, may facilitate seawater intrusion into the aquifer. Most of the areas along the coast have been adversely affected by saline water intrusion whereas only a few areas in a lagoonal side are affected by saline intrusion. Sea level raise may cause an upward movement of saline water in coastal aquifers.

GEOPHYSICAL METHODS

Although theoretical approaches allow determination of the position of the saline/fresh water interface, they can only be applied with confidence where the vertical component of flow is zero or close to zero i.e. inland and distant

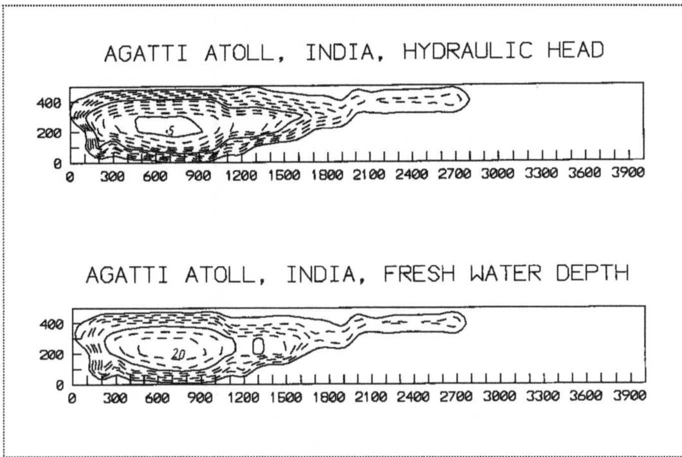

Figure 19a and 19b: Simulated hydraulic head and freshwater depth of Agati Island due to 0.1 m sea level rise in two dimensions (Bobba et al., 2000).

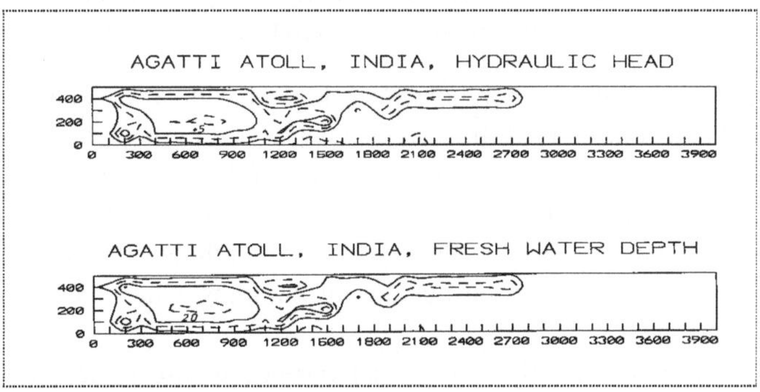

Figure 20a and 20b: Simulated hydraulic head and freshwater depth of Agati Island due to 0.5 m sea level rise in two dimensions (Bobba et al., 2000).

from the effects of abstraction. Various geophysical techniques have been used to identify the morphology of underground saltwater bodies (Tellam et al., 1986).

Borehole and surface geophysical approaches are available to investigate the geometry of saltwater bodies. They rely on the increase in conductivity between fresh and saltwater.

Borehole Logging

Borehole fluid conductivity logs may be used to determine the position of the interface within the borehole. Provided there is not a major component

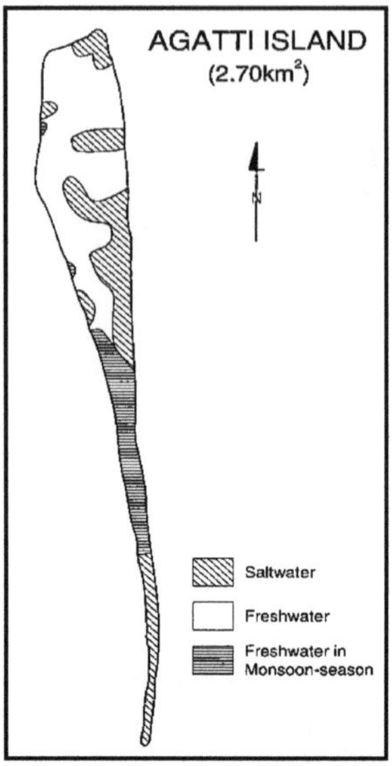

Figure 21. Simulated freshwater availability of Agati Island in different seasons; comparison with Sundaresan (1993).

of vertical flow within the borehole, such logs provide a reasonably accurate indication (to within 1 m) of the interface depth within the formation (Tellam et al., 1986). However, if the borehole and aquifer are not in equilibrium, errors may be significant. Correlating conductivity with flow meter logs can identify strata that contribute saltwater to the borehole. This is particularly useful if the logging is carried out during pumping when the normal flow regime of the well exists.

Various types of resistivity logs can provide a direct indication of the salinity of the water within a formation. However, their applicability is limited by the fact that boreholes generally have to be screened. If the screening material is metal, resistivity methods are not applicable.

Surface Geophysical Methods

Electrical resistivity and electromagnetic methods have been investigated as a means of delineating saltwater bodies (Arora and Rose, 1981; Prasad et al., 1983). All methods rely on the fact that saltwater has different electrical properties to fresh water. They have advantages in that: (i) geophysical surveys are low cost compared with drilling boreholes; and (ii) data can be directly related to water quality.

Electrical Resistivity

In areas of relatively homogeneous sediments, electrical resistivity surveys can be used to indicate anomalies attributable to saltwater bodies. Depth probing can be used to estimate the thickness of the saltwater zone, and resistivity traversing to delineate the surface of the saltwater body across an area.

Tellam et al. (1986) predicted a slight overestimation of the depth to the saltwater interface using this approach. This would be due to the fact that the real resistivity change is gradual, but is represented by the model as a sharp interface, with a large resistivity step. They suggested prediction accuracies of within 10 m at 30 m depth.

The effectiveness of the technique in detecting intrusion may be limited because of the similarity in electrical conductivity between clay, salt water in sandy aquifers, and brackish water in sandy aquifers. The occurrence of clay layers may also confuse interpretation (Prasad et al., 1983). Realistic evaluation is generally only possible if the results of such surveys are complemented by geological or geochemical data.

Electromagnetic Methods

Electromagnetic methods use two wire coils, one to generate an alternating magnetic field, and the other to detect the resultant magnetic field generated in the sub-surface (Anthony, 1992). The method may be used to obtain general information at shallow depths over large areas (frequency domain electromagnetic profiling), or to collect more detailed quantitative data about the electrical profile of the subsurface (time domain electromagnetic sounding).

Frequency domain (FDEM) profiling is most suitable for shallow depths (up to 30-40 m). It does not require any contact between the transmitter and receiver antennae and the ground surface, and is therefore rapid. However, its main disadvantage results from the fact that the receiver measures the bulk conductivity of the subsurface, and results are therefore only qualitative.

Time domain electromagnetic (TDEM) sounding or profiling has advantages over both electrical resistivity and FDEM methods. It has good lateral and vertical resolution, it is minimally influenced by near surface in-homogeneities, and non-uniqueness of data interpretation is reduced compared with electrical resistivity (Goldman et al., 1991). The method has been applied to the coastal aquifer of Israel (calcareous sandstone) and was found to be a feasible approach for detecting the saline/fresh water interface (Goldman et al., 1991).

SUMMARY

The groundwater protection is a very complex and difficult issue, which will require sustained effort at all levels of government over a long period of time before the resource is adequately protected. The basic components of strategy are: (1) to strengthen the groundwater programmes; (2) to cope with currently un-addressed groundwater problems; and (3) to create internal groundwater organizations. This strategy should come about in response to the national recognition of the serious problem posed by the present and possible future contamination of the nation's groundwater.

Here is presented a four-point groundwater protection strategy: (1) establishment of planning, management and/or preventative practices; (2) improvement of government coordination, policy and legal authorities related to environmental protection, water resources development, production and distribution; (3) improvement of utility system management and operation; and (4) provision of information for problem identification, informed decision-making and public education.

To implement this strategy, recommendations have been identified for the following categories: water resources policy, water system management policy and procedures, waste water system management policy and procedures, pollution control policy and procedures, land use policy, sanitation practices and procedures and public education. Taken together, these different elements make up a total groundwater protection programme. The next step is for the responsible agencies of government to work together and select the course of action, which best fits the requirements of each island. The central government needs to establish funding levels and priorities for groundwater protection and to create incentives for local governments to implement local strategies. It is important that elected officials recognize that groundwater protection is a continuous ongoing necessity. The groundwater protection becomes more critical as population and economy grow and the supply limits of surface water resources are reached. With adequate funding, policy support and supervision, the responsible agencies should be in a better position to address the issue of groundwater protection in a comprehensive and sustained manner. Management must be cognizant of agency responsibilities and ensure that concerned personnel are competent, trained and diligent in their jobs. Numerical simulation results will give the information to sustainable saltwater intrusion due to climate change effects. The results of simulation were represented according to their three main characteristics, that is, water table levels, the position of interface between salt and fresh waters and the thickness of freshwater and saltwater available in the aquifer.

This research has provided numerical simulation of the influence of surface flows, coming from the water management of the projected reservoir, on the regional groundwater behaviour on the island aquifer. A single-phase two-dimensional finite element model, considering open boundary conditions for steep coasts and a sharp interface between freshwater and salt water, was applied for steady-state conditions to the phreatic aquifer. When recharges of saltwater occur at the coastline, essentially of freshwater deficits, a hypothesis of mixing for the freshwater-saltwater transition zone allows the model to calculate the resulting seawater intrusion in the aquifer. Hence, an adequate treatment and interpretation of the hydrogeological data that are available for the coastal aquifer were of main concern in satisfactorily applying the proposed numerical model. Results of the steady state simulations showed reasonable calculations of the water table levels and the freshwater and saltwater thickness, as well as, the extent of the interface and seawater intrusion into the aquifer for the total discharges or recharges along the coastline. As a result of the present hydrogeological simulations on the phreatic aquifer, a considerable advance in seawater intrusion would be expected in the coastal aquifer if current rates of groundwater exploitation continue and an important part of the fresh water from the river is annually channelled from the reservoir for irrigation purposes.

Providing safe drinking water to the people of the island is a reachable goal. A detailed study of water resources and environmental study is necessary for future development. Of course, it will take time, money and dedicated effort, but it can be done. In combination with other efforts to improve overall sanitation, health care and public education, the provision of safe drinking water will greatly improve public health and the quality of life. Adequate and safe drinking water is something every islander can and should have. It will take a major, long-term effort on the part of the local governments to improve the condition of water supplies. The effort must take into account the many factors that influence the condition of these supplies and it must proceed in an orderly, co-ordinated manner.

Managing our surface waters and groundwater is complicated. In order to do so effectively, we must consider many issues, including water quality, freshwater supply, flood protection and floodplain management, and natural systems management. Managing our water resources involves the challenge of balancing varying water uses and demands with availability. Conflicting priorities often must be addressed to ensure that there are sufficient water supplies for human needs while maintaining water quality and viable, functioning natural systems. While it may be difficult to set management guidelines, our coastal waters have a direct connection with inland aquifers and should therefore be considered as an integral part in any management plan. Current studies of coastal hydrology by researchers at universities and elsewhere are helping land managers and planners protect these areas for future generations. It is important that we, as individuals, take an active role in protecting our limited surface and groundwater resources as well. By reducing the amount of fertilizer applied to lawns, responsibly disposing off our petroleum and hazardous waste materials, reducing the amount of water used at home and the office, and other common sense actions, we can help in protecting the water that we depend on in our daily lives.

APPENDIX

FINITE ELEMENT MODELS

A powerful numerical method is the finite element method. In groundwater flow problems one could imagine that a region is subdivided into small elements, each with its own physical properties, in such a way that for each element the flow is described in terms of the head in the nodal points, using Darcy's law in each element, and that a system of algebraic equations is obtained from the condition that the flow must be continuous at each node.

The usual way of presenting the finite element method does not employ such a physical reasoning, however. Instead, a mathematical argument is used, in which the system of equations is obtained by requiring that the differential equation be satisfied 'on the average', using certain weight functions. The system of equations obtained in the finite element method has the same structure as in the finite difference method. Actually, the two methods are very similar, and, for certain problems, it has been shown that they can be considered as two representations of the same mathematical model. However, the usual way of deriving and developing the equations reveals certain differences. For instance, the natural and simplest type of element is the triangle, which enables a flexible representation of the field, whereas the simplest and most natural mesh in the finite difference method is a mesh of squares or rectangles, which is less flexible. It can be seen from Fig. A.1 that in a mesh of triangles it is easily possible to have elements of different sizes, and also to follow smooth curved boundaries. Another advantage of the finite element method is that in its standard formulation it has the immediate property that each element can have its own values for the physical parameters, such as transmissivity and storativity.

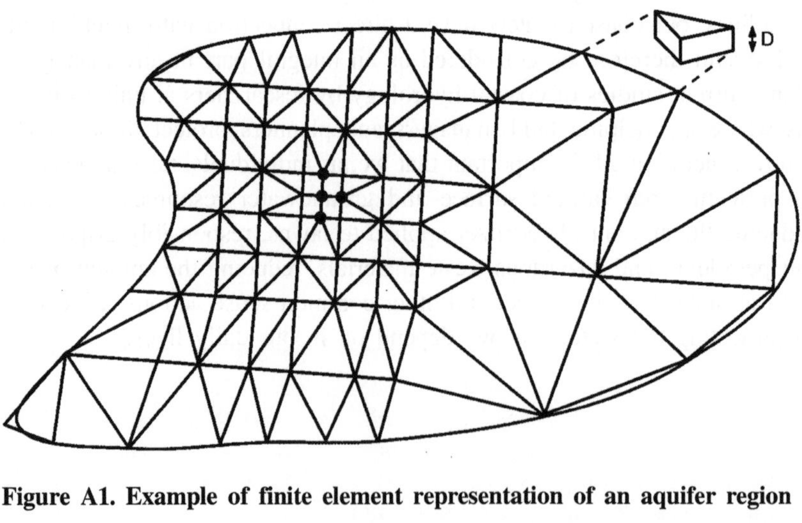

Figure A1. Example of finite element representation of an aquifer region

Galerkin's Approximation

To solve a differential equation

$$L(U) = 0 \tag{1}$$

by Galerkin's method, a trial solution of the form

$$U \approx \hat{U}(x, y, t) = \sum_{n=1}^{n} C_i(t)\, N_i(t)\; N_i(x, y) \tag{2}$$

is assumed. Here N_i (x, y) $(i = 1, 2, \ldots n)$ is system of functions (referred to as basis functions) selected beforehand and which satisfy the essential boundary conditions of the original differential equations. The functions N_i (x,y) $(i = 1, 2, 3, \ldots n)$ are assumed to be linearly independent and to represent the first N functions of some system of functions that is complete in the given region. The functions $C_i(t)$ are the undetermined coefficients.

The approximating functions $\hat{U}(x, y, t)$ will be an exact solution to Equation (1) only if $L(U)$ is identically equal to zero, otherwise a residual $R_E = L(\hat{U})$ will result. In the method of weighted residuals an attempt is made to force this residual to zero in an average sense through the selection of constants C_i $(i = 1, 2, \ldots n)$. To do this the residual is distributed over the spatial domain Ω by using a weighting function W_i (x,y) such that

$$\iint_{\Omega} R_E\, W_i\, (x, y)\, dxdy = 0 \qquad i = 1, 2 \ldots, n \tag{3}$$

or recalling the definition of the inner product

$$< R_E(X, Y), W_i(X, Y) > = 0 \qquad i = 1, 2, \ldots, n \tag{4}$$

i.e., the residual is made orthogonal to the weighting function. In Galerkin's method, the weighting function is chosen to be the same as basis functions N_i (x, y). As n basis functions are selected and as there are n undetermined coefficients $C_i(t)$ $(i = 1,2,3,\ldots n)$, only n conditions of orthogonality can be satisfied. These are

$$\iint_{\Omega} (\hat{U})\, N_i(x, y) dxdy = \iint_{\Omega} L\left[\sum_{j=1}^{n} C_j(t)\, N_j(x, y)\right] N_i(x, y) dxdy = 0 \tag{5}$$

$$i = 1, 2, 3, \ldots n$$

Performing appropriate integrations of the basis functions, and solving the resulting set of simultaneous equations, the coefficients $C_j(t)$ could be determined. Substituting these values in Equation (2), the solution to Equation (1) could be obtained.

$$\lim_{n\rightarrow} \hat{U} = U \tag{6}$$

Steps in the Finite Element Method Solution

The following basic steps are involved in obtaining the solution to the differential equation using the finite element method.

(1) The domain of interest is divided into a number of finite elements described by node points. (2) Suitable basis (shape) functions are selected to approximate the dependent variable at each nodal point. Values between these points are calculated using piecewise continuous interpolating functions defined over each element. (3) The weighted integrals of the residuals are set to zero to obtain a set of simultaneous equations. (4) The resulting set of equations are solved to obtain the values of the dependent variable at all nodal points. (5) Partial integration can be carried out to reduce the continuity required and also satisfy only the kinematic boundary conditions.

Choice of the Element Shape and Basis Functions

Ergatoudis et al. (1968) introduced the curved isoparametric quadrilateral element and used it mainly in structural engineering applications. The basic shape of the element is quadrilateral, but the sides may be distorted in a prescribed way. Eragatoudis et al. (1968) developed basis functions using the same order of approximations. Pinder et al. (1972) developed this deformed, mixed isoparametric quadrilateral element. The advantage of this element is that all four sides of the quadrilateral need not be of the same order. For each side of the element linear, quadratic, or cubic approximation could be used depending upon the shape of the boundaries or the expected shape of the unknown function.

The basis functions $N_i(x,y)$ in Equation (2) are selected such that $N_i(x,y)$ is unity at node i and zero at all other nodes. Because of the fact that the basis functions are chosen with this constraint, the undetermined coefficients $C_i(t)$ $(i = 1, 2,...n)$ in Equation (2) are equal to the required function $\hat{U}$ at the n node points in the domain.

To facilitate integration of the basis functions over each element a dimensionless and curvilinear local coordinate system (ξ, η) is introduced. All elements appear as squares in the local coordinate system with no side nodes (linear side), one side at the mid point (Quadratic node), or two side nodes at one-third points (cubic side). The functions $N_i(x,y)$ referred to as basis functions are also sometimes known as shape functions.

The mapping from local coordinates to cartesian coordinates x,y is by the following shape function mapping

$$X = N_1X_1 + N_2X_2 + ... = \sum_{j=1}^{n} N_jX_j \quad (7)$$

$$Y = N_1Y_1 + N_2Y_2 + ... = \sum_{j=1}^{n} N_jY_j \quad (8)$$

The summation is over each element and N_j are functions of ξ and η. The unknown functions $\hat{U}_i$ are prescribed by shape functions as

$$\hat{U}_i = \sum_{i=1}^{n} N_i U_i \tag{9}$$

As a point is moved around in the parent element its local coordinates in that element will generate the cartesian coordinates through Equations (7) and (8) and the corresponding U_i values through Equation (9). The functions $N_i(x,y)$ written in terms of ξ and η are selected such that in addition to satisfying the basic requirements of a basis function they also relate the global and local co-ordinate systems.

Application of Galerkin's Method

The objective is to obtain the values of the dependent variable H (hydraulic head) and C (concentration) at the n nodal points in the system. The trial solutions for H and C are now assumed in the form

$$H \approx \hat{H} = \sum_{i=1}^{n} H_i(t) N_i(x,y) \tag{10}$$

and

$$C \approx \hat{C} = \sum_{i=1}^{n} C_i(t) N_i(x,y) \tag{11}$$

where $N_i(x,y)$ (i = 1, 2, 3....n) are the basis functions (N_i need not necessarily be the same for approximating H and C). The basis functions satisfy the essential boundary conditions of the original partial differential equations. $H_i(t)$ and $C_i(t)$ are a set of undetermined coefficients and at the specified points (referred to as nodes) in the domain Ω.

Define

$$L_h(H) \equiv \frac{\partial}{\partial_x}\left(K_x \frac{\partial H}{\partial x}\right) + \frac{\partial}{\partial y}\left(K_y \frac{\partial H}{\partial y}\right) + q = 0 \tag{12}$$

and

$$L_c(C) = \frac{\partial}{\partial x}\left(D_{xx}\frac{\partial C}{\partial x}\right) + \frac{\partial}{\partial y}\left(D_{yy}\frac{\partial C}{\partial y}\right) + \frac{\partial}{\partial x}\left(D_{xy}\frac{\partial C}{\partial y}\right) + \frac{\partial}{\partial y}\left(D_{yx}\frac{\partial C}{\partial x}\right)$$

$$-\frac{\partial}{\partial x}(V_{xx}C) - \frac{\partial}{\partial y}(V_{yy}C) + S(x,y) - \theta\frac{\partial C}{\partial t} = 0 \tag{13}$$

The approximating functions and when substituted in $L_h(H)$ and $L_c(C)$ will yield the residual. The residual is now minimized by using the orthogonality condition with respect to the basis functions.

$$\iint_{\Omega} L_h(\hat{H}) N_i \, d\Omega = 0, \quad i = 1, 2, ...n \tag{14}$$

and $$\iint_{\Omega} L_c(\hat{C})\, N_i \, d\Omega = 0, \quad i = 1, 2, ...n \tag{15}$$

The n orthogonality conditions can be satisfied as n basis functions have been selected. These are

$$\iint_{\Omega} L_h\left[\sum_{j=1}^{n} H_j N_j(x,y)\right] N_i(x,y)\, d\Omega = 0, \quad i = 1, 2, ...n \tag{16}$$

and $$\iint_{\Omega} L_c\left[\sum_{j=1}^{n} C_j N_j(x,y)\right] N_i(x,y)\, d\Omega = 0, \quad i = 1, 2, ...n \tag{17}$$

Substituting $L_h(H)$ and $L_c(C)$ in Equations (16) and (17) the following equations are obtained:

$$\iint_{\Omega}\left\{\left[\frac{\partial}{\partial x}\left(K_x\frac{\partial}{\partial x}\right)+\frac{\partial}{\partial y}\left(K_y\frac{\partial}{\partial y}\right)\right]\Sigma_{j=1}^{n} H_j N_j - q\right\} N_i \, d\Omega = 0,$$

$$i = 1, 2, ...n \tag{18}$$

and $$\iint_{\Omega}\left\{\left[\frac{\partial}{\partial x}\left(D_{xx}\frac{\partial}{\partial x}\right)+\frac{\partial}{\partial y}\left(D_{yy}\frac{\partial}{\partial y}\right)+\frac{\partial}{\partial x}\left(D_{xy}\frac{\partial}{\partial y}\right)+\frac{\partial}{\partial y}\left(D_{yx}\frac{\partial}{\partial x}\right)\right]\right.$$

$$\Sigma_{j=1}^{n} C_j n_j - \frac{\partial}{\partial x}(\Sigma_{j=1}^{n} C_j N_j V_x) - \frac{\partial}{\partial y}(\Sigma_{j=1}^{n} C_j N_j V_y)$$

$$\left. -\frac{\partial}{\partial t}(\theta \Sigma_{j=1}^{n} C_j N_j) + S\right\} N_i \, d\Omega = 0 \quad i = 1, 2, 3.....n \tag{19}$$

Application of Green's Theorem

The second derivative terms appearing in Equations (18) and (19) would require continuity of the first derivative at all element interfaces. To avoid this restriction the second derivative terms are transformed to first derivatives by using Green's theorem in the following form

$$\iint_{\Omega} \Psi \frac{\partial^2 H}{\partial x_\alpha \partial x_\beta} d\Omega = -\iint_{\Omega} \frac{dH}{\partial x_\alpha} \cdot \frac{d\Psi}{dx_\beta} d\Omega + \int_s \frac{\partial H}{\partial n} ds, \quad \alpha, \beta = 1, 2 \tag{20}$$

where $$\frac{\partial H}{\partial n} = \frac{\partial}{\partial x_\alpha} l_\alpha ; \frac{\partial H}{\partial n}$$

being the outward normal derivatives of the function $H(x_\alpha)$and l_α's are the direction cosines.

Application of Green's theorem is made element by element. Using Green's theorem, Equations (18) and (19) are now transformed as follows:

$$\iint_\Omega \Sigma_{j=1}^{n}\left(K_x \frac{\partial N_i}{\partial x}\frac{\partial N_j}{\partial x} + K_y \frac{\partial N_i}{\partial y}\frac{\partial N_j}{\partial y}\right) H_j \, dx\, dy +$$

$$\iint_\Omega q\, N_i \, dx\, dy - \int_s N_i \sum_{j=1}^{n}\left(K_x \frac{\partial N_j}{\partial x} l_x + K_y \frac{\partial N_j}{\partial y} l_y\right) H_j \, ds \tag{21}$$

$i = 1, 2, 3, \ldots n$

and

$$\iint_\Omega \Sigma_{j=1}^{n}\left(D_{xx} \frac{\partial N_i}{\partial x}\frac{\partial N_j}{\partial x} + D_{yy} \frac{\partial N_i}{\partial y}\frac{\partial N_j}{\partial y} + D_{xy} \frac{\partial N_i}{\partial x}\frac{\partial N_j}{\partial y} + D_{yx} \frac{\partial N_i}{\partial x}\frac{\partial N_j}{\partial y} +\right.$$

$$\left. N_j V_x \frac{\partial N_i}{\partial x} + N_j V_y \frac{\partial N_i}{\partial y} + N_i N_j \frac{\partial v_x}{\partial x} + N_i N_j \frac{\partial v_y}{\partial y}\right)$$

$$C_j \, dx\, dy + \iint_\Omega \theta N_i N_j \, dx\, dy \frac{\partial C}{\partial t} - \iint_\Omega S N_i \, dxdy -$$

$$\int_s N_i \left(D_{xx} \frac{\partial C}{\partial x} l_x + D_{xy} \frac{\partial C}{\partial y} l_x + D_{yx} \frac{\partial C}{\partial x} l_y + D_{yy} \frac{\partial C}{\partial y} l_y\right) ds = 0$$

$i = 1, 2, 3, \ldots n$ (22)

The above equations in matrix form are:

$$[A]\{H\} + \{B\} = 0 \tag{23}$$

for groundwater flow and

$$[K] + [L]\{C\} + [M]\{\partial C/\partial t\} + \{F\} = 0 \tag{24}$$

for solute transport.

Typical elements of different matrices are

$$A_{ij} = \iint_\Omega \left(K_x \frac{\partial N_i}{\partial x}\frac{\partial N_j}{\partial x} + K_y \frac{\partial N_i}{\partial x}\frac{\partial N_j}{\partial x}\right) dx\, dy \tag{25}$$

$$B_i = \iint \Omega q\, N_i \, dx\, dy - \int_S N_i \sum_{j=1}^{n}\left(K_x \frac{\partial N_i}{\partial x} l_x + K_y \frac{\partial N_j}{\partial x} l_y\right) N_j \, ds \tag{26}$$

$$K_{ij} = \iint_\Omega D_{xx} \frac{\partial N_i}{\partial x}\frac{\partial N_j}{\partial x} + D_{yy} \frac{\partial N_i}{\partial y}\frac{\partial N_j}{\partial y} + D_{xy} \frac{\partial N_i}{\partial x}\frac{\partial N_j}{\partial y} + D_{yx} \frac{\partial N_i}{\partial y}\frac{\partial N_j}{\partial x} \tag{27}$$

$$L_{ij} = \iint_{\Omega} N_j \left(V_x \frac{\partial Ni}{\partial x} + V_y \frac{\partial Ni}{\partial y} \right) + N_i N_j \left(\frac{\partial V_x}{\partial x} + \frac{\partial V_y}{\partial y} \right) \tag{28}$$

$$M_{ij} = \iint_{\Omega} \theta N_i N_j \; dx \; dy$$

$$F_i = -\iint_{\Omega} SN_i \, dxdy - \int_{\Omega} N_i \left(D_{xx} \frac{\partial}{\partial x} l_x + D_{yy} \frac{\partial}{\partial y} l_y + D_{xy} \frac{\partial}{\partial x} l_x + D_{yx} \frac{\partial}{\partial y} l_y \right)$$

$$\sum_{j=1}^{n} C_j N_j \; ds \qquad i = 1, 2, 3....n; \; j = 1, 2, 3, ...n \tag{30}$$

Integration of the Approximating Equations

An analysis of Equations (23) to (28) indicates that integration of the basis functions and their derivatives has to be done. As the basis functions are selected in such a way that they are non-zero only over the elements over which they are located, the domain Ω at any time consists of one or more elements depending upon the location of a particular node.

As stated earlier, each element appears as a square in the local coordinate system. The integration is done on the square element in the local coordinate system and the limits of integration are +1 and –1. The basis function derivatives $\partial N_i/\partial x$ and $\partial N_i/\partial y$ are obtained using the relationship.

$$\begin{Bmatrix} \dfrac{\partial N_i}{\partial \zeta} \\ \dfrac{\partial N_i}{\partial \zeta} \end{Bmatrix} = \begin{bmatrix} \dfrac{\partial X}{\partial \zeta} & \dfrac{\partial Y}{\partial \zeta} \\ \dfrac{\partial X}{\partial \eta} & \dfrac{\partial Y}{\partial \eta} \end{bmatrix} \begin{Bmatrix} \dfrac{\partial N_i}{\partial x} \\ \dfrac{\partial N_i}{\partial y} \end{Bmatrix} = [J] \begin{Bmatrix} \dfrac{\partial N_i}{\partial x} \\ \dfrac{\partial N_i}{\partial y} \end{Bmatrix} \tag{31}$$

[J] is the Jacobian matrix evaluated numerically (Ergatoudis et al., 1968) noting that

$$[J] = \begin{bmatrix} \dfrac{\partial N_1}{\partial \zeta}, & \dfrac{\partial N_2}{\partial \zeta}, & .... \dfrac{\partial N_n}{\partial \zeta} \\ . & . & . \\ . & . & . \\ \dfrac{\partial N_1}{\partial \zeta}, & \dfrac{\partial N_2}{\partial \zeta}, & ... \dfrac{\partial N_n}{\partial \zeta} \end{bmatrix} \begin{bmatrix} X_1 & Y_1 \\ X_2 & Y_2 \\ . & . \\ . & . \\ . & . \end{bmatrix} \tag{32}$$

The element area in the global coordinate system is also changed to local-coordinate system using the relationship:

$$dx.dy = \det[J]\ d\eta\ d\xi \tag{33}$$

Numerical integration is done over each element using a Gaussian quadrature. In this technique a polynomial $f(x)$ of a degree $2N–1$ could be integrated exactly as a weighted mean of its N partial values at specified points referred to as Gauss points. If cubic elements are used a sixth degree polynomial in ξ and η is generated and 16 Gauss points (4 × 4 rules) are required. For all elements, generally nine Gauss points (3 × 3) are used.

Boundary Conditions

The hydraulic head and transport equations are subject to certain boundary conditions and they should be incorporated in the calculations. For both the flow and transport equations two types of boundary conditions could be described. These are: (a) Dirichlet type wherein the hydraulic head or concentration at certain nodes is specified, and (b) Neumann type wherein the flux of water or contaminant entering the aquifer is specified.

The second term in the $\{B\}$ matrix (Equation (26)) for flow equation and the second term in the $\{F\}$ matrix (Equation (30)) for the transport equation are generated only if Neumann type boundary conditions are prescribed. For nodes where Dirichlet type boundary conditions are prescribed the elements of the matrices $[A]$ in the flow equation and $([K] + [L])$ and $[M]$ in the transport equation are not generated in the simultaneous set of equations resulting from Equations (23) and (24). The nodes for which the Dirichlet conditions are prescribed are known values. In the computational procedure the $\{B\}$ or $\{F\}$ matrix depending on whether the flow equation or the transport equation is being considered has to be modified as follows. Let the elements of $\{B\}$ and $\{F\}$ be denoted by b_i^* (i =1, 2, 3...n) and f_i^* (1, 2, 3,...n) respectively. Then modify $\{B\}$ or $\{F\}$ as per rule

$$b_i = b_i^* + \Sigma_{\text{allj}}\ A_{ij}\ H_j \tag{34}$$

and

$$f_i = f_i^* + \Sigma_{\text{allj}}\ ([K] + [L]_{ij})\ C_j \tag{35}$$

where j is the number of nodes for which the Dirichlet type boundary conditions are prescribed in each case. The computational steps are shown as follows:

Consider Equation (24) in the form

$$[N]\ \{C\} + [M]\ \left\{\frac{\partial C}{\partial t}\right\} + \{F\} = 0 \tag{36}$$

The partitioning of the matrices $[N]$, $[M]$ and $\{F\}$ for a four-node element using the Dirichlet boundary conditions is as follows. Write the coefficient matrices in the following form.

$$\begin{bmatrix} N_{11} & N_{12} & N_{13} & N_{14} \\ N_{21} & N_{22} & N_{23} & N_{24} \\ N_{31} & N_{32} & N_{33} & N_{34} \\ N_{41} & N_{42} & N_{43} & N_{44} \end{bmatrix} \{C\} + \begin{bmatrix} M_{11} & M_{12} & M_{13} & M_{14} \\ M_{21} & M_{22} & M_{23} & M_{24} \\ M_{31} & M_{32} & M_{33} & M_{34} \\ M_{41} & M_{42} & M_{43} & M_{44} \end{bmatrix} \left\{ \frac{\partial C}{\partial t} \right\} + \{F\} = 0 \tag{37}$$

and assume the Dirichlet boundary conditions are prescribed for node 1, first the rows and columns of [*N*] and [*M*] matrices for node 1, and made zero.

Next, applying the backward finite difference approximation to dC/dt, Equation (37) will be as follows:

$$\begin{bmatrix} N_{22} + \frac{M_{22}}{\Delta t} & N_{23} + \frac{M_{23}}{\Delta t} & N_{24} + \frac{M_{24}}{\Delta t} \\ N_{32} + \frac{M_{32}}{\Delta t} & N_{33} + \frac{M_{33}}{\Delta t} & N_{34} + \frac{M_{34}}{\Delta t} \\ N_{42} + \frac{M_{42}}{\Delta t} & N_{43} + \frac{M_{43}}{\Delta t} & N_{44} + \frac{M_{44}}{\Delta t} \end{bmatrix} \{C\}_{t_0+\Delta t}$$

$$= \begin{bmatrix} \frac{M_{22}}{\Delta t} & \frac{M_{23}}{\Delta t} & \frac{M_{24}}{\Delta t} \\ \frac{M_{32}}{\Delta t} & \frac{M_{33}}{\Delta t} & \frac{M_{34}}{\Delta t} \\ \frac{M_{42}}{\Delta t} & \frac{M_{43}}{\Delta t} & \frac{M_{44}}{\Delta t} \end{bmatrix} \{C\}_{t_0} - \begin{Bmatrix} N_{21} \bullet BC_1 \\ N_{31} \bullet BC_1 \\ N_{41} \bullet BC_1 \end{Bmatrix} \tag{38}$$

The right side matrices are now combined into one matrix as given by

$$\begin{bmatrix} \frac{M_{22}}{\Delta t} C_2 + \frac{M_{23}}{\Delta t} C_3 + \frac{M_{24}}{\Delta t} C_4 - N_{21} \bullet BC_1 \\ \frac{M_{32}}{\Delta t} C_2 + \frac{M_{33}}{\Delta t} C_3 + \frac{M_{34}}{\Delta t} C_4 - N_{31} \bullet BC_1 \\ \frac{M_{42}}{\Delta t} C_2 + \frac{M_{43}}{\Delta t} C_3 + \frac{M_{44}}{\Delta t} C_4 - N_{41} \bullet BC_1 \end{bmatrix} \tag{39}$$

where C_2, C_3 and C_4 are concentrations at nodes 2, 3 and 4 respectively at time t_0, and BC_1 is the value of the concentration at node 1.

Calculation of Velocity Vectors

Apart from the dispersion coefficients, the main input to the transport equation is the seepage velocity coefficients. The two-dimensional horizontal groundwater flow equation is solved to obtain the hydraulic head values at the nodes. To obtain the seepage velocity components at the nodes of each element, the basis functions are differentiated at each Gauss point and the functional coefficient approach is used (Pinder and Frind, 1972).

$$V_x = K_x \frac{\partial H}{\partial x} = K_x \sum_{i=1}^{n} \frac{\partial N_i}{\partial x} H_i \tag{40}$$

$$V_y = K_y \frac{\partial H}{\partial y} = K_y \sum_{i=1}^{n} \frac{\partial N_i}{\partial y} H_i \tag{41}$$

Again, for calculating the gradient of the seepage velocity components a similar approach is used.

$$\frac{\partial V_x}{\partial x} = \sum_{i=1}^{n} \frac{\partial N_i}{\partial x} V_{x_i} \tag{42}$$

$$\frac{\partial V_y}{\partial x} = \sum_{i=1}^{n} \frac{\partial N_i}{\partial y} V_{y_i} \tag{43}$$

The gradients of the seepage velocity components are calculated at the centre of the element ($\xi = 0$; $\eta = 0$). For calculating the dispersion coefficients using the velocity relationships, the velocity components calculated at the centre of the element are used. For each element the components of the hydrodynamic dispersion coefficient are assumed to be the same.

Calculation of the Time Derivative

The set of Equations (24) are a set of first order differential equations. To solve these equations a finite difference approximation to the time derivative *dC/dt* is adopted. The time derivative could be approximated either by central, forward or backward differences. The backward difference scheme, which is fully implicit, is found to be stable and accurate (Pinder and Frind, 1972). Using the backward difference method, Equation (24) can be written as:

$$([K] + [L]) \{C\}_{t+\Delta t} + \frac{[M](\{C\}_{t+\Delta t} - \{C\}_t)}{\Delta t} + \{F\} = 0 \tag{44}$$

By rearranging, the following recurrence relationship is obtained:

$$([K] + [L]) + \frac{[M]}{\Delta t}\{C\}_{t+\Delta t} = \left(\frac{[M]}{\Delta t}\right)\{C\}_t - \{F\} \tag{45}$$

Starting with the initial, known concentration distribution for all the nodes, this recurrence relationship calculates the concentration for the subsequent time steps.

Nature of the Matrices

Equations (23) and (24) as stated earlier are a set of simultaneous equations. The matrices $[A]$, $([K] + [L])$ and $[M]$ are $(n - m) \times (n - m)$ size where m is the number of passive nodes due to Dirichlet boundary conditions. While the matrix $[A]$ is symmetric, the matrix $([K] + [L])$ is non-symmetric, the non-symmetry being due to the non-equality of the components of the seepage velocity. This lack of symmetry requires an increased computational effort as the entire matrix is to be considered in computations, whereas in case of symmetric matrices only the upper triangular part needs to be considered. Even though the matrices are nonsymmetric they are sparse and banded. The bandwidth can be kept to a minimum by proper numbering of the nodes in the system.

Considering any two of the sixteen nodes which describe an element, let the differences in their node numbers be denoted by N_d. For a given element let the maximum value of N_d be denoted by NE_d. Considering all elements in the system let the maximum value of NE_d be denoted as NE_{max}. For minimization of the bandwidth and consequently the computational cost for a given analysis, the node points should be numbered so as to minimize the value of NE_{max}.

Comments on Finite Element Method

The finite element method, in its standard form, can be considered to be somewhat more flexible than the standard form of the finite difference method. This may explain its popularity. It must be noted, however, that there exist variants of the finite difference method that are very similar to the finite element method. Thus, the two methods may be considered as more or less equivalent. The method to be used for a given class of problems may be determined for a large part by such considerations as the convenience of a programme, its availability and cost, and the level of experience of the user.

A disadvantage of the finite element method, as compared to the finite difference method, is that the memory requirement is often considerably larger. The formulation of a finite element programme leads to a system of linear equations, just as the finite difference method, but the general procedure is such that these equations are constructed by generating a system matrix, the coefficients of which have to be stored in the computer's memory. In the most elementary form this means that for a system with N unknowns a matrix of $N \times N$ coefficients must be stored. Fortunately, the matrix is sparse, due to the fact that connections between nodes happen only when they are in one element. Thus, by careful programming, it is possible to reduce the memory requirement to about $5 \times N$, taking into account the symmetry of the system matrix also. When these properties of the system matrix are used, the finite element method becomes a real competitor for the finite difference method. Of course, the advantage, that each element, at least in principle, may have its own values for all the physical parameters, also involves a certain penalty, in that all these values have to be stored in the computer's memory. In general one may conclude that, as a rule, the memory requirements of a finite element programme are larger than those of a finite difference programme. It is worthwhile to pay this price only when the flexibility of the finite element method is really needed.

Accuracy of Numerical Models

Numerical models can, of course, never give more than approximate representation of reality. In general these models use a mesh of discrete points, in which the physical values are calculated, and it can be expected that this creates a lower bound for the fluctuations that the model can represent. In addition to this there are some more difficulties inherent to numerical modeling of transport processes. These are discussed in what follows.

Numerical Dispersion

The first difficulty is that there appears to be a contradiction in the basics of numerical modeling of transport by dispersion. As mentioned before, the dispersivities are of the order of magnitude of the pore size, or, more generally, of the size of the random irregularities in the porous medium. The size will be denoted by d. The transition to a continuum model involves an averaging process over some representative elementary volume (Bear, 1972), which must be an order of magnitude larger than the size of the irregularities, d. This means that the smallest physically relevant dimension in the continuum formulation is about $10d$,

$$x >> d \tag{46}$$

On the other hand, in a numerical solution a spatial discretization is used. The characteristic mesh size is denoted by x. As a result of the discretization various numerical errors are introduced. One of these is due to the approximation of the first order derivative which describes the advective transport. It can be shown that the resulting error is an additional dispersion-like term (numerical dispersion), having a dispersion coefficient $v\ x$. In order to suppress this error the numerical dispersion effect should be small compared to the physical dispersion. This leads to the requirement that the mesh size x should be small compared to the dispersivities, hence

$$x << d \tag{47}$$

This is in contradiction with the earlier conclusion that the smallest dimension in the problem should be large compared to d.

For the solution of this problem of numerical dispersion various procedures have been suggested, and are currently being investigated. Most of these are based upon the analysis of the transport problem by using the method of characteristics. This can also be viewed as a process of tracing particles, as the pore fluid is transporting them. In this way the error in modeling the advective transport is eliminated. The dispersive transport now can be modeled using discrete particles also, and a convenient method to do so is to use a random walk simulation. A disadvantage of such a simulation is that it does not immediately give quantitative results in the form of the concentration at a point as a function of time. The results give a good insight into the real behaviour.

REFERENCES

Anthony, S.S. (1992). Electromagnetic methods for mapping freshwater lenses on Micronesian atoll islands. *Journal of Hydrology*, **137:** 99-111.

Arora C.L. and Rose, R.N. (1981). Demarcation of fresh- and saline-water zones, using electrical methods (Abohar area, Ferozepur district, Punjab). *Journal of Hydrology*, **49:** 75-86.

Bear, J. (1972). Dynamics of fluids in porous media. American Elsevier Publishing Company, Inc.

Bear, J. (1979). Hydraulics of groundwater. McGraw Hill, New York.

Bobba, A.G. (1993a). Mathematical models for saltwater intrusion in coastal aquifers-Literature review. *Water Resources Management,* **7:** 3-37.

Bobba, A.G. (1993b). Field validation of "SUTRA" groundwater flow model to Lambton County, Ontario, Canada. *Water Resources Management*, **7:** 289-310.

Bobba, A.G. (1998). Application of numerical model to predict freshwater depth in islands due to climate change effect: Agati Island, India. *Journal of Environmental Geology*, **6:** 1-13.

Bobba, A.G. (2000). Numerical simulation of saltwater intrusion into east coastal basin of Indian sub-continent due to anthropogenic effects. ICIWRM-2000, Proceedings of International Conference on Integrated Water Resources Management for Sustainable Development. National Institute of Hydrology, 323-340.

Bobba, A.G. (2002). Numerical modelling of saltwater intrusion due to human activities and sea level change in the Godavari delta. *Hydrological Sciences Journal*, **47(S):** S67-S80.

Bobba, A.G. and Joshi, S.R. (1989). Application of an inverse approach to a Canadian radioactive waste disposal site. *Ecological Modelling*, **46:** 195-211.

Bobba, A.G. and Joshi, S.R. (1988). Groundwater transport of Radium-226 and uranium from Port Granby Waste management site to Lake Ontario. *Nuclear and Chemical Waste Management*, **8:** 199-209.

Bobba, A.G. and Singh, V.P. (1995). Groundwater contamination modelling. *In:* Environmental Hydrology. V.P. Singh (Ed.), Kluwer Academic Publishers. pp. 225-319.

Bobba, A.G., Singh, V.P., Berndtsson, R. and Bengtsson, L. (2000). Numerical simulation of saltwater intrusion into Laccadive island aquifers due to climate change. *Journal of Geological Society of India*, **55:** 589-612.

Bokuniewicz, H.J. (1980). Groundwater seepage into Great South Bay, New York. *Estuarine and Coastal Marine Science*, **10:** 437-444.

Custodio, E. (1985). Saline intrusion. *In:* Hydrogeology in the Service of Man. IAHS Publication No. 154, I, IAHS, Wallingford, 65-90.

Ergatoudies, I., Irons, B.M. and Zienkiewicz, O.C. (1968). Curved isoparametric "Quadrilateral" elements for finite element analysis. *International Journal of Solids and Structures,* **4:** 31-42.

Falkland, A. (1991). Hydrology and Water Resources of Small islands—A Practical Guide. UNESCO, Paris.

Freeze, R.A. and Cherry, J.A. (1979). Groundwater. Prentice-Hall, Englewood Cliffs.

Geirnaret W. and Laveven, M.P. (1992). Composition and history of groundwater in the western Nile Delta. *Journal of Hydrology*, **138:** 169-189.

Goldman M., Gilad, D., Ronen, A. and Melloul, A. (1991). Mapping of seawater intrusion into the coastal aquifer of Israel by the time domain electromagnetic method. *Geoexploration*, **28:** 153-174.

Hem, J.D. (1970). Study and interpretation of the chemical characteristics of natural water. USGS Water Supply Paper 1473, USGS.

Kashef, A.I. (1977). Management and control of salt-water intrusion in coastal aquifers. *Critical Reviews in Environmental Control*, **7:** 217-275.

Kashef, A.I. 1986. Groundwater Engineering, McGraw-Hill, New York.

Piggott, A.R., Bobba, A.G and Novakowski, K.S. 1996. Regression and inverse analysis in regional groundwater modeling. *Journal of Water Resources Planning and Management*, ASCE, **122:** 1-10.

Pinder, G.F. and Frind, E.O. (1972). Application of Galerkin procedure to aquifer analysis. *Water Resources Research*, **8:** 108-120.

Pinder, G.F., Frind, E.O. and Papadopules, S.S. (1972). Fundamental coefficients in the analysis of groundwater flow using finite elements. *Water Resources Research*, **9:** 222-226.

Prasad, P.R., Pekdeger, A. and Ohse, W. (1983). Geochemical and geophysical studies of salt water intrusion in coastal regions. *In:* Relation of Groundwater Quantity and Quality (Proceedings of the Hamburg Symposium, August 1983). IAHS Publ. No. 146, F.X. Dunn, G. Matthess and R.A. Gras (Eds.), IAHS Press, Wallingford, Oxon, 209-218.

Reddel, D.I. and Sunada, D.K. (1970). Numerical simulation of dispersion in groundwater aquifers. *Hydrology*, paper 41, Colorado State University, Fort Collins.

Revelle, R. (1941). Criteria for recognition of sea water in ground-waters. *Transactions of the American Geophysical Union*, **22:** 593-597.

Rushton, K.R. (1980). Differing positions of saline interfaces in aquifers and observation boreholes. *Journal of Hydrology*, **48:** 185-189.

Sundaresan, J. (1993). Impact of sea level rise on aquifer system of Agatti atoll. *Environmental Geology*, **21:** 51-54.

Tellam, J.H., Lloyd, J.W. and Walters, M. (1986). The morphology of a saline groundwater body: Its investigation, description and possible explanation. *Journal of Hydrology*, **83:** 1-21.

Valiela, I., Foreman, K., LaMontagne, M., Hersh, D., Costa, J., Peckol., DeMeo-Anderson., D'Avanzo, C., Babione, M., Sham, C., Brawley, J., Lajyha, K. (1992). Coupling of watersheds and coastal waters: Sources and consequences of nutrient enrichment in Waquoit Bay, Massachusetts. *Estuaries*, **15:** 443-457.

Valiela, L., Costa, J., Foreman, K., Teal, J.M., Howes, B. and Aurrey, D. (1990). Transport of groundwater borne nutrients from watersheds and their effects on coastal waters. *Biogeochemistry*, **10:** 177-197.

Voss, C.I. (1984). SUTRA—A finite element simulation model for saturated-unsaturated fluid density dependent groundwater flow with energy transport or chemically – reactive single species solute transport. U.S. Geol. Surv., Water Resources Invest. Rep., 84-4239.

11

Management of Groundwater Resources

M. Thangarajan

Retired Scientist-G
National Geophysical Research Institute
Hyderabad, India

PREAMBLE

Groundwater plays a major life support to mankind. It is the major source to meet the domestic, irrigation and industrial demands. Groundwater occurs in a wide range of rock types and usually requires little or no treatment; therefore, it is often the cheapest and simplest water supply option.

In India too, groundwater is the main source both for irrigation and drinking purposes. In early days, abstraction from the shallow aquifer has been limited, mainly because water-lifting devices were animal-powered. However, since the 1950s groundwater abstraction has increased substantially, both as a consequence of the increase in the number of wells and of progressive replacement of the animal-powered lifting devices by energized pumps capable of much higher yields. This rapid development of groundwater resources in India and other developing countries had resulted in the declining of water table/levels rapidly in many parts of the country causing shallow wells to dry up with a particular impact on those rural poor farmers unable to deepen their wells to chase the declining water levels. In coastal areas, declining water levels are also associated with the ingress of saline water, leading to reduced crop yields, loss of drinking water supplies and ultimately loss of both fertile land and water supply wells.

Monitoring water level over three decades (from 1970), in many parts of the globe, have provided clear evidence of a long-term water-level decline, as a result of increased groundwater abstraction. This has resulted in the deterioration of water quality and the widespread drying-up of wells following a 'failure' of the monsoon. Deepening of wells does not appear to be a viable option as most wells already fully penetrate the shallow weathered aquifer. This has resulted only in debt trap of farmers particularly from the monsoon climatic countries such as India and African continent. Therefore, a detailed discussion on the management aspects have been given in this article.

COMMUNITY MANAGEMENT OF GROUNDWATER RESOURCES

The livelihood of rural population mainly depends on the sustainable development and management of groundwater resources and more than 70% of India's population live in rural areas. In India and elsewhere, however, groundwater is considered to be the personal property of the landholders and the abnormal growth of population and spurt in industrial activities for the last four decades and conversion of vast waste lands into cropping land forced more abstraction of groundwater. This has resulted in socio-economic conflicts. The land has been divided in to small land-holdings and each holder started developing new wells (shallow and deep bore wells) without any management approach. Today, in India millions of wells are not in operation as they are dried up. Competitive spirit in deepening wells is going on without understanding its consequences. During drought years entire rural folk suffers for their day-to-day food. Thus, groundwater plays a major role in the livelihood of entire rural sector.

Addressing the problem of poverty and over-exploitation of groundwater in India and other African countries is a complex phenomenon. If there is sustainable source of water available both for drinking and irrigation, it may be possible to alleviate the rural poverty by 75% and this is possible only by effective groundwater management. If one answers the question to whom and how much the groundwater resources are to be distributed as depicted in Fig. 1, then one can resolve the management of groundwater resources.

REQUIREMENTS FOR EFFECTIVE GROUNDWATER MANAGEMENT

The following are some of the requirements for effective groundwater management in rural area:

- good understanding of the aquifer system,
- practical measures to control abstraction to be identified including education to rural people about the aquifer and its behaviour, and
- augment groundwater resource through artificial recharge.

Water resources management means intervention in matters concerning water, which may be planning, design and operation. Water resources management needs a number of steps carried out in sequence as given below:

- assessment of water demands/requirements,

- assessment of water resources through geophysical and other methods,
- evaluation of biochemical processes in the soil environment with special reference to the stability of soil organic matter and the carbon/nitrogen balance,
- selection of suitable plants and determination of the rotation system,
- determination of irrigation schemes and design of drainage pattern, if necessary,
- monitoring of water quality and pollution sources, vertical profiling of the unsaturated zone and aquifer,
- study of wash-out and leaching of salts from the soil into the unsaturated zone,
- study of solute transport and transformation processes in the unsaturated zone,
- modeling of groundwater flow and mass transport in the unsaturated and saturated aquifer system,
- evaluation of economic and social consequences of recommended changes in the agro-eco system, and
- identification of problems and constraints in decision-making.

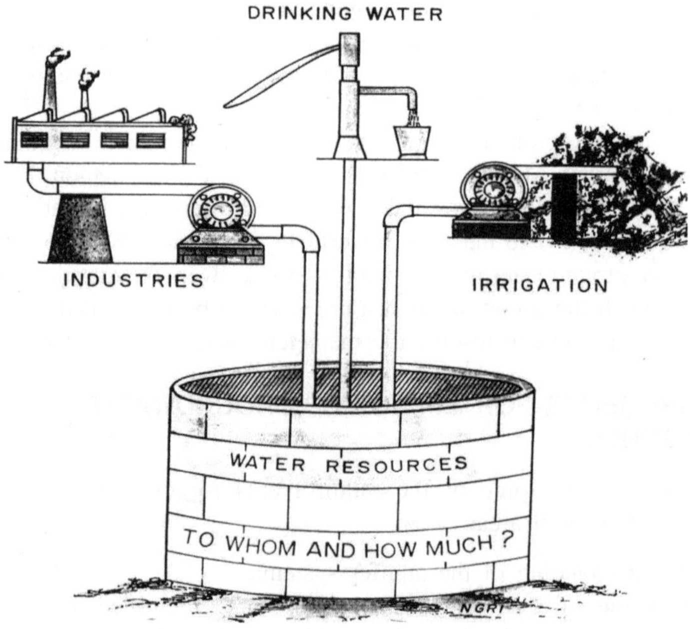

Figure 1. Groundwater resources distribution

Controlling abstraction is the most sensitive one and *could not be achieved by mere legislation* and it needs to educate the rural folk to share the available groundwater by themselves. This again needs some sort of technical approach, which can induce them to understand the problem. Implementing such reforms is problematic as monitoring (controlling) the activities of individual owners and stopping subsidies, is very difficult. Against this background, the application of *simple groundwater models* and the development of *user group based institutions (community participation)* for groundwater management, could be a viable and innovative idea and it has to be tested (BGS, 2000).

AUGMENTATION OF GROUNDWATER RESOURCES THROUGH ARTIFICIAL RECHARGE

Augmentation of groundwater resources through artificial recharge is becoming popular slogan among the developing countries. Even in early days, the spreading (ponding) of surface water in minor and medium irrigation tanks had helped to infiltrate water into shallow and unconfined aquifers. The rapid decline of water levels in semi-arid and arid regions due to over-exploitation had forced to augment the groundwater either through surface spreading (ponding) or direct injection into the deeper aquifer through boreholes. The artificial augmentation of groundwater otherwise called derived recharge is defined as a process by which excess surface water is directed into the ground, either by spreading on the surface, using recharge wells or by altering natural conditions to increase infiltration to replenish an aquifer. Pyne (1995), Pavelic and Dillon (1997) and Barber (2002) had reviewed the various techniques used to augment the groundwater resources. Aquifer storage and recovery (ASR) method is becoming more common nowadays wherein a single bore well is used both for injection and recovery. By using this method, the injection wells can be developed regularly during recovery phases and it helps for self-cleaning the clogging of bore wells. Surface spreading infiltration can augment only the shallow aquifer system whereas the ASR and recharge wells can be used to augment the deeper aquifer. The problem of clogging with solids and bio-films in the injection well is still a major bottleneck in ASR and recharge wells, which has to be rectified. Reducing the level of suspended solids to the level of 10 mg/l by pre-treatment can reduce this problem to some extent.

Surface spreading on the other hand attenuates the suspended solids in vadose zone while the water infiltrates vertically downward before reaching the water table. Here, the vadose (unsaturated) zone acts as a natural filter for suspended solids and other bacteria etc. This process is otherwise called soil-aquifer treatment (SAT) or geo-purification. Check dams and percolation dams constructed across the rivulets and small streams in hard rock region has proved to be a viable technique to improve the derived recharge close to the vicinity of the rivulet/stream and it is also proved that it will transmit the water very fast to a far place if it encounters a fracture of good length say a few kilometres.

GROUNDWATER CONTAMINATION

Sources of Contamination

The sources of groundwater contamination are many and the contaminants are numerous in the developed and developing countries. Common industrial solvents such as trichloro-ethylene, 1,1,1-trichloroethene, benzene and carbon tetrachloride found in widespread areas, with all indications being there, are multiple sources (Fusillo et al., 1985). Suburban areas have groundwater with high levels of nitrate due to the use of lawn fertilizers found in groundwater (Pionke and Urban, 1985), with specialized synthetic organic agricultural chemicals as well (Rothschild et al., 1982). Landfills in urban and semi-urban areas are known sources of contamination (Noss and Johnson, 1984; Leod, 1984). Groundwater contamination is also identified due to leakage from underground petroleum products (Kramer, 1982) and synthetic organic chemicals (Oliveira and Sitar, 1985). The urban aquifers are found to be contaminated in many folds than the rural aquifers. Contaminated groundwater will in most cases not travel more than a few hundred metres from the source and in many cases not more than a few tens of metres. If there is a single source, the contamination may be localized whereas the contamination will widespread, if there are multiple sources, or if the contamination is a result of widespread land-use practices (fertilizers and pestcides). Natural barriers in the soil acts as attenuator and prevents the contamination to become irreversible process. Attenuation mechanisms include dilution, dispersion, mechanical filtration, volatilization, biological activity, ion exchange and adsorption on soil particle surfaces, chemical reactions and radioactive decay. Even synthetic organic compounds can undergo biological decay (Cline and Viste, 1984), although the decay products may also be toxic. In recent years a number of cleaning technologies have been developed for restoring the quality of groundwater that has been contaminated. This is discussed in following sections.

Septic Tanks and Cesspools

The disposal of domestic wastewater through septic tanks and drain tile fields is a common practice in many semi-urban areas. Anaerobic decomposition of wastes takes place in the septic tank. The liquid waste is carried to a drain tile field, where it seeps to the water table through the vadose zone.

Septic tank effluent contains bacteria and viruses. It is a major factor in the incidences of waterborne disease from wells all over the world. The most important factor that influences the development of groundwater contamination from septic tanks is the density of septic tank systems in the area. Documented cases of widespread groundwater contamination from septic tank systems have been reported in areas where the sizes range from less than one-quarter of an acre to three acres (Yatas, 1985).

Septic tanks are most likely to contribute to groundwater contamination in areas where there is a high density of houses with septic tanks, thin soil layer exists over permeable bedrock, the soil is extremely permeable, such as gravel, and shallow water table exists within a couple of metres from the land surface. Areas with high population density should not be served with septic tanks, and areas with thin soils, extremely permeable soils, and shallow water tables should be avoided.

Landfills

Burial in a landfill is the most common means of disposing off municipal refuse, ashes, garbage, leaves, demolition debris and sludges from municipal and industrial wastewater treatment facilities. Radioactive, toxic and hazardous wastes have also been subjected to land burial as a means of disposal. Precipitation that infiltrates the waste can mix with liquids already present in the waste and leach compounds from the solid waste. The result is a liquid known as leachate. Leachate can move downward from the landfill into the water table and cause groundwater contamination. If the waste is buried below the water table, moving groundwater can leach compounds from the waste and become contaminated. Landfill leachate can contain very high concentrations of both inorganic and organic compounds.

When leachate from a landfill mixes with groundwater, it forms a plume that spreads in the direction of the flowing groundwater (advection). As one goes away from the source, the concentration decreases owing to hydrodynamic dispersion and retardation. The volume of leachate that is produced is a function of the amount of water percolating through the refuse. Land disposal of solid waste in humid areas is more likely to produce large volumes of leachate than land disposal in arid zones. The vadose zone in arid regions may receive little or no charge. Under such conditions, solid-waste disposal is not likely to result in groundwater contamination.

One has to look into the hydro-geological factor in selecting alternative sites so that there is a very good isolation (separation) between the leachate production site and the water table. Landfills may be designed to minimize the formation of leachate as well as to minimize the amount of leachate that escapes from the landfill. Leachate could also be collected and treated.

It is desirable to construct landfills to accommodate the solid waste above the water table. Vadose zone plays a major role in attenuating the leachate produced in the refuse site. A *natural-attenuation* landfill is one that relies totally on natural processes to attenuate any leachate formed. Such landfills should lie above the water table to promote maximum attenuation in the vadose zone. Clays in the soil have the highest attenuation factor because they have the most ion exchange and adsorption sites. Unfortunately, in humid areas, the water table in clay soils tends to be close to the land surface; this means that much of the landfill should be above grade. Capping the landfill with one to two metre of compacted clay soil or a synthetic membrane may reduce leachate generation. If large amounts of leachate are generated in a natural-attenuation landfill located in low-permeability clay, there is a tendency for leachate to come to the land surface and form leachate springs. Natural-attenuation landfill design is shown in Fig. 2.

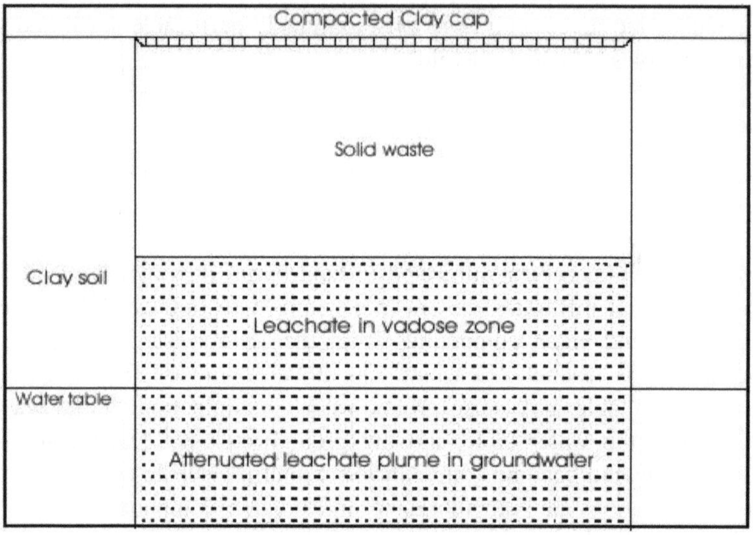

Figure 2. Sketch of natural-attenuation landfill

A lined landfill is one designed to capture part or all of the leachate generated. Landfill liners are typically constructed of 1 to 3 metres of compacted clay soils. The permeability of the liner should not be greater than 1 x 10^{-7} cm/sec. A synthetic membrane such as HDPE (high density polyethylene) could also be used as the liner. Because leachate will collect on the liner, a leachate-collection system is also needed. The leachate-collection system consists of a blanket of sand or gravel, with perforated drainage lines, lying on the liner. The base of the liner is sloped towards the drain tiles. Leachate drains through the leachate-collection system to a holding tank or sewer and is ultimately removed and treated. Clay-lined systems can be designed to collect about 70 to 90 percent of the leachate produced (Kmet et al., 1981). The remainder of the leachate will seep through the liner. A double liner and secondary leachate-collection system installed beneath the primary liner can be constructed to capture the leakage through the primary liner. A membrane or clay liner could be used for the secondary liner. A double-lined system with leachate collection is shown in Fig. 3.

In areas with a high water table and low-permeability soils (10^{-6} cm/sec), a zone-of-saturation landfill could be constructed (Fig. 4). An excavation below the water table is made. In clay soils this can easily be done because groundwater seeps into the excavation at a very slow rate and evaporates;

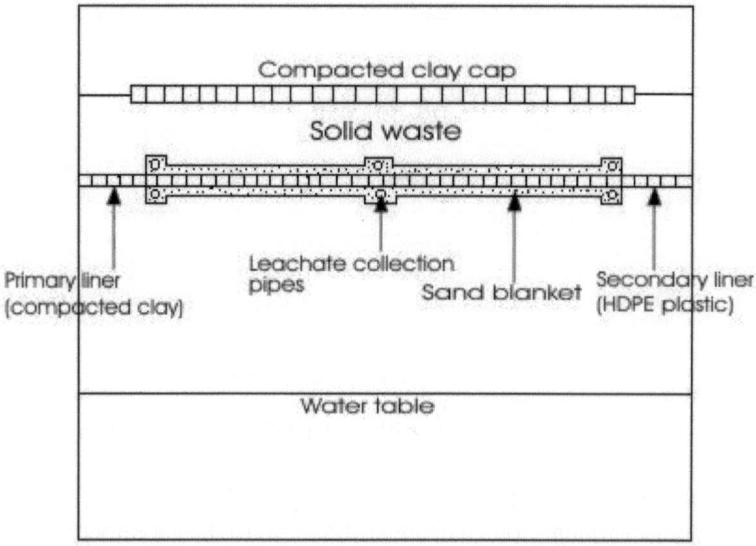

Figure 3. Double-lined landfill with leachate collection system.

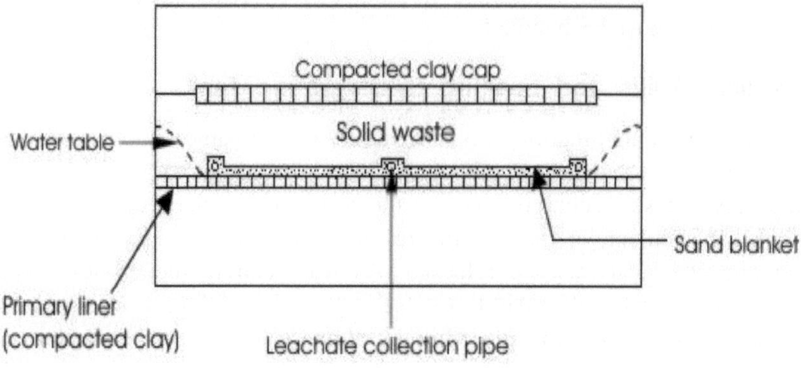

Figure 4. Zone-of-saturation landfill

it does not accumulate. A re-compacted clay liner with re-compacted clay sidewalls is installed to further reduce the amount of seepage into the excavation. A leachate-collection system is installed. This collects not only the leachate that forms, but any groundwater that seeps in as well. The zone-of-saturation landfill is more efficient than the lined landfill because no leachate escapes. However, it will be necessary to collect the leachate long after the landfill is closed because the waste will be below the water table.

As an alternative to the zone-of-saturation landfill, a hydraulic-gradient control landfill can be constructed in areas of a high water table (Fig. 5). Under-drains are placed beneath the liner to lower the water table so that it

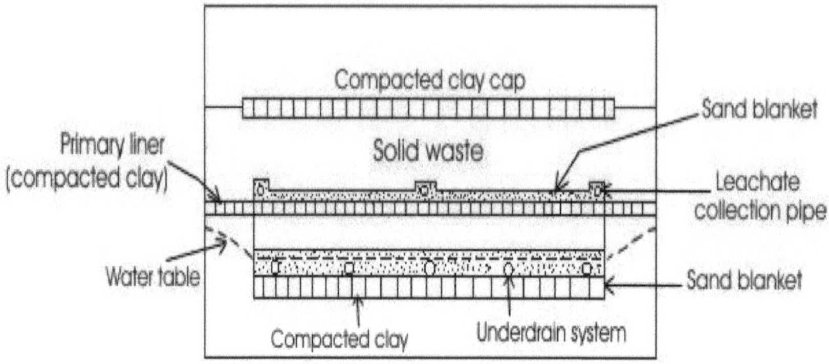

Figure 5. Lined landfill with water-table control system

is below the liner. A leachate-collection system is installed above the liner. The water discharging from the under-drains must be monitored to determine whether the portion of the leachate that drains through the liner is impacting it. If it is, then it must be treated before being discharged. As with all of the systems, a well-designed, well-constructed cap is essential in reducing the amount of leachate formed, thereby reducing the amount of leachate that must be handled.

Chemical Spills and Leaking Underground Tanks

Groundwater contamination due to a variety of inorganic and organic compounds has occurred as a result of spills and leaks of toxic and hazardous chemicals. These discharges may be the result of a sudden action, such as a tank car accident, or may be the result of slow leakage. Typically more than one chemical may be released. As was seen in the earlier section different chemicals will travel through the ground at different rates owing to retardation effects. As a result, complex plumes of contaminated water may result.

If the contaminant dissolves in the water, it will flow along with the groundwater. However, if a liquid discharged into ground has a specific gravity less than that of water, it can float on the water table. This is what happens when a petroleum product leaks into the ground.

Mining

Extraction and processing of metallic ore and coal has been the source of both surface and groundwater contamination. Groundwater moving through mineralized rock zones may contain excessive amounts of heavy metals (Klusman and Edwards, 1977). Mining and milling expose overburden and waste rock to oxidation. Oxidation of pyrite, a common mineral, can produce sulfuric acid. Uranium and thorium mining and milling operations can release radioactive isotopes to the atmosphere, surface water and groundwater (Kaufman et al., 1976).

Other Sources of Groundwater Contamination

Waste-disposal practices include liquid industrial waste disposal in lagoons and injection wells, oil-field brine disposal in lagoons and wells, land-spreading of sewage and industrial sludges, leakage from municipal waste-water sewers and lagoons, and land disposals of animal waste from feed lots. Other major causes of groundwater contamination could be spills and leaks; mine drainage; salt-water intrusion; poorly constructed or abandoned water, oil and gas wells; infiltration of contaminated surface water; agricultural activities; highway deicing salts; and atmospheric contaminants.

GROUNDWATER RESTORATION

Many researchers have reported that once groundwater has been contaminated, it may take many years to remove the contaminants through natural process from an aquifer after the source of contamination has been eliminated. During the 1980's methods of restoring groundwater quality by implementing various types of remedial measures were developed. However, most of these methods are time-consuming and extremely expensive. There are two broad categories of remedial measures; one must remove or isolate the source and/or pump and treat the groundwater (JRB Associates Inc. Book, 1982; Canter and Knox, 1985; Canter, 1982).

Source-Control Measures

One extreme method of source control is to excavate and remove the source. The Pollution Control of Tamilnadu on Tamilnadu Chrome and Chemical Ltd., at Ranipet, Tamilnadu, India, did this, where the solid waste containing hexavalent chromium (a toxic chemical) was dumped on a landfill for years and it had contaminated the groundwater. The Pollution Control had ordered to remove the solid waste from the factory site (Thangarajan, 1999a). A less drastic approach is to isolate the waste in place. If the waste is entirely above the water table, this is much more easily done than if the waste extends below the water table.

Modifying the surface drainage pattern so that runoff from upland areas does not cross the land surface above the waste will reduce the amount of surface water that infiltrates into the waste and produces leachate. The construction of a low-permeability cap above the waste can also be very effective in reducing the amount of infiltration through the waste. Caps can be constructed of compacted clay, synthetic membranes, concrete, asphalt and other types of materials. The most effective caps have several layers and include coarse granular layers between fine-grained layers to act as drains to divert infiltration away from the waste (Herzog et al., 1982). A multi-media cap design is shown in Fig. 6.

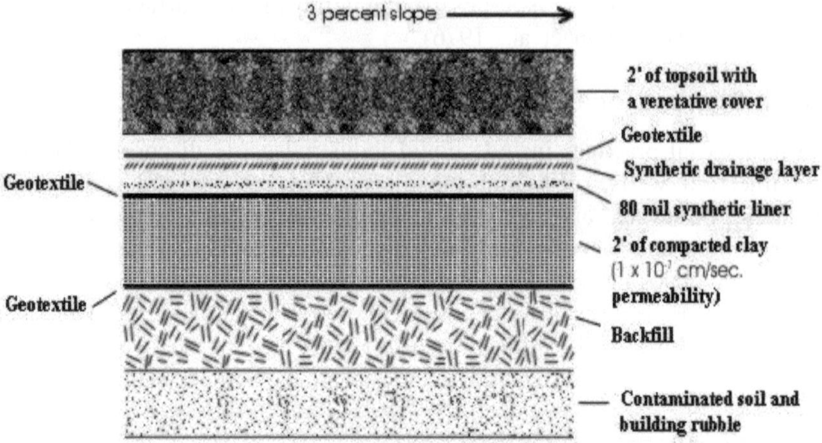

Figure 6. Design of a low-permeability multimedia cap to cover waste.

If the waste extends below the water table, then it is necessary to keep the groundwater from flowing through it. This can be accomplished by installing a low-permeability vertical barrier around the waste body (Ayres et al., 1983; Brunsing and Cleary, 1983; Fitzwater et al., 1983; Druback and Arlotta, 1985; Lynch et al., 1984). Vertical barriers can be constructed by digging the trench and backfilling it with a slurry-type mixture of water, soil and bentonitic clay. This is called slurry wall. Slurry wall construction is limited to the depths that the trench can be constructed. A grout curtain can be installed by injecting any one of a number of compounds into boreholes around the site. The materials fill the pore spaces in the rock or soil and harden. Grout curtains can be installed to great depths. Interlocking metal sheet piling can also be driven into soil to form a cutoff wall.

One can see from Fig. 7a, a waste material buried beneath the water table. On the other hand, Fig. 7b indicates the installation of a low-permeability slurry wall to lower the water table and divert flowing groundwater from the waste source. In this case, one wall-installed up gradient of the waste is sufficient. In other cases, the waste could be surrounded with walls. Grout can even be injected through the waste to form a bottom seal. Used in conjunction with walls around the waste, the waste can be totally isolated and this aspect is depicted in Fig. 7c.

Hydraulic gradient control measures can be used to lower the water table where it is in contact with the waste (Keely, 1984; Schafer, 1984; Campbell et al., 1984; Poulos and Laws, 1985). Figure 7d shows a pumping well installed in the upgradient of buried waste material and by doing this the water table is lowered so that it no longer is in contact with the waste.

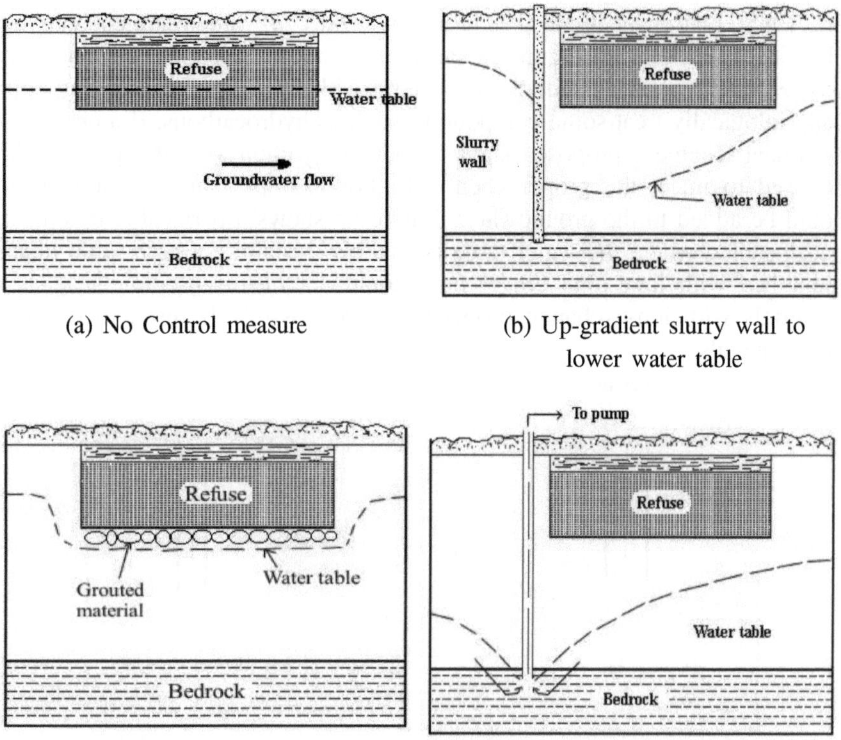

(a) No Control measure

(b) Up-gradient slurry wall to lower water table

(c) Grouting to create seal on sides and bottom

(d) Gradient control through pumping well

Figure 7. Techniques used to isolate the waste from water table (Source: U.S. Environmental Protection Agency)

Plume Treatment

Once the contamination source is isolated, the task remains to restore the quality of the groundwater that has become contaminated. One option could be to take no action and let the contaminants be flushed from the aquifer by natural recharge. This process could be made more efficient in some hydrogeologic settings by artificially increasing the amount of water that enters the aquifer and hence accelerating the natural process. This approach is usually not desirable if the aquifer is used as a source of drinking water. Additionally, if the plume eventually discharge to a surface-water body, it might cause contamination of the surface water. The time for natural restoration might be tens to hundreds of years. Even if the aquifer is not presently a drinking-water source, in the future it could be put to such a use. If the waste source were no longer present, future generations could unsuspectingly drill wells into the contaminated aquifer.

Plume treatment can occur *in situ* or by extraction of the water via wells (Lenzo, 1984; Flathman et al., 1984; Brenoel and Brown, 1985; Yaniga et al., 1985; Flathman et al., 1985; Fetter, 1988). *In situ* treatment can be chemical or biological. Figure 8a shows a method of injecting nutrients and oxygen into a plume. Natural soil bacteria with the proper mix of nutrients can biologically treat some compounds, such as hydrocarbons. If a chemical treatment scheme is proposed, the up gradient injection well of Fig. 8a could be used to inject the proper chemicals. For example, an oxidizing agent could be added to the groundwater. Figure 8b shows a permeable treatment bed installed where the water table is shallow. Installing a permeable treatment bed with limestone gravel in it could neutralize an acidic leachate. Materials with ion-exchange properties could possibly be used to remove heavy metals.

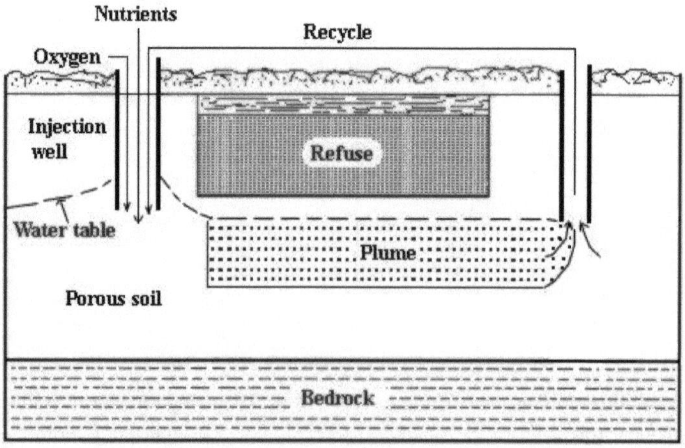

A. Injecting nutrients and oxygen into an aquifer to promote bioreclamation.

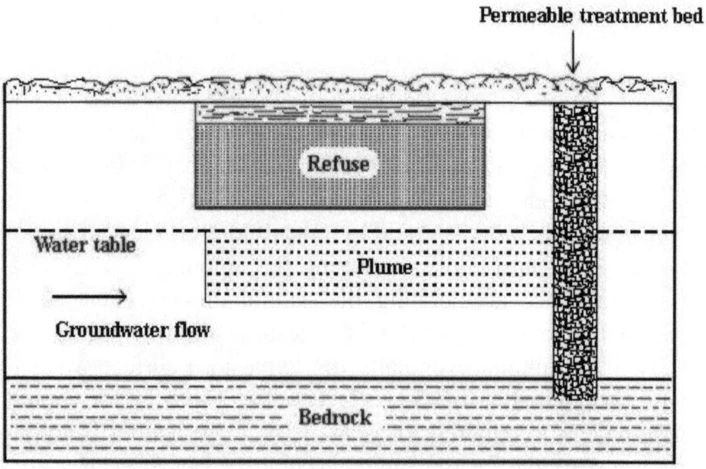

B. Installation of a permeable treatment bed in path of plume to provide contact-type treatment such as ion exchange or neutralization.

Figure 8. In situ treatment methods. Source: Modified from U.S. Environmental Protection Agency.

Where the plume contains water contaminated with chlorinated solvents, many of which are believed to be carcinogenic at the low parts per billion ranges, the best option in most cases could be to use extraction wells to remove the contaminated water. Figure 9 shows plume removal by means of shallow gradient-controlled wells. Mathematical models can play a vital role in deciding the pumping well spacing and its rate. The extraction wells are designed to capture the plume while at the same time removing as little of the uncontaminated water as possible. Extraction wells can also be planned as plume-stabilization wells. In this case they are located somewhere within the plume and sized to reverse the hydraulic gradient beyond the edge of the plume. They then prevent further movement of the plume. Locating an extraction well outside the plume will tend to expand the plume boundaries.

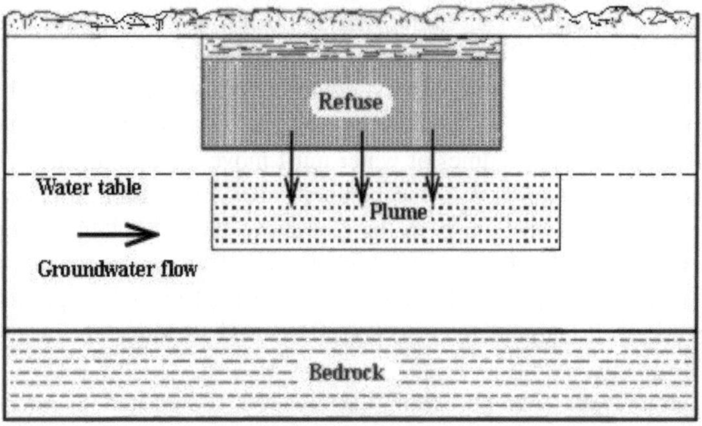

Before pumping

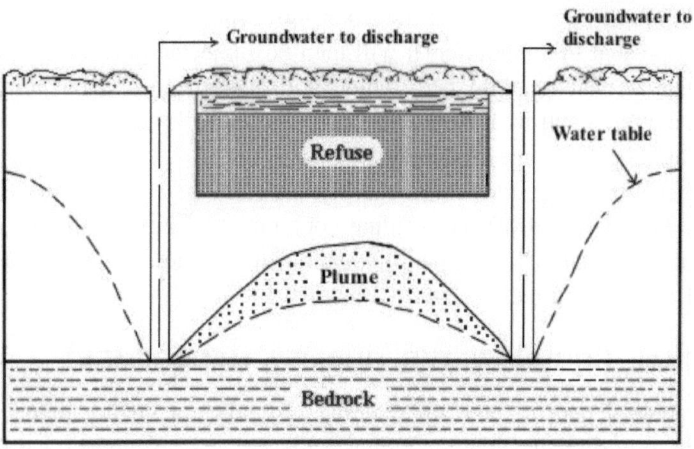

After pumping

Figure 9. Use of extraction wells to remove contaminated groundwater. Source: U.S. Environmental Protection Agency.

Contaminated water that has been removed from the ground is usually treated before being discharged. Naturally, the treatment will depend upon the types and concentrations of the contaminants. One or more of the following methods usually removes synthetic organic compounds from contaminated water: volatilization to the atmosphere in an air-stripping column, adsorption on activated carbon, and biological treatment.

Remedial Action Plan

Private water-supply wells were located to the north of the site in the shallow aquifer. As a part of the remedial effort at the site, the U.S. Environmental Protection Agency has extended public water supply to the residents of this area. While this alleviated an obvious danger, as the shallow aquifer discharges into surface water and there is no institutional control over new shallow wells being drilled, restoration of the shallow aquifer was necessary.

The long period of pumpage is required for plume treatment because dilution and dispersion, as well as the low rate of groundwater movement, mean that many pore volumes of water must move through the area in order to reduce the contaminants to the planned level. In addition, a reservoir of organic compounds still exists in the soil above the water table. In order to reduce the strength of the source, a vapour-extraction system for the shallow soil zone and upper part of the shallow aquifer is planned. The groundwater withdrawal wells around the site could be used to lower the water table beneath the site.

All groundwater pumped from the area will be treated to remove the organic compounds. Treatment will consist of spraying the water through an air-stripping column to volatilize part of the organic matter. The water will then be put through a sand filter to remove precipitated iron, which is naturally present in the aquifer. It will then be subjected to filtration through a granular activated-carbon bed, which will absorb most of the remaining organic matter. Final treatment will occur biologically as the water is sent to the municipal sewage treatment plant.

As a final measure, a low-permeability cap (cf. Fig. 6) will be placed over the site. This will seal off the residual contamination in the soil and greatly reduce the amount of precipitation that infiltrates through the unsaturated zone to the water table.

REGIONAL GROUNDWATER QUALITY MANAGEMENT

It is suggested here to take up regional groundwater quality management approach to ensure sustainable and hazard-free water resources for drinking, agriculture and industries. The following can be considered for any regional groundwater basin:

- Remediation
- Planning to stop it happening again or getting worse (prevention is better than cure)

Possible Remedial Measures of Industrial Effluents

The following are some of the remedial measures to be considered for treating the contaminated groundwater system:

- Treatment of effluents and cautious disposal of residues
- Improving lining of flow channels and discharge of effluents in low permeability zones
- Pumping of polluted groundwater, treating and recharging into the system
- Use of treated or partially treated groundwater in industry/agriculture

Treatment of effluents through common effluent treatment plants (CETP) is found to be economically and technologically viable one (Rajamani, 2002). Central Leather Research Institute (CLRI), Chennai, India, has designed a model plant to treat the tannery effluents and is shown in Figs. 10 and 11. The collection of effluent from group of industries is made to be treated in a single plant and disposing the sludge in a low permeable zone.

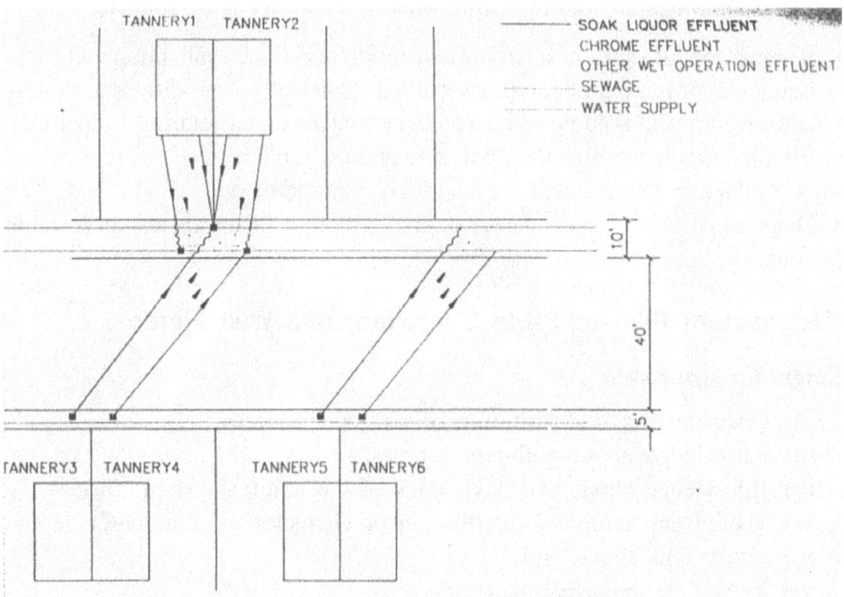

Figure 10. Proposed effluent collection systems by CLRI (Chennai)

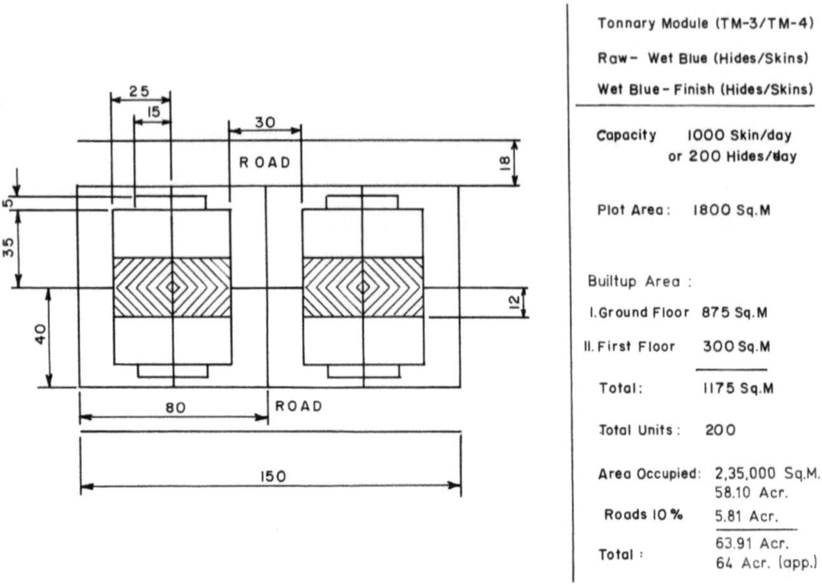

Figure 11. Proposed Tannery treatments Module (TM-3/TM-4) by CLRI (Chennai)

The technological options recommended by CLRI is as follows:

- Segregation of certain sectional waste streams like soak liquor, chrome liquor etc. or mixing of suitable sectional wastewater from different process and cleaner technologies like recovery and reuse in tanning industries;
- Primary treatment in individual unit or in a CETP;
- Secondary biological anaerobic/aerobic treatment; and
- Disposal of sludge from the treatment sites to a far away low permeable zone.

Management Plan to Protect Aquifers and Well Fields

Scientific Approach

(i) Occurrence and distribution of pollutant sources,
(ii) Knowledge about pollutant sources,
(iii) Knowledge about hydrodynamics of the aquifer system,
(iv) Spatial and temporal distribution of characteristic parameters,
(v) Cause and effect, and
(vi) Impact on management action etc.

Data Network and Monitoring

Integrated geophysical, geohydrological and geo-chemical approach should be followed to assess the groundwater pollution. This needs a systematic data, which have to be monitored through a network of stations.

Regulations

Adequate legislation to prevent pollution and enforce clean up through pump and treat technology, dilution through artificial recharge, soil reclamation etc.

Changing Practices and Changing Technology

Industries with out-dated technology should be closed down and opt for modern eco-friendly technology. The industries also should be provided incentives.

Relocation of Industries

If it is found very difficult to remediate the contaminant aquifer, it is preferred to relocate the industries after studying the vulnerability of the area to the contaminants. One such attempt was made at Calcutta, where a number of tannery industries have been brought under one roof and all effluents are collected and treated.

A schematic block diagram on the concept of water quality management is shown in Fig. 12.

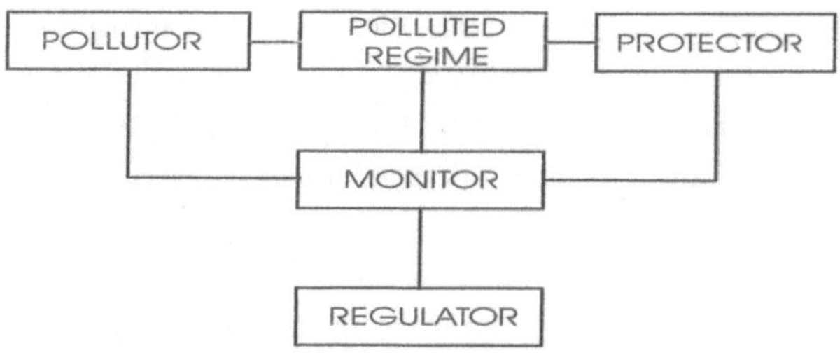

Figure 12. Concept of water quality management

Solid Waste Management

The garbage generated in the urban and industrial solid wastes disposed on the land surface also gives rise to groundwater pollution. The most serious risks occur where uncontrolled tipping (refuse dumping), as opposed to controlled sanitary landfill, is practiced and where hazardous industrial wastes, including drums of liquid effluents are disposed off at inappropriate sites. The municipal solid wastes are disposed in low-lying grounds without any record on the quantity of wastes and its quality. Even if the disposal is stopped, the site will represent a potential hazard to groundwater for decades together. The solid wastes in many developing countries are generally less toxic, having a high content of water and decaying vegetable matter, compared to typical solid wastes from over developed countries which may contain significant levels of heavy metals (chromium, lead, mercury and cadmium) and diverse synthetic organic compounds (solvents, phenols and PCB's). However, most municipal wastes generally contain only small quantities of hazardous materials. But the plume migration from the refuse site to the down gradient site will contaminate the drinking water supply thereby creating a serious health hazard.

Metros and other municipal towns will continue to grow rapidly and this, in turn, will impose enhanced stress on groundwater resources. The rapid development of industry, often small scale units and highly dispersed, is likely to produce widespread contamination of groundwater by metals, phenols and solvents. This is because (i) these contaminants frequently occur in industrial effluents and (ii) the industrial effluents itself is often disposed to the ground or to surface water channels, where it may directly pollute the groundwater. The scale of this problem is not yet fully appreciated in developing countries as very little monitoring for solid waste disposal contamination is undertaken. As mentioned above, the solution to the problem would require provisions for the collection, treatment and proper disposal of such solid wastes and liquid effluents.

Rural Drinking Water Management

We have discussed in this article, that there is an urgent need to manage the water resources for its sustainability in coming years. Getting adequate both in quantity and quality should be the main concern. Groundwater pollution is one of the hindrance to have adequate fresh-water. It is, therefore, suggested to evolve a comprehensive groundwater management plan so that an adequate supply of drinking water is assured to the rural poor. A schematic block diagram of rural water supply management scheme is illustrated in Fig. 13. The schematic is based on the idea of World Health Organization (WHO) and it is explanatory one. It is the need of the hour that those who are involved in water affairs should unite together and take the community at the center-stage and evolve a pragmatic management plan so that sustainable development and management will ensure an adequate and potable water supply to the rural folk who are really in need of it.

It is suggested that for sustainable development of groundwater resources, exploitation of deeper aquifers for irrigation should be avoided, except where it can be proved that recharge to these aquifers is considerable and does not occur as leakage from shallow aquifer. Here afterwards, deeper aquifers should be reserved for emergency drinking water supplies. Groundwater model can play a major role in the assessment and management of groundwater system in an area.

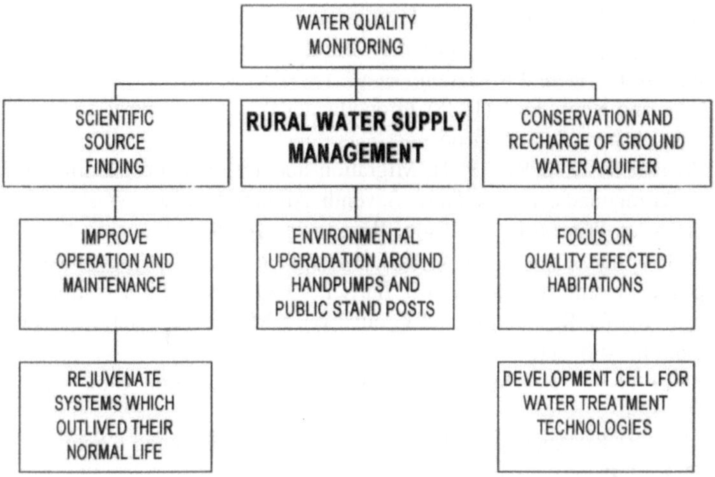

Figure 13. Rural drinking water management schemes

REFERENCES

Ahmed, M.U. (1974). Coal mining and its effect on Water quality. *American Water Resources Proceedings*, **18:** 138-148.

Aller, L., Bennett, T., Lehr, J.H. and Petty, R.J. (1985). A standardized system for evaluating groundwater pollution potential using hydrogeologic settings. United States Environmental Protection Agency Report EPA/600/2-87/035, 182 pp.

Ayres, J.E., Lager, D.C. and Barvenik, M. J. (1983). The first EPA superfund cutoff wall: Design and Specifications. *Proceedings*, Third National Symposium on Aquifer Restoration and Groundwater Monitoring. National Water Well Association, 13-22.

Barber, C. (2002). Artificial recharge of groundwater and aquifer storage and recovery of water and wastewater: Approaches for evaluation of potential water quality impacts. Proceedings of IGC-2002. Oxford & IBH, New Delhi, 117-128.

Brenoel, M. and Brown, R.A. (1985). Remediation of a Leaking Underground Storage Tank with Enhanced Bioreclamation. *Proceedings*, Fifth National Symposium on Aquifer Restoration and Groundwater Monitoring. National Water Well Association, 527-537.

British Groundwater Survey (2000). Technical report WC/95/52. Wallingford, UK.

Brunsing, T.P. and Cleary, J. (1983). Isolation of Contaminated Groundwater by Slurry-Induced Ground Displacement. *Proceedings*, Third National Symposium on Aquifer Restoration and Groundwater Monitoring. National Water Well Association, 28-36.

Campbell, P., Bost, R.C. and Jacobson, R.W. (1984). Subsurface Organic recovery and contaminant migration Simulation. *Proceedings*, Fourth National Symposium on Aquifer Restoration and Groundwater Monitoring. National Water Well Association, 82-91.

Canter, L.W. (1982). Overview of Aquifer Restoration. *Proceedings*, Second National Symposium on Aquifer Restoration and Groundwater Monitoring. National Water Well Association, vii-x.

Canter, L.W. and Knox, R.C. (1985). Groundwater Pollution Control. Chelsea, Mich.: Lewis Publishers, Inc., 526 pp.

Cline, P.V. and Viste, D.R. (1984). Migration and Degradation patterns of volatile organic compounds. *Proceedings*, Seventh Annual Madison Waste Conference, University of Wisconsin, Madison, Wisconsin, 14-29.

Druback, G.W. and Arlotta, S.V. (1985). Subsurface Pollution Contaminant using a Composite system vertical Cut-off Barrier. *Proceedings*, Fifth National Symposium on Aquifer Restoration and Groundwater Monitoring. National Water Well Association, 400-411.

Farquhar, G.J. (1988). Leachate: Production and characterization. *Can. J. Civil, Eng.*, **16:** 317-325.

Fetter, C.W. (1988). Applied Hydro-Geology. Merrill Publishing Company, Columbus, Ohio, 548 pp.

Fitzwater, P.L., Brassow, C.L. and Fetter, C.W. Jr. (1983). Assessment of Groundwater Contamination and Remedial action for a hazardous waste facility in the Gulf Coast. *Proceedings*, Third National Symposium on Aquifer Restoration and Groundwater Monitoring. National Water Well Association, 135-141.

Flathman, P.W., Quince, J.R. and Bottomley, L.S. (1984). Biological treatment of Ethylene Glycol-Contaminated Groundwater at the Naval Air Engineering Center, Lakehurst, New Jersey. *Proceedings*, Fourth National Symposium on Aquifer Restoration and Groundwater Monitoring. National Water Well Association, 111-119.

Flathman, P.W. et al. (1985). *In-Situ* Physical/Biological Treatment of Methylene Chloride (Dichloromenthene) Contaminated Groundwater. *Proceedings*, Fifth National Symposium on Aquifer Restoration and Groundwater Monitoring. National Water Well Association, 571-597.

Fusillo, T.V., Horchreiter, J.J. Jr. and Lord, D.G. (1985). Distribution of volatile organic compounds in a New Jersey Coastal Plain Aquifer System. *Groundwater*, **23:** 354-360.

Gupta, C.P., Thangarajan, M., Rao, V.V.S.G., Ramachandra, Y.M. and Sharma, M.R.K. (1994). Preliminary study of groundwater pollution in the Upper Palar basin and feasibility of mass transport modeling to predict pollutant migration. NGRI Tech. Rept., no. 94-GW-168, 45p.

Herzog, B.L., Cartwright, K., Johnson, T.M. and Harris, H.J.H. (1982). A Study of Trench Covers to Minimize Infiltration of Waste Disposal sites. Illinois State Geological Survey Contract Report no. 1981-5, Nuclear Regulatory Commission, NUREG/CR-2478, 245 pp.

Johnson, T.J. et al. (1983). Hydraulic Investigations of Failure mechanisms and migration of organic chemicals at Wilsonville, Illinois. *Proceedings*, Third National Symposium on Aquifer Restoration and Groundwater Monitoring. National Water Well Association, 413-420.

JRB Associates, Inc. (1982). Handbook: Remedial Actions at waste disposal sites. U.S. Environmental Protection Agency, EPA-625/6-82-006, 497 pp.

Kaufman, R.F., Eadie, G.G. and Russell, C.R. (1976). Effects of Uranium mining and milling on Groundwater in the grants Uranium Belt, New Mexico. *Groundwater*, **14:** 296-308.

Keely, J.F. (1984). Optimizing Pumping strategies for Contaminant studies and Remedial actions. *Proceedings*, Fourth National Symposium on Aquifer Restoration and Groundwater Monitoring. National Water Well Association, 34-42.

Klusman, R.W. and Edwards, K.W. (1977). Toxic Metals in Groundwater of the Front Range, Colorado. *Groundwater*, **15:** 160-169.

Kmet, P., Quinn, K.J. and Slavik, C. (1981). Analysis of Design Parameters affecting the collection efficiency of Clay lined landfills. *Proceedings*, Fourth Annual Madison Waste Conference. Madison, Wis.: University of Wisconsin.

Kramer, W.H. (1982). Groundwater Pollution from Gasoline. *Groundwater Monitoring Review*, **2(2):** 18-22.

Lee, F.F. and Jones, R.A. (1991). Landfills and groundwater quality. *Ground Water*, **29(4):** 482-486.

Lenzo, F.C. (1984). Air-Stripping for VOCs in Water: Pilot, Design, Construction. *Proceedings*, Fourth National Symposium on Aquifer Restoration and Groundwater Monitoring. National Water Well Association, 100-110.

Lynch, E.R. et al. (1984). Design and Evaluation of In-Place structures utilizing Groundwater cutoff walls. *Proceedings*, Fourth National Symposium on Aquifer Restoration and Groundwater Monitoring. National Water Well Association, 1-7.

McLeod, R.S. (1984). Evaluation of 'Superfund' sites for Control of Leachate and Contaminant Migration. *Proceedings*, Fifth National Conference on Management of Uncontrolled Hazardous Waste Sites, Hazardous Materials Control Research Institute, 114-121.

McWhorter, D.B., Skogerboe, R.K. and Skogerboe, G.V. (1974). Potential of mine and mill spoils for water quality degradation. *American Water Resources Association Proceedings*, **18:** 123-137.

Norbeck, P.N., Mink, L.L. and Williams, R.E. (1974). Groundwater Leaching of Jig Tailing deposits in the Coeur d'Alene District of Northern Idaho. *American Water Resources Association Proceedings*, **18:** 149-157.

Noss, R.R. and Johnson, E.T. (1984). Field monitoring of the Adams, Massachusetts, Landfill Leachate Plume. *Proceedings*, Fourth National Symposium and exposition on Aquifer Restoration and Groundwater Monitoring. National Water Well Association, 356-362.

Oliveira, D.P. and Sitar, N. (1985). Groundwater Contamination from Underground Solvent Storage Tanks, Santa Clara, California. *Proceedings*, Fifth National Symposium and exposition on Aquifer Restoration and Groundwater Monitoring. National Water Well Association, 691-708.

Pavelic, P. and Dillon, P.J. (1997). Review of international experience in injecting natural and reclaimed waters into aquifers for storage and reuse. Centre for Groundwater Studies report No 74, May 1997.

Pionke, H.B. and Urban, J.B. (1985). Effect of Agricultural land use on Groundwater quality in a small Pennsylvania Watershed. *Groundwater*, **23:** 68-80.

Poulos, S.J. and Laws, A.C. (1985). Gradient control for Contaminants and Pollutants. *Proceedings*, Fifth National Symposium on Aquifer Restoration and Groundwater Monitoring. National Water Well Association, 390-399.

Pyne, R.D.G. (1995). Groundwater recharge and wells: A guide to aquifer storage and recovery. Lewis Publishers, Florida, USA.

Rajamani, S. (2002). Environmental management of industrial waste discharges to prevent groundwater pollution. *Proceedings*, International Groundwater Conference (IGC-2002) held at Dindigul, South India, 255-260.

Ralston, D.R. and Morilla, A.G. (1974). Groundwater movement through an abandoned Tailings Pile. *American Water Resources Association Proceedings*, **18:** 174-183.

Rothschild, E.R., Manser, R.J. and Anderson, M.P. (1982). Investigation of Aldicarb in Groundwater in selected areas of the Central sand plain of Wisconsin. *Groundwater*, **20:** 437-445.

Roux, P.H. and Althoff, W.F. (1980). Investigation of Organic Contamination of Groundwater in South Brunswick Township, New Jersey. *Groundwater*, **18:** 464-471.

Schafer, J.M. (1984). Determining Optimum Pumping rates for creation of Hydraulic barriers to Groundwater Pollutant Migration. *Proceedings*, Fifth National Symposium on Aquifer Restoration and Groundwater Monitoring. National Water Well Association, 50-63.

Schroeder, P.R., Morgan, J.M., Walski, T.M and Gibson, A.C. (1984). Hydrologic Evaluation of Landfill Performance (HELP) Model; USEPA, Washington.

UNDP. 1971. Groundwater investigations in Tamilnadu (Phase 1), Tech. Rep., New York, 88 pp.

U.S. Environmental Protection Agency (1977). The report to Congress—Waste Disposal Practices and their effects on Groundwater. Washington, D.C.: U.S. Environmental Protection Agency, 512 pp.

Yaniga, P.M., Matson, C. and Demko, D.J. (1985). Restoration of Water Quality in a Multi-aquifer System via *In-Situ* Biodegradation of the Organic Contaminants. *Proceedings*, Fifth National Symposium on Aquifer Restoration and Groundwater Monitoring. National Water Well Association, 510-526.

Yates, M.V. (1985). Septic Tank Density and Groundwater Contamination. *Groundwater*, **23:** 586-591.

Index

activated alumina, 135
abstraction, 191
activated-carbon, 348
adsorb nutrients, 118
adsorption, 338
advective migration, 192
Agati Island, 312
algae (chlorophylls), 12
alluvial aquifer, 192
analog models, 194
analytical element method, 206
analytical solution, 156, 200
anisotropic, 156
apparent resistivity, 30
aquifer parameters (T & S), 61
aquifer response, 190
aquifer storage and recovery (ASR), 112, 337
aquifer, 285
aquitard, 286
aquiclude, 62
archaean ages, 157
arid climate, 157
array, 29
arsenic contamination, 128
arsenicism, 128
arsenate, 142
arsenic, 137
arsenite, 143
artificial recharge, 3, 191, 337
asymptotic value, 89
augmentation, 112, 191, 337
automatic inversion, 44

Badon Ghyleen-Herzberg, 286, 287
barometric effects, 302
Barrett's indexing technique, 17
basement rock, 53
basic functions, 206, 322
basin hydrology, 6
bedrock ridge, 58
bedrock topography, 57
bedrock, 158
benzene, 115
betex, 115
bias, 252
biodegradation, 245
biotite granite, 161
bipole-bipole array, 29
block estimation, 101
bootstrapping, 264
borehole logging, 314
borehole silting, 59
Botswana, 214
Boulton, 173
boundary conditions, 77, 220, 290

cables, 32
calculated variogram, 84
calibration, 8
chemical analysis, 289
chemical spills, 342
chemicals, 12
clays, 31
climate change, 294
clogging of bore wells, 337
clogging, 113
cloud characteristics, 13
coastal watershed, 295
colluvial sediments, 129
common effluent treatment plant (CETP), 349
community management, 335
compacted clay, 343
computer modeling, 44
concentration gradient, 192
conceptual models, 238
conditional simulation, 80
conductive fracture zone, 156, 165
conductive sphere, 38
conductivity, 301
confined aquifer, 62, 205
connectivity, 159
courant numbers, 246
contaminant transport, 115
contaminants, 243

coral atolls, 313
coralline atolls, 291
correlation coefficient, 243
covariance, 82
cross covariance, 80
cross validation, 84
cryolite, 136
current, 29

Darcy's law, 67, 294, 307
Darcian velocity, 194
data acquisition, 10
data error, 253
Death valley, 271
debt trap, 191
defluordiation, 135
delayed yield, 173
demographic pressures, 298
dental enamel, 129
deterministic methods, 80
dielectric properties, 14
digital data bases, 10
digital elevation models (DEM), 7
digital image processing, 9
digital techniques, 4
Dindigul, 55
dipole-dipole array, 30
dipping beds, 39
discontinuity, 159
dissolved organic matter (DOM), 12
double continuum, 169
double logarithmic scale, 45
double porosity model, 156, 168
drainage density, 5
drawdowns, 67
drought years, 335
Dupuit formula, 165

earth resistivity, 28, 316
eco-friendly technology, 351
effective porosity, 292
effluent, 113
electric analog, 190
electrical imaging, 50
electrical potential, 27
electrode spacing, 33, 36
electrodes, 32
electromagnetic profiling, 54, 317
electromagnetic radiation, 11
electronegative, 129
electronic conduction, 31
empirical techniques, 44
energy balance equation, 11
expected value (ensemble average), 81
erosion, 158
error variance, 253
evapo-transpiration, 13, 253
experimental variogram, 83

finite difference method, 53, 76, 201
finite element models, 320
finite element, 190, 204
first-order reactions, 245
flood control, 10
flood plain mapping, 10
flow meter, 156
fluoride, 103, 129
four electrode array, 30
Frac3DVS software, 212
FracMan software, 211
fractional dimension flow solution, 160
fracture density, 158
fracture properties, 156
fractures, 53
frequency domain electromagnetic method (FDEM), 317

Galerkin method, 206, 321, 323
Gauss-Newton method, 241
generalized radial flow, 177
generic models, 189
geochemical process, 136
geoelectrical sounding, 50
Geographic Information System (GIS), 1
geological structure, 4, 31
geology, 3
geometric transformation, 9
geomorphic features, 6
geomorphological processes, 158
geomorphology, 3
geophysical methods, 314
geo-purification, 114
geostationary satellites, 13
geostatistics, 78
Ghyben-Herzberg, 313
Godavari Delta, 310

Green's theorem, 324
Gringarten method, 176
groundwater management, 191
groundwater modeling, 1, 157
groundwater pollution, 191
groundwater restoration, 343
Groundwater Vistas, 210
Grout curtains, 344

hand pump, 58
hard rock aquifer, 156
hard rock, 35
hazard-free, 352
hazardous wastes, 339, 352
heavy metals (metalloids), 137
heavy metals, 351
hexavalent chromium, 343
Hele-Shaw physical model, 190
high islands, 291
histogram, 164
homogeneous half-space, 33
horizontal traversing, 34
hydraulic conductivity, 5
hydraulically conductive fractures, 164
hydrodynamic properties, 156, 157
hydrodynamic dispersion, 192, 290
hydrogeology, 2
hydrofluoric acid, 136
hydrological modeling, 3
hydrological variables, 80
hypertension, 144

impedance, 31
inductive coupling, 32
infiltration, 112
infrared images, 13
ingress of saline water, 190
initial conditions, 220
injection test, 165
injection, 112
interactive computer modeling, 44
interpolation, 79
interpretation techniques, 93
interpretive models, 189
intrinsic hypothesis, 82
inverse methods, 238
inverse modeling, 239
ionic conduction, 31
irregular function, 80

Jack knifing, 264

Kalahari beds, 216
Kavaratti Island, 313
keratosis, 138
Kodaganar, 55
Kriging, 78
Kulandaipatty, 56
Kumyere river valley, 214

Laccadive Islands, 312
leachate, 339
lag, 83
lagoons, 313
Lagrangian multiplier, 87
Lake Ontario, 309
Lambton county, 307
land fills, 339
land use, 11
landsat, 17
landscape roughness, 11
latent heat flux, 11
lineament, 58
linear confidence, 253
linear drift, 96
logarithmic drift, 79
longitudinal conductance, 48
longitudinal dispersivity, 306
low islands, 291

macroscopic dispersion, 289
Maheswaram, 161
Maheswaram watershed, 90
management models, 303
management of groundwater resources, 334
mass transport, 303
mathematical modeling, 200
Mathinipatty, 57
mechanical filtration, 338
melanosis, 138
metamorphic rocks, 104
method of characteristics, 247, 331
metrosat data, 14
micro-biological pollution, 191
microwave radiometer, 14
mineralization of bones, 129

mineralogical analysis, 122
miscible fluids, 298
MMOC, 247
MOC, 207
model calibration, 221, 237
model error, 253
model predictions, 238, 254
MODFLOW, 201, 239
MODPATH, 207
molecular diffusion, 289
molecular diffusivity, 304
monsoon climatic regions, 191
moving neighbourhood, 89
MT3D, 207
MT3DMS, 247
multi component reactive transport, 244
multicore cables, 32
multi-layer, 206
multiple stakes, 32
multivariate, 79

NAPSAC, 213
Navier-Stokes equation, 174
nested models, 84
Neuman technique, 174
nitrification, 114
non-conductors, 31
non-continuum, 190
non-linear regression, 238
non-point sources, 296
non-randomness, 252
non-symmetric, 330
nugget effect, 86
numerical dispersion, 204
numerical modeling, 80
numerical models, 302
numerical simulation, 190
nutrients, 118

objective function, 239
offset Wenner electrode, 38
Ohm's law, 195
Okavango Delta, 215
ordinary kriging, 103
organic pollution, 191
over-exploitation, 191, 335

packers, 300
parameter correlation coefficients (PCC), 257
parameter uncertainty, 256
partial differential equation, 204
particle tracking, 245
percolation tanks, 4
permeability, 17, 156
PEST, 239
phosphorous, 141
physical models, 115, 193
physiography, 292
plume treatment, 345
point sources, 296
pollution sources, 296
porosity, 31
porous medium, 158, 190
post-audits, 256, 265
post-monsoon, 91
potential electrode, 41
pre-cambrian, 157
predictor-corrector (P-C), 247
prediction scaled sensitivities (PSS), 257
prediction scenario runs, 222
prediction sensitivities, 256
predictive model, 189
pre-monsoon, 91
prevalence, 144
probabilistic framework, 81
pseudo-depth, 51
pseudo-section, 51
pulsed DC instruments, 32
pumping test, 64, 159

quadrilateral elements, 307

radar remote sensing, 14
radial flow, 164, 165
radio-active decay, 338
rainfall, 295
rainfall recharge, 91
random errors, 47
random function, 81
random walk model, 247
randomness, 9
range, 83
reaction kinetics, 115
realizations, 81

recharge wells, 113
recovery, 122
redox reactions, 117
regional scale, 238
regionalized variables, 78
remote sensing, 1
representative elementary volume (REV), 330
resistance meters, 32
resistance-capacitance (R-C) analog model, 195
resistivity techniques, 59
retardation, 243
reverse osmosis, 135
river Kodaganar, 55
RMS error, 221
rock matrix, 159
runoff models, 16
runoff, 295
rural drinking water management, 352

saltwater intrusion, 296, 302
satellite imagery, 59
scale effect, 179
Scatter plot, 221
Schlumberger, 29
seawater intrusion, 193, 288
secondary fracture network, 157
semi-arid climate, 157
semi-confined aquifer, 63
sensitivity analysis, 237
septic tanks, 338
silicate mineral, 129
sill, 83
simultaneous equations, 330
site-scale, 238
skeletal fluorosis, 132
slug tests, 157
slurry wall, 344
soil aquifer treatment (SAT), 114, 337
soil cover, 292
soil heat flux, 11
soil moisture, 11, 13
solar radiation, 11, 14
solid waste management, 351
sounding curve interpretation, 44
space, 99
spatial data, 10
spatial parameters, 80
spatial variability, 80
specific yield, 174
spectral properties, 12
stationarity, 82
statistical distribution, 156
steady state, 64
steady state calibration, 221
stochastic models, 9
storativity, 17, 61
submarine groundwater discharge, 286
surface parameters, 11
surface runoff, 295
surface spreading, 337
Sutra model, 304
SUTRA, 210

Taylor's series, 196
tectonic activity, 158
tectonically fractured areas, 159
tectono-geomorphic model, 159
Theis type curve, 70
thematic layers, 4
theoretical variogram, 84
thermal properties, 12
thermal radiation, 12
thermodynamic properties, 244
tidal effects, 302
time, 9
time domain electromagnetic method (TDEM), 317
toluene, 115
tomography, 50
topography, 295
total dissolved solids (TDS), 191
total suspended solids, 191
total variation diminishing (TVD), 247
toxicological, 133
transient solution, 160
transition zone, 289
transmissive function, 158
transmissivity, 18, 61
transverse dispersivity, 306
transverse resistance, 48
traversing, 34
triangular elements, 306
true resistivity, 36

turbidity, 12
two-layer curve, 36

Ucode, 239
unbiased condition, 86
unbounded variogram, 92
uncertainties, 10
uncertainty evaluation, 237
unconfined aquifer, 285
unconsolidated weathered mantle, 159
uncropped dry fields, 58
un-fractured, 54
unique neighbourhood, 89
universal kriging, 89
unsaturated soil zone, 192, 238
unsaturated zone, 296
unsteady state condition, 64

vadose zone, 114, 337
Vaippar river basin, 85
validation, 237
value of improved information (*voii*) statistic 262
v-anomalies, 52
vapour-extraction, 348
variance, 78
variographic analysis, 85
variogram, 79
vector format, 9
velocity, 164
vertical electrical sounding, 30, 35
Virudalaipatty, 57
virus concentration, 266
viscous fluid models, 194
volatilization, 338
volcanic rocks, 291
voltage, 29
voltmeter, 31

waste water, 118
water borne disease, 338
water interception, 294
water related issues, 10
water resources, 1, 6
watershed, 156, 295
weathered fractured medium, 156
weathering, 54
weighted residual, 241
well storage, 76
Wenner array, 29
Wenner sounding, 37
Winflow, 211

Plate 1

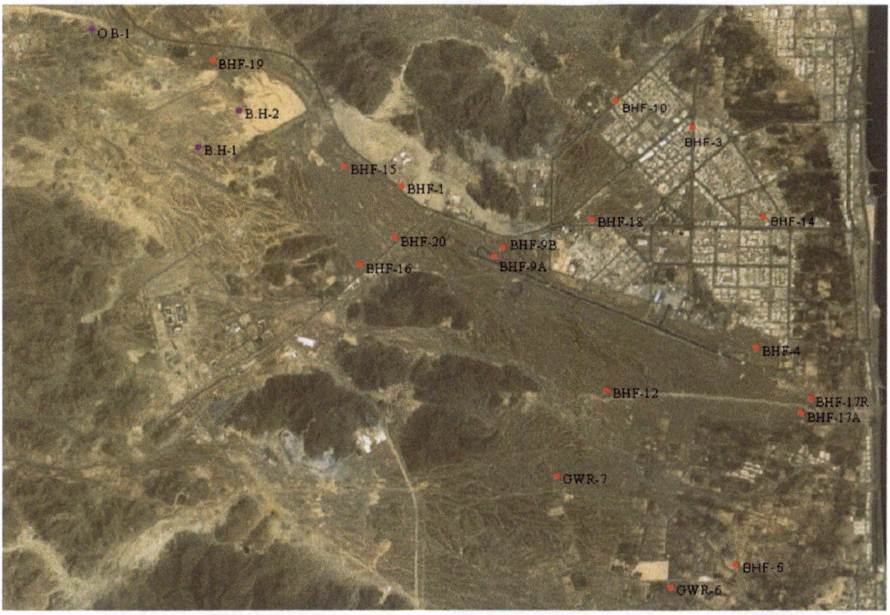

RS image of Wadi Ham in the UAE

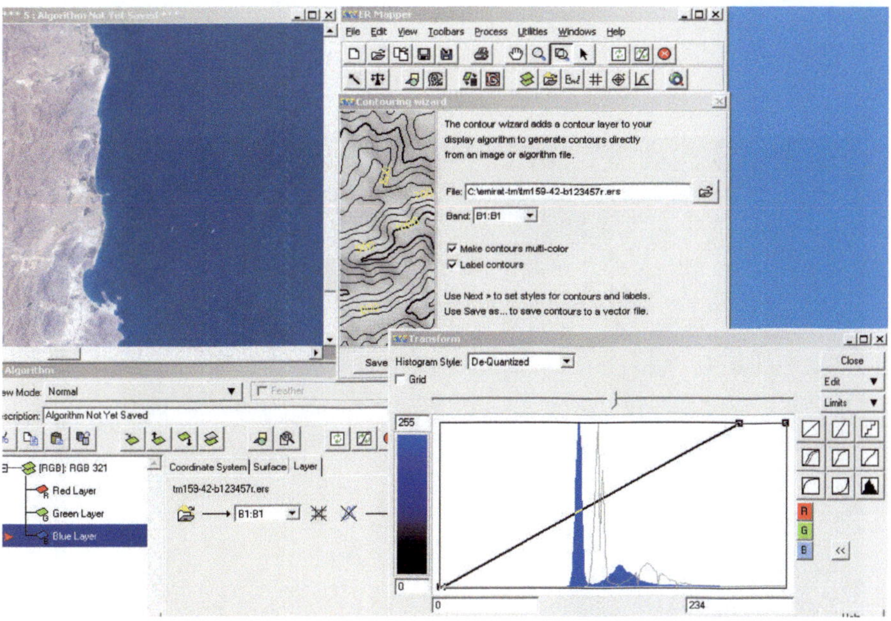

Image processing activities to improve feature sharpness and contrasts

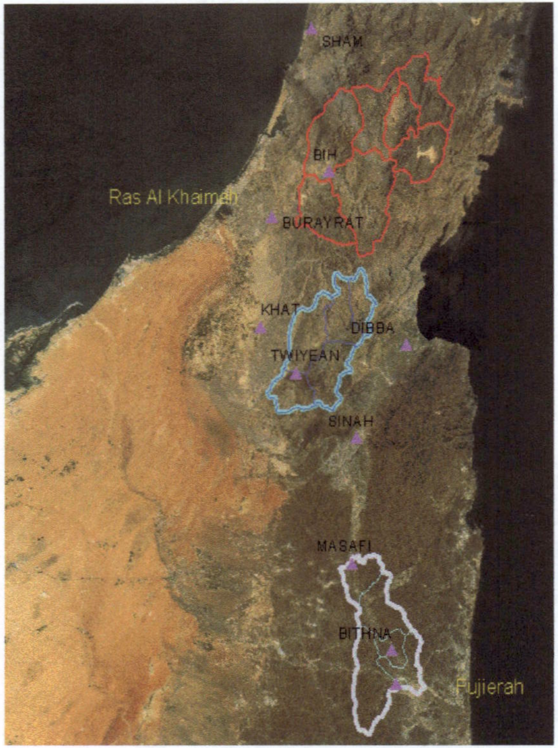

Remote sensing image for three locations in Wadi along with rain gauge stations

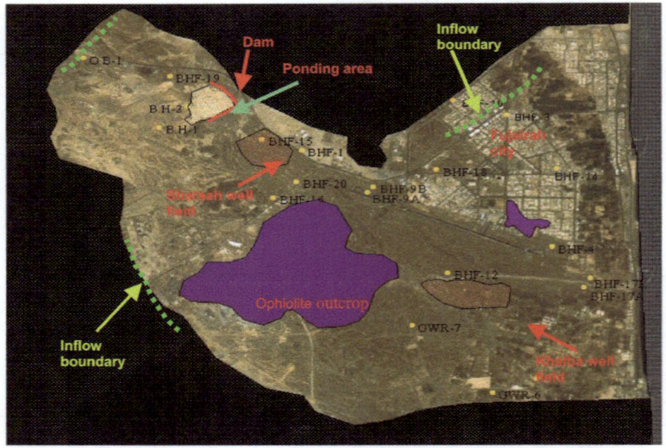

RS image including the main features of the study area in Wadi Ham, UAE

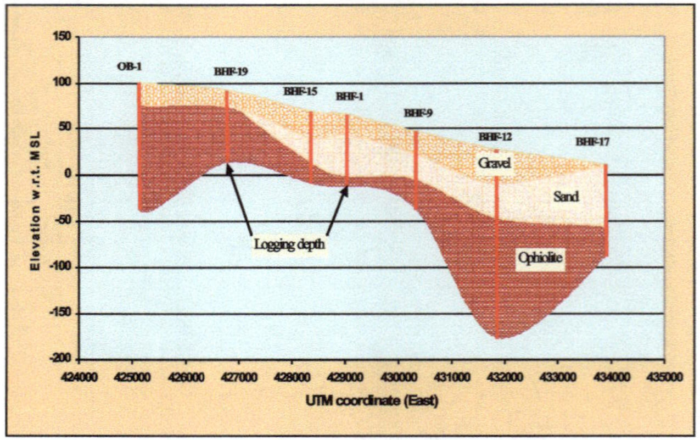

Typical geological cross-section along the course of Wadi Ham

Plate 3

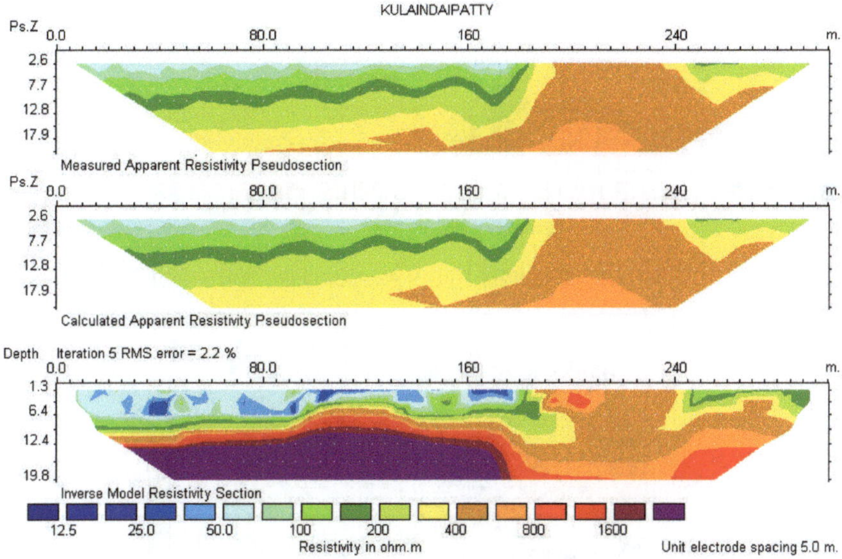

Electrical image over weathered basement in India. (a) is observed data plotted as a coloured pseudosection (Ps.z =pseudo-depth), (b) is the pseudosection computed from the model and (c) is the image or model showing true depth and true formation resistivity.

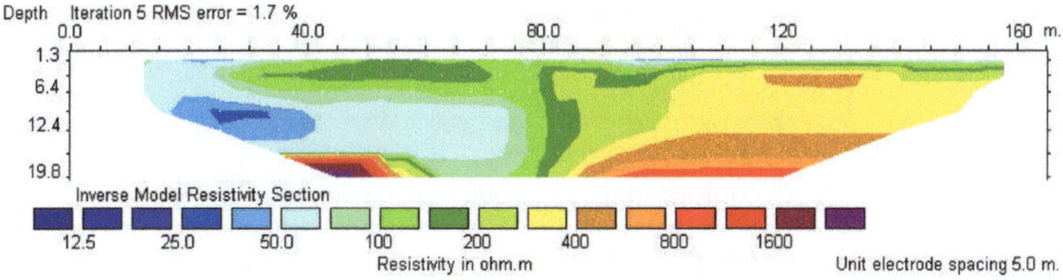

Electrical image along Line 2 at Mathinipatty

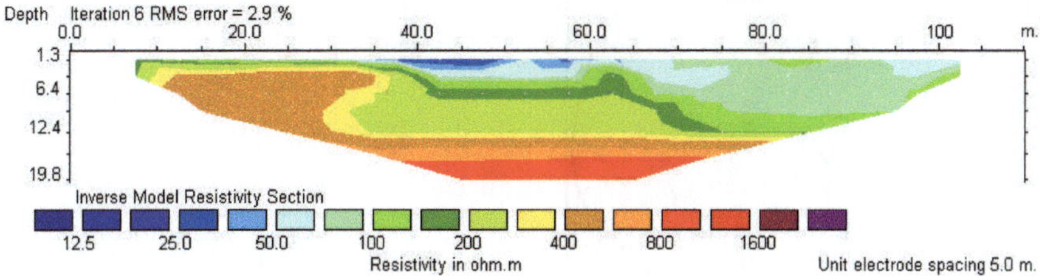

Electrical image along Line 3 at Virudalaipatty

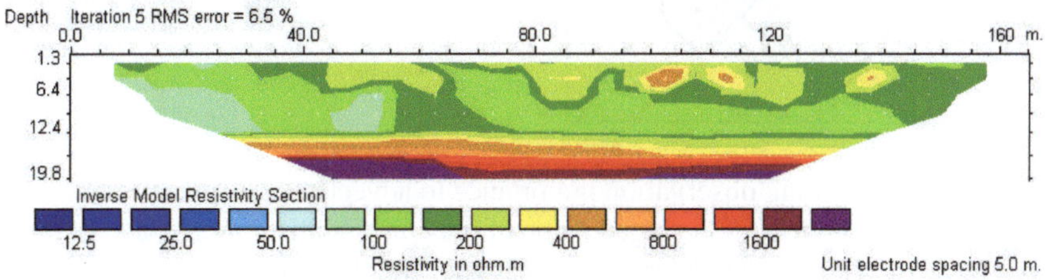

Electrical image along Line 4 at Rajagoundanoor

Plate 4

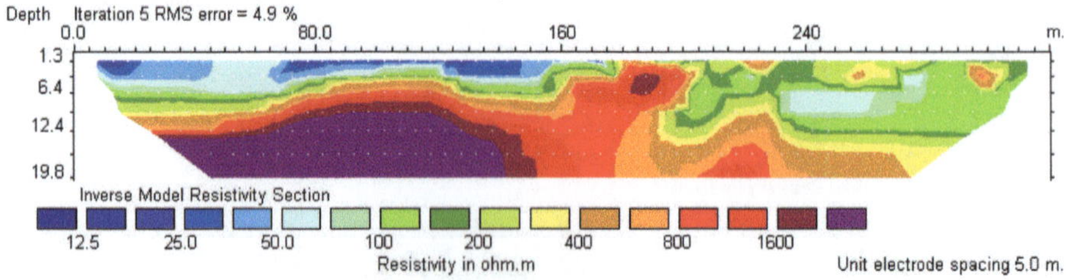

Electrical image along Line 5 at Giriyappa Naiyakanoor

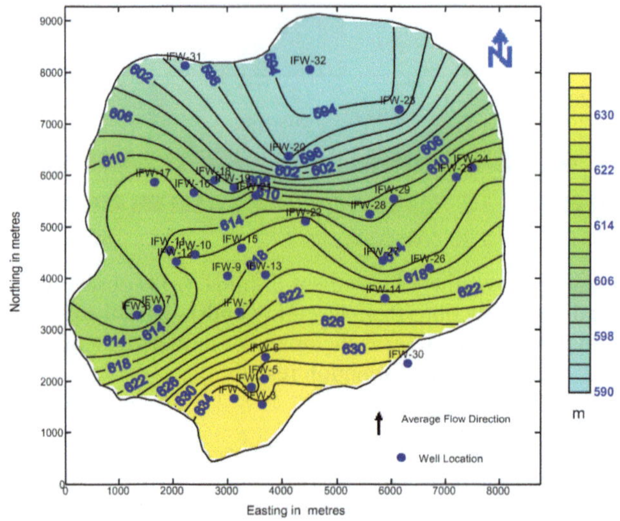

Water level contour map, Jan.2000 (Surfer without kriging)

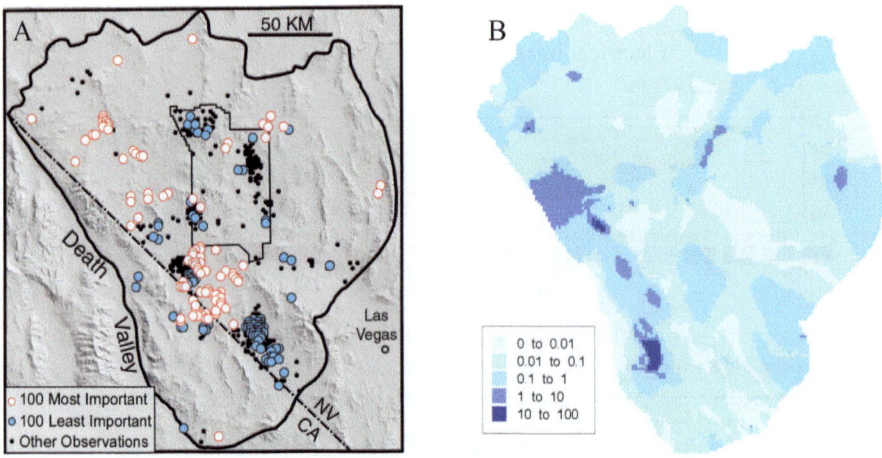

Results of determining observation importance to advective transport predictions in the DVRFS model. An averaged value of the $opr_{i(-i)}$ statistic was used to rank (A) the 501 existing head observation locations and (B) potential new observation locations in each cell of model layer 1

MIX
Papier aus verantwortungsvollen Quellen
Paper from responsible sources
FSC® C105338

If you have any concerns about our products,
you can contact us on
ProductSafety@springernature.com

In case Publisher is established outside the EU,
the EU authorized representative is:
Springer Nature Customer Service Center GmbH
Europaplatz 3, 69115 Heidelberg, Germany

Printed by Libri Plureos GmbH
in Hamburg, Germany